Student Solutions Manual

for

Applied Calculus for the Managerial, Life, and Social Sciences

Fifth Edition

S. T. Tan
Stonehill College

BROOKS/COLE

THOMSON LEARNING

Australia • Canada • Mexico • Singapore • Spain • United Kingdom • United States

Assistant Editor: *Ann Day*
Editorial Assistant: *Suzannah Alexander*
Marketing Manager: *Karin Sandberg*

Ancillary Coordinator: *Rita Jaramillo*
Print Buyer: *Christopher Burnham*
Permissions Editor: *Robert Kauser*

For more information, contact
Wadsworth/Thomson Learning
10 Davis Drive
Belmont, CA 94002-3098
USA

For more information about our products, contact:
Thomson Learning Academic Resource Center
1-800-423-0563
http://www.wadsworth.com

International Headquarters
Thomson Learning
International Division
290 Harbor Drive, 2nd Floor
Stamford, CT 06902-7477
USA

UK/Europe/Middle East/South Africa
Thomson Learning
Berkshire House
168-173 High Holborn
London WC1V 7AA
United Kingdom

Asia
Thomson Learning
60 Albert Complex, #15-01
Singapore 189969

Canada
Nelson Thomson Learning
1120 Birchmount Road
Toronto, Ontario M1K 5G4
Canada

ISBN 0-534-38788-8

CONTENTS

CHAPTER 9 DIFFERENTIAL EQUATIONS

CHAPTER 10 PROBABILITY AND CALCULUS

CHAPTER 11 TAYLOR POLYNOMIALS AND INFINITE SERIES

CHAPTER 12 TRIGONOMETRIC FUNCTIONS

CHAPTER 1

EXERCISES 1.1, page 12

1. The statement is false because -3 is greater than -20. (See the number line that follows).

    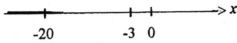

2. The statement is false because 2/3 [which is equal to (4/6)] is less than 5/6.

    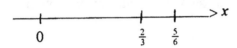

5. The interval (3,6) is shown on the number line that follows. Note that this is an open interval indicated by **(** and **)**

    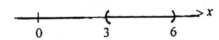

7. The interval [-1,4) is shown on the number line that follows. Note that this is a half-open interval indicated by **[** (closed) and **)** (open).

9. The infinite interval (0,∞) is shown on the number line that follows.

11. First, $2x + 4 < 8$ (Add -4 to each side of the inequality.)
    Next, $2x < 4$ (Multiply each side of the inequality by 1/2)
    and $x < 2.$
    We write this in interval notation as $(-\infty,2)$.

13. We are given the inequality $-4x \geq 20$.
   Then $x \leq -5$.　　　　　　　　　　(Multiply both sides of the inequality by $-1/4$
   　　　　　　　　　　　　　　　　　　and reverse the sign of the inequality.)
   We write this in interval notation as $(-\infty, -5]$.

15. We are given the inequality $-6 < x - 2 < 4$.
   First　　$-6 + 2 < x < 4 + 2$　　　　(Add $+2$ to each member of the inequality.)
   and　　　$-4 < x < 6$,
   so the solution set is the open interval $(-4, 6)$.

17. We want to find the values of x that satisfy the inequalities
   $$x + 1 > 4 \text{ or } x + 2 < -1.$$
   Adding -1 to both sides of the first inequality, we obtain
   $$x + 1 - 1 > 4 - 1,$$
   or　　　　　　　$x > 3$.
   Similarly, adding -2 to both sides of the second inequality, we obtain
   $$x + 2 - 2 < -1 - 2,$$
   or　　　　$x < -3$.
   Therefore, the solution set is $(-\infty, -3) \cup (3, \infty)$.

19. We want to find the values of x that satisfy the inequalities
   $$x + 3 > 1 \text{ and } x - 2 < 1.$$
   Adding -3 to both sides of the first inequality, we obtain
   $$x + 3 - 3 > 1 - 3,$$
   or　　　　　　　　$x > -2$.
   Similarly, adding 2 to each side of the second inequality, we obtain
   $$x < 3,$$
   and the solution set is $(-2, 3)$.

21. $|-6 + 2| = 4$.

23. $\dfrac{|-12 + 4|}{|16 - 12|} = \dfrac{|-8|}{|4|} = 2.$

25. $\sqrt{3}|-2| + 3|-\sqrt{3}| = \sqrt{3}(2) + 3\sqrt{3} = 5\sqrt{3}$.

27. $|\pi - 1| + 2 = \pi - 1 + 2 = \pi + 1$.

29. $\left|\sqrt{2}-1\right|+\left|3-\sqrt{2}\right|=\sqrt{2}-1+3-\sqrt{2}=2.$

31. False. If $a > b$, then $-a < -b$, $-a+b < -b+b$, and $b - a < 0$.

33. False. Let $a = -2$ and $b = -3$. Then $a^2 = 4$ and $b^2 = 9$, and $4 < 9$. Note that we only need to provide a counterexample to show that the statement is not always true.

35. True. There are three possible cases.

> **Case 1** If $a > 0$, $b > 0$, then $a^3 > b^3$, since $a^3 - b^3 = (a - b)(a^2 + ab + b^2) > 0$.

> **Case 2** If $a > 0$, $b < 0$, then $a^3 > 0$ and $b^3 < 0$ and it follows that $a^3 > b^3$.

> **Case 3** If $a < 0$ and $b < 0$, then $a^3 - b^3 = (a - b)(a^2 + ab + b^2) > 0$, and we see that $a^3 > b^3$. (Note that $(a - b) > 0$ and $ab > 0$.)

37. False. Take $a = -2$, then $|-a| = |-(-2)| = |2| = 2 \neq a$.

39. True. If $a - 4 < 0$, then $|a - 4| = 4 - a = |4 - a|$. If $a - 4 > 0$, then
$$|4 - a| = a - 4 = |a - 4|.$$

41. False. Take $a = 3$, $b = -1$. Then $|a + b| = |3 - 1| = 2 \neq |a| + |b| = 3 + 1 = 4$.

43. $27^{2/3} = (3^3)^{2/3} = 3^2 = 9.$

45. $\left(\dfrac{1}{\sqrt{3}}\right)^0 = 1$. Recall that any number raised to the zero power is 1.

47. $\left[\left(\dfrac{1}{8}\right)^{1/3}\right]^{-2} = \left(\dfrac{1}{2}\right)^{-2} = (2^2) = 4.$

49. $\left(\dfrac{7^{-5} \cdot 7^2}{7^{-2}}\right)^{-1} = (7^{-5+2+2})^{-1} = (7^{-1})^{-1} = 7^1 = 7.$

51. $(125^{2/3})^{-1/2} = 125^{(2/3)(-1/2)} = 125^{-1/3} = \dfrac{1}{125^{1/3}} = \dfrac{1}{5}$.

53. $\dfrac{\sqrt{32}}{\sqrt{8}} = \sqrt{\dfrac{32}{8}} = \sqrt{4} = 2$.

55. $\dfrac{16^{5/8}16^{1/2}}{16^{7/8}} = 16^{(5/8+1/2-7/8)} = 16^{1/4} = 2$.

57. $16^{1/4} \cdot 8^{-1/3} = 2 \cdot \left(\dfrac{1}{8}\right)^{1/3} = 2 \cdot \dfrac{1}{2} = 1$.

59. True.

61. False. $x^3 \times 2x^2 = 2x^{3+2} = 2x^5 \neq 2x^6$.

63. False. $\dfrac{2^{4x}}{1^{3x}} = \dfrac{2^{4x}}{1} = 2^{4x}$.

65. False. $\dfrac{1}{4^{-3}} = 4^3 = 64$.

67. False. $(1.2^{1/2})^{-1/2} = (1.2)^{-1/4} \neq 1$.

69. $(xy)^{-2} = \dfrac{1}{(xy)^2}$.

71. $\dfrac{x^{-1/3}}{x^{1/2}} = x^{(-1/3)-(1/2)} = x^{-5/6} = \dfrac{1}{x^{5/6}}$

73. $12^0(s+t)^{-3} = 1 \cdot \dfrac{1}{(s+t)^3} = \dfrac{1}{(s+t)^3}$.

75. $\dfrac{x^{7/3}}{x^{-2}} = x^{(7/3)+2} = x^{(7/3)+(6/3)} = x^{13/3}$.

77. $(x^2y^{-3})(x^{-5}y^3) = (x^{2-5}y^{-3+3}) = x^{-3}y^0 = x^{-3} = \dfrac{1}{x^3}$.

79. $\dfrac{x^{3/4}}{x^{-1/4}} = x^{(3/4)-(-1/4)} = x^{4/4} = x$.

81. $\left(\dfrac{x^3}{-27y^{-6}}\right)^{-2/3} = x^{3(-2/3)}\left(-\dfrac{1}{27}\right)^{-2/3}y^{6(-2/3)} = x^{-2}\left(-\dfrac{1}{3}\right)^{-2}y^{-4} = \dfrac{9}{x^2y^4}$.

83. $\left(\dfrac{x^{-3}}{y^{-2}}\right)^2\left(\dfrac{y}{x}\right)^4 = \dfrac{x^{-3(2)}y^4}{y^{-2(2)}x^4} = \left(\dfrac{y^{4+4}}{x^{4+6}}\right) = \dfrac{y^8}{x^{10}}$.

85. $\sqrt[3]{x^{-2}} \cdot \sqrt{4x^5} = x^{-2/3} \cdot 4^{1/2} \cdot x^{5/2} = x^{-(2/3)+(5/2)} \cdot 2 = 2x^{11/6}$.

87. $-\sqrt[4]{16x^4y^8} = -(16^{1/4} \cdot x^{4/4} \cdot y^{8/4}) = -2xy^2$.

89. $\sqrt[6]{64x^8y^3} = (64)^{1/6} \cdot x^{8/6}y^{3/6} = 2x^{4/3}y^{1/2}$.

91. $2^{3/2} = (2)(2^{1/2}) = 2(1.414) = 2.828$.

93. $9^{3/4} = (3^2)^{3/4} = 3^{6/4} = 3^{3/2} = 3 \cdot 3^{1/2} = 3(1.732) = 5.196$.

95. $10^{3/2} = 10^{1/2} \cdot 10 = (3.162)(10) = 31.62$.

97. $10^{2.5} = 10^2 \cdot 10^{1/2} = 100(3.162) = 316.2$.

99. $\dfrac{3}{2\sqrt{x}} \cdot \dfrac{\sqrt{x}}{\sqrt{x}} = \dfrac{3\sqrt{x}}{2x}$.

101. $\dfrac{2y}{\sqrt{3y}} \cdot \dfrac{\sqrt{3y}}{\sqrt{3y}} = \dfrac{2y\sqrt{3y}}{3y} = \dfrac{2}{3}\sqrt{3y}$.

103. $\dfrac{1}{\sqrt[3]{x}} \cdot \dfrac{\sqrt[3]{x^2}}{\sqrt[3]{x^2}} = \dfrac{\sqrt[3]{x^2}}{\sqrt[3]{x^3}} = \dfrac{\sqrt[3]{x^2}}{x}$.

105. $\dfrac{2\sqrt{x}}{3} \cdot \dfrac{\sqrt{x}}{\sqrt{x}} = \dfrac{2x}{3\sqrt{x}}$.

107. $\sqrt{\dfrac{2y}{x}} = \dfrac{\sqrt{2y}}{\sqrt{x}} \cdot \dfrac{\sqrt{2y}}{\sqrt{2y}} = \dfrac{2y}{\sqrt{2xy}}$.

109. $\dfrac{\sqrt[3]{x^2z}}{y} \cdot \dfrac{\sqrt[3]{xz^2}}{\sqrt[3]{xz^2}} = \dfrac{\sqrt[3]{x^3z^3}}{y\sqrt[3]{xz^2}} = \dfrac{xz}{y\sqrt[3]{xz^2}}$.

111. If the car is driven in the city, then it can be expected to cover

$(18.1)(20) = 362$ \hspace{1cm} (miles/gal · gal)

or 362 miles on a full tank. If the car is driven on the highway, then it can be expected to cover

$(18.1)(27) = 488.7$ \hspace{1cm} (miles/gal · gal)

or 488.7 miles on a full tank. Thus, the driving range of the car may be described by the interval [362, 488.7].

113. \hspace{1cm} $6(P - 2500) \leq 4(P + 2400)$

$6P - 15000 \leq 4P + 9600$

$2P \leq 24600$, or $P \leq 12300$.

Therefore, the maximum profit is $12,300.

115. Let x represent the salesman's monthly sales in dollars. Then
$$0.15(x - 12000) \geq 3000$$
$$15(x - 12000) \geq 300000$$
$$15x - 180000 \geq 300000$$
$$15x \geq 480000$$
$$x \geq 32000.$$
We conclude that the salesman must attain sales of at least $32,000 to reach his goal.

117. The rod is acceptable if $0.49 < x < 0.51$ or $-0.01 < x - 0.5 < 0.01$. This gives the required inequality $|x - 0.5| \leq 0.01$.

119. We want to solve the inequality
$$-6x^2 + 30x - 10 \geq 14. \qquad \text{(Remember } x \text{ is expressed in thousands.)}$$
Adding -14 to both sides of this inequality, we have
$$-6x^2 + 30x - 10 - 14 \geq 14 - 14,$$
or $\qquad -6x^2 + 30x - 24 \geq 0.$

Dividing both sides of the inequality by -6 (which reverses the sign of the inequality), we have
$$x^2 - 5x + 4 \leq 0.$$
Factoring this last expression, we have
$$(x - 4)(x - 1) \leq 0.$$
From the following sign diagram,

Sign of $(x - 4)$ - - - - - - - - - - - - - - - -0+ + + + + + + + +

Sign of $(x - 1)$ - - - - - -0 + + + + + + + + + + + + + + + +

$$\xrightarrow{\qquad\quad}\; x$$
$$\quad 0 \quad 1 \qquad\qquad 4$$

we see that x must lie between 1 and 4. (The inequality is only satisfied when the two factors have opposite signs.) Since x is expressed in thousands of units, we see that the manufacturer must produce between 1000 and 4000 units of the commodity.

121. False. Take $a = 1, b = 2,$ and $c = 3$. Then $a < b$, but
$$a - c = 1 - 3 = -2 \ngtr 2 - 3 = -1 = b - c.$$

123. True. $|a - b| = |a + (-b)| \le |a| + |-b| = |a| + |b|$.

EXERCISES 1.2, page 26

1. $(7x^2 - 2x + 5) + (2x^2 + 5x - 4) = 7x^2 - 2x + 5 + 2x^2 + 5x - 4$
    $$= 9x^2 + 3x + 1.$$

3. $(5y^2 - 2y + 1) - (y^2 - 3y - 7) = 5y^2 - 2y + 1 - y^2 + 3y + 7$
    $$= 4y^2 + y + 8.$$

5. $x - \{2x - [-x - (1 - x)]\} = x - \{2x - [-x - 1 + x]\}$
    $$= x - \{2x + 1\}$$
    $$= x - 2x - 1$$
    $$= -x - 1.$$

7. $(\dfrac{1}{3} - 1 + e) - (-\dfrac{1}{3} - 1 + e^{-1}) = \dfrac{1}{3} - 1 + e + \dfrac{1}{3} + 1 - \dfrac{1}{e}$
    $$= \dfrac{2}{3} + e - \dfrac{1}{e}$$
    $$= \dfrac{3e^2 + 2e - 3}{3e}.$$

9. $3\sqrt{8} + 8 - 2\sqrt{y} + \dfrac{1}{2}\sqrt{x} - \dfrac{3}{4}\sqrt{y} = 3\sqrt{4 \cdot 2} + 8 + \dfrac{1}{2}\sqrt{x} - \dfrac{11}{4}\sqrt{y}$
    $$= 6\sqrt{2} + 8 + \dfrac{1}{2}\sqrt{x} - \dfrac{11}{4}\sqrt{y}.$$

11. $(x + 8)(x - 2) = x(x - 2) + 8(x - 2) = x^2 - 2x + 8x - 16 = x^2 + 6x - 16.$

13. $(a + 5)^2 = (a + 5)(a + 5) = a(a + 5) + 5(a + 5) = a^2 + 5a + 5a + 25$
    $$= a^2 + 10a + 25.$$

15. $(x + 2y)^2 = (x + 2y)(x + 2y) = x(x + 2y) + 2y(x + 2y)$
    $$= x^2 + 2xy + 2yx + 4y^2 = x^2 + 4xy + 4y^2.$$

1 Preliminaries

17. $(2x + y)(2x - y) = 2x(2x - y) + y(2x - y) = 4x^2 - 2xy + 2xy - y^2$
$$= 4x^2 - y^2.$$

19. $(x^2 - 1)(2x) - x^2(2x) = 2x^3 - 2x - 2x^3 = -2x.$

21. $2\left(t + \sqrt{t}\right)^2 - 2t^2 = 2(t + \sqrt{t})(t + \sqrt{t}) - 2t^2$
$$= 2(t^2 + 2t\sqrt{t} + t) - 2t^2$$
$$= 2t^2 + 4t\sqrt{t} + 2t - 2t^2$$
$$= 4t\sqrt{t} + 2t = 2t(2\sqrt{t} + 1).$$

23. $4x^5 - 12x^4 - 6x^3 = 2x^3(2x^2 - 6x - 3).$

25. $7a^4 - 42a^2b^2 + 49a^3b = 7a^2(a^2 - 6b^2 + 7ab).$

27. $e^{-x} - xe^{-x} = e^{-x}(1 - x).$

29. $2x^{-5/2} - \frac{3}{2}x^{-3/2} = \frac{1}{2}x^{-5/2}(4 - 3x).$

31. $6ac + 3bc - 4ad - 2bd = 3c(2a + b) - 2d(2a + b) = (2a + b)(3c - 2d).$

33. $4a^2 - b^2 = (2a + b)(2a - b).$ (Difference of two squares)

35. $10 - 14x - 12x^2 = -2(6x^2 + 7x - 5) = -2(3x + 5)(2x - 1).$

37. $3x^2 - 6x - 24 = 3(x^2 - 2x - 8) = 3(x - 4)(x + 2).$

39. $12x^2 - 2x - 30 = 2(6x^2 - x - 15) = 2(3x - 5)(2x + 3).$

41. $9x^2 - 16y^2 = (3x)^2 - (4y)^2 = (3x - 4y)(3x + 4y).$

43. $x^6 + 125 = (x^2)^3 + (5)^3 = (x^2 + 5)(x^4 - 5x^2 + 25).$

45. $(x^2 + y^2)x - xy(2y) = x^3 + xy^2 - 2xy^2 = x^3 - xy^2.$

47. $2(x - 1)(2x + 2)^3[4(x - 1) + (2x + 2)]$

$$= 2(x - 1)(2x + 2)^3[4x - 4 + 2x + 2]$$
$$= 2(x - 1)(2x + 2)^3[6x - 2]$$
$$= 4(x - 1)(3x - 1)(2x + 2)^3.$$

49. $4(x - 1)^2(2x + 2)^3(2) + (2x + 2)^4(2)(x - 1)$

$$= 2(x - 1)(2x + 2)^3[4(x - 1) + (2x + 2)]$$
$$= 2(x - 1)(2x + 2)^3(6x - 2)$$
$$= 4(x - 1)(3x - 1)(2x + 2)^3.$$

51. $(x^2 + 2)^2[5(x^2 + 2)^2 - 3](2x) = (x^2 + 2)^2[5(x^4 + 4x^2 + 4) - 3](2x)$
$$= (2x)(x^2 + 2)^2(5x^4 + 20x^2 + 17).$$

53. $x^2 + x - 12 = 0$, or $(x + 4)(x - 3) = 0$, so that $x = -4$ or $x = 3$. We conclude that the roots are $x = -4$ and $x = 3$.

55. $4t^2 + 2t - 2 = (2t - 1)(2t + 2) = 0$. Thus, $t = 1/2$ and $t = -1$ are the roots.

57. $\frac{1}{4}x^2 - x + 1 = (\frac{1}{2}x - 1)(\frac{1}{2}x - 1) = 0$. Thus $\frac{1}{2}x = 1$, and $x = 2$ is a double root of the equation.

59. Here we use the quadratic formula to solve the equation $4x^2 + 5x - 6 = 0$. Then, $a = 4$, $b = 5$, and $c = -6$. Therefore,

$$x = \frac{-b \pm \sqrt{b^2 - 4ac}}{2a} = \frac{-(5) \pm \sqrt{(5)^2 - 4(4)(-6)}}{2(4)} = \frac{-5 \pm \sqrt{121}}{8}$$
$$= \frac{-5 \pm 11}{8}.$$

Thus, $x = -\frac{16}{8} = -2$ and $x = \frac{6}{8} = \frac{3}{4}$ are the roots of the equation.

61. We use the quadratic formula to solve the equation $8x^2 - 8x - 3 = 0$. Here $a = 8$, $b = -8$, and $c = -3$. Therefore,

$$x = \frac{-b \pm \sqrt{b^2 - 4ac}}{2a} = \frac{-(-8) \pm \sqrt{(-8)^2 - 4(8)(-3)}}{2(8)} = \frac{8 \pm \sqrt{160}}{16}$$

$$= \frac{8 \pm 4\sqrt{10}}{16} = \frac{2 \pm \sqrt{10}}{4}.$$

Thus, $x = \frac{1}{2} + \frac{1}{4}\sqrt{10}$ and $x = \frac{1}{2} - \frac{1}{4}\sqrt{10}$ are the roots of the equation.

63. We use the quadratic formula to solve $2x^2 + 4x - 3 = 0$. Here, $a = 2$, $b = 4$, and $c = -3$. Therefore

$$x = \frac{-b \pm \sqrt{b^2 - 4ac}}{2a} = \frac{-(4) \pm \sqrt{(4)^2 - 4(2)(-3)}}{2(2)} = \frac{-4 \pm \sqrt{40}}{4}$$

$$= \frac{-4 \pm 2\sqrt{10}}{4} = \frac{-2 \pm \sqrt{10}}{2}.$$

Thus, $x = -1 + \frac{1}{2}\sqrt{10}$ and $x = -1 - \frac{1}{2}\sqrt{10}$ are the roots of the equation.

65. $\dfrac{x^2 + x - 2}{x^2 - 4} = \dfrac{(x+2)(x-1)}{(x+2)(x-2)} = \dfrac{x-1}{x-2}.$

67. $\dfrac{12t^2 + 12t + 3}{4t^2 - 1} = \dfrac{3(4t^2 + 4t + 1)}{4t^2 - 1} = \dfrac{3(2t+1)(2t+1)}{(2t+1)(2t-1)} = \dfrac{3(2t+1)}{2t-1}.$

69. $\dfrac{(4x-1)(3) - (3x+1)(4)}{(4x-1)^2} = \dfrac{12x - 3 - 12x - 4}{(4x-1)^2} = -\dfrac{7}{(4x-1)^2}.$

71. $\dfrac{2a^2 - 2b^2}{b-a} \cdot \dfrac{4a+4b}{a^2 + 2ab + b^2} = \dfrac{2(a+b)(a-b)4(a+b)}{-(a-b)(a+b)(a+b)} = -8.$

73. $\dfrac{3x^2 + 2x - 1}{2x+6} \div \dfrac{x^2-1}{x^2+2x-3} = \dfrac{(3x-1)(x+1)}{2(x+3)} \cdot \dfrac{(x+3)(x-1)}{(x+1)(x-1)} = \dfrac{3x-1}{2}.$

75. $\dfrac{58}{3(3t+2)} + \dfrac{1}{3} = \dfrac{58 + 3t + 2}{3(3t+2)} = \dfrac{3t + 60}{3(3t+2)} = \dfrac{t+20}{3t+2}.$

77. $\dfrac{2x}{2x-1} - \dfrac{3x}{2x+5} = \dfrac{2x(2x+5) - 3x(2x-1)}{(2x-1)(2x+5)} = \dfrac{4x^2 + 10x - 6x^2 + 3x}{(2x-1)(2x+5)}$

$$= \frac{-2x^2 + 13x}{(2x-1)(2x+5)} = -\frac{x(2x-13)}{(2x-1)(2x+5)}.$$

79. $\dfrac{4}{x^2-9} - \dfrac{5}{x^2-6x+9} = \dfrac{4}{(x+3)(x-3)} - \dfrac{5}{(x-3)^2}$

$$= \frac{4(x-3)-5(x+3)}{(x-3)^2(x+3)} = -\frac{x+27}{(x-3)^2(x+3)}.$$

81. $\dfrac{1+\dfrac{1}{x}}{1-\dfrac{1}{x}} = \dfrac{\dfrac{x+1}{x}}{\dfrac{x-1}{x}} = \dfrac{x+1}{x} \cdot \dfrac{x}{x-1} = \dfrac{x+1}{x-1}.$

83. $\dfrac{4x^2}{2\sqrt{2x^2+7}} + \sqrt{2x^2+7} = \dfrac{4x^2 + 2\sqrt{2x^2+7}\sqrt{2x^2+7}}{2\sqrt{2x^2+7}} = \dfrac{4x^2+4x^2+14}{2\sqrt{2x^2+7}}$

$$= \frac{4x^2+7}{\sqrt{2x^2+7}}.$$

85. $\dfrac{2x(x+1)^{-1/2} - (x+1)^{1/2}}{x^2} = \dfrac{(x+1)^{-1/2}(2x-x-1)}{x^2} = \dfrac{(x+1)^{-1/2}(x-1)}{x^2}$

$$= \frac{x-1}{x^2\sqrt{x+1}}.$$

87. $\dfrac{(2x+1)^{1/2} - (x+2)(2x+1)^{-1/2}}{2x+1} = \dfrac{(2x+1)^{-1/2}(2x+1-x-2)}{2x+1}$

$$= \frac{(2x+1)^{-1/2}(x-1)}{2x+1} = \frac{x-1}{(2x+1)^{3/2}}.$$

89. $\dfrac{1}{\sqrt{3}-1} \cdot \dfrac{\sqrt{3}+1}{\sqrt{3}+1} = \dfrac{\sqrt{3}+1}{3-1} = \dfrac{\sqrt{3}+1}{2}.$

91. $\dfrac{1}{\sqrt{x}-\sqrt{y}}\cdot\dfrac{\sqrt{x}+\sqrt{y}}{\sqrt{x}+\sqrt{y}}=\dfrac{\sqrt{x}+\sqrt{y}}{x-y}.$

93. $\dfrac{\sqrt{a}+\sqrt{b}}{\sqrt{a}-\sqrt{b}}\cdot\dfrac{\sqrt{a}+\sqrt{b}}{\sqrt{a}+\sqrt{b}}=\dfrac{(\sqrt{a}+\sqrt{b})^{2}}{a-b}.$

95. $\dfrac{\sqrt{x}}{3}\cdot\dfrac{\sqrt{x}}{\sqrt{x}}=\dfrac{x}{3\sqrt{x}}.$

97. $\dfrac{1-\sqrt{3}}{3}\cdot\dfrac{1+\sqrt{3}}{1+\sqrt{3}}=\dfrac{1^{2}-(\sqrt{3})^{2}}{3(1+\sqrt{3})}=-\dfrac{2}{3(1+\sqrt{3})}.$

99. $\dfrac{1+\sqrt{x+2}}{\sqrt{x+2}}\cdot\dfrac{1-\sqrt{x+2}}{1-\sqrt{x+2}}=\dfrac{1-(x+2)}{\sqrt{x+2}(1-\sqrt{x+2})}=-\dfrac{x+1}{\sqrt{x+2}(1-\sqrt{x+2})}.$

101. True. The two real roots are $\dfrac{-b\pm\sqrt{b^{2}-4ac}}{2a}.$

103. False. Take $a=2,\ b=3,\ $ and $\ c=4.$ Then

$\dfrac{a}{b+c}=\dfrac{2}{3+4}=\dfrac{2}{7}.$ But $\dfrac{a}{b}+\dfrac{a}{c}=\dfrac{2}{3}+\dfrac{3}{4}=\dfrac{8+9}{12}=\dfrac{17}{12}.$

EXERCISES 1.3, page 34

1. The coordinates of A are (3,3) and it is located in Quadrant I.

3. The coordinates of C are (2,-2) and it is located in Quadrant IV.

5. The coordinates of E are (-4,-6) and it is located in Quadrant III.

7. A   9.   E, F, and G.   11. F

For Exercises 13-19, refer to the following figure.

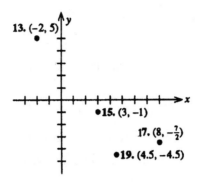

13. (-2, 5)

15. (3, -1)

17. $(8, -\frac{7}{2})$

19. (4.5, -4.5)

21. Using the distance formula, we find that $\sqrt{(4-1)^2 + (7-3)^2} = \sqrt{3^2 + 4^2} = \sqrt{25} = 5$.

23. Using the distance formula, we find that
$$\sqrt{(4-(-1))^2 + (9-3)^2} = \sqrt{5^2 + 6^2} = \sqrt{25+36} = \sqrt{61}.$$

25. The coordinates of the points have the form $(x, -6)$. Since the points are 10 units away from the origin, we have
$$(x - 0)^2 + (-6 - 0)^2 = 10^2$$
$$x^2 = 64,$$
or $x = \pm 8$. Therefore, the required points are $(-8, -6)$ and $(8, -6)$.

27. The points are shown in the diagram that follows.

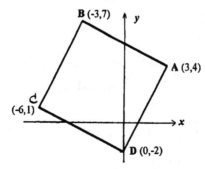

B (-3,7)

A (3,4)

(-6,1)

D (0,-2)

To show that the four sides are equal, we compute the following:
$$d(A,B) = \sqrt{(-3-3)^2 + (7-4)^2} = \sqrt{(-6)^2 + 3^2} = \sqrt{45}$$

$$d(B,C) = \sqrt{[(-6-(-3)]^2 + (1-7)^2} = \sqrt{(-3)^2 + (-6)^2} = \sqrt{45}$$
$$d(C,D) = \sqrt{[0-(-6)]^2 + [(-2)-1]^2} = \sqrt{(6)^2 + (-3)^2} = \sqrt{45}$$
$$d(A,D) = \sqrt{(0-3)^2 + (-2-4)^2} = \sqrt{(3)^2 + (-6)^2} = \sqrt{45}.$$

Next, to show that $\triangle ABC$ is a right triangle, we show that it satisfies the Pythagorean Theorem. Thus,

$$d(A,C) = \sqrt{(-6-3)^2 + (1-4)^2} = \sqrt{(-9)^2 + (-3)^2} = \sqrt{90} = 3\sqrt{10}$$

and $[d(A,B)]^2 + [d(B,C)]^2 = 90 = [d(A,C)]^2$. Similarly, $d(B,D) = \sqrt{90} = 3\sqrt{10}$, so $\triangle BAD$ is a right triangle as well. It follows that $\angle B$ and $\angle D$ are right angles, and we conclude that $ADCB$ is a square

29. The equation of the circle with radius 5 and center (2,-3) is given by
$$(x-2)^2 + [y-(-3)]^2 = 5^2$$
or $\quad (x-2)^2 + (y+3)^2 = 25.$

31. The equation of the circle with radius 5 and center (0, 0) is given by
$$(x-0)^2 + (y-0)^2 = 5^2$$
or $\quad x^2 + y^2 = 25$

33. The distance between the points (5,2) and (2,-3) is given by
$$d = \sqrt{(5-2)^2 + (2-(-3))^2} = \sqrt{3^2 + 5^2} = \sqrt{34}.$$
Therefore $r = \sqrt{34}$ and the equation of the circle passing through (5,2) and (2,-3) is
$$(x-2)^2 + [y-(-3)]^2 = 34$$
or $\quad (x-2)^2 + (y+3)^2 = 34.$

35. Referring to the diagram on page 35 of the text, we see that the distance from A to B is given by $d(A,B) = \sqrt{400^2 + 300^2} = \sqrt{250,000} = 500.$ The distance from B to C is given by
$$d(B,C) = \sqrt{(-800-400)^2 + (800-300)^2} = \sqrt{(-1200)^2 + (500)^2}$$
$$= \sqrt{1,690,000} = 1300.$$

The distance from C to D is given by
$$d(C,D) = \sqrt{[-800-(-800)]^2 + (800-0)^2} = \sqrt{0 + 800^2} = 800.$$

The distance from D to A is given by

$$d(D, A) = \sqrt{[(-800) - 0]^2 + (0 - 0)} = \sqrt{640000} = 800.$$

Therefore, the total distance covered on the tour, is

$$d(A, B) + d(B, C) + d(C, D) + d(D, A) = 500 + 1300 + 800 + 800$$
$$= 3400, \quad \text{or } 3400 \text{ miles.}$$

37. Referring to the following diagram,

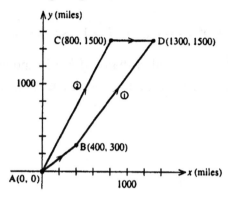

we see that the distance he would cover if he took Route (1) is given by

$$d(A, B) + d(B, D) = \sqrt{400^2 + 300^2} + \sqrt{(1300 - 400)^2 + (1500 - 300)^2}$$
$$= \sqrt{250,000} + \sqrt{2,250,000} = 500 + 1500 = 2000,$$

or 2000 miles. On the other hand, the distance he would cover if he took Route (2) is given by

$$d(A, C) + d(C, D) = \sqrt{800^2 + 1500^2} + \sqrt{(1300 - 800)^2}$$
$$= \sqrt{2,890,000} + \sqrt{250,000} = 1700 + 500 = 2200,$$

or 2200 miles. Comparing these results, we see that he should take Route (1).

39. Calculations to determine VHF requirements:

$$d = \sqrt{25^2 + 35^2} = \sqrt{625 + 1225} = \sqrt{1850} \approx 43.01.$$

Models B through D satisfy this requirement.

Calculations to determine UHF requirements:

$$d = \sqrt{20^2 + 32^2} = \sqrt{400 + 1024} = \sqrt{1424} = 37.74$$

Models C through D satisfy this requirement. Therefore, Model C will allow him to receive both channels at the least cost.

41. a. Let the position of ship A and Ship B after t hours be $A(0, y)$ and $B(x, y)$, respectively. Then $x = 30t$ and $y = 20t$. Therefore, the distance between the two ships is $D = \sqrt{(30t)^2 + (20t)^2} = \sqrt{900t^2 + 400t^2} = 10\sqrt{13}t$.

   b. The required distance is obtained by letting $t = 2$ giving $D = 10\sqrt{13}(2)$ or approximately 72.11 miles.

43. True. Plot the points.

45. True. $kx^2 + ky^2 = a^2$; $x^2 + y^2 = \dfrac{a^2}{k} < a^2$ if $k > 1$. So the radius of the circle with equation (1) is a circle of radius smaller than a if $k > 1$ (and centered at the origin). Therefore, it lies inside the circle of radius a with equation $x^2 + y^2 = a^2$.

EXERCISES 1.4, page 48

1. e   3. a   5. f

7. Referring to the figure shown in the text, we see that $m = \dfrac{2-0}{0-(-4)} = \dfrac{1}{2}$.

9. This is a vertical line, and hence its slope is undefined.

11. $m = \dfrac{y_2 - y_1}{x_2 - x_1} = \dfrac{8-3}{5-4} = 5.$ 13. $m = \dfrac{y_2 - y_1}{x_2 - x_1} = \dfrac{8-3}{4-(-2)} = \dfrac{5}{6}.$

15. $m = \dfrac{y_2 - y_1}{x_2 - x_1} = \dfrac{d-b}{c-a}.$

17. Since the equation is in the slope-intercept form, we read off the slope $m = 4$.
   a. If x increases by 1 unit, then y increases by 4 units.
   b. If x decreases by 2 units, y decreases by $4(-2) = -8$ units.

19. The slope of the line through A and B is $\dfrac{-10-(-2)}{-3-1} = \dfrac{-8}{-4} = 2$.

   The slope of the line through C and D is $\dfrac{1-5}{-1-1} = \dfrac{-4}{-2} = 2$.

Since the slopes of these two lines are equal, the lines are parallel.

21. The slope of the line through A and B is $\dfrac{2-5}{4-(-2)} = -\dfrac{3}{6} = -\dfrac{1}{2}$.

    The slope of the line through C and D is $\dfrac{6-(-2)}{3-(-1)} = \dfrac{8}{4} = 2$. Since the slopes of these two lines are the negative reciprocals of each other, the lines are perpendicular.

23. The slope of the line through the point $(1, a)$ and $(4,-2)$ is $m_1 = \dfrac{-2-a}{4-1}$ and the

    slope of the line through $(2,8)$ and $(-7, a+4)$ is $m_2 = \dfrac{a+4-8}{-7-2}$. Since these two

    lines are parallel, m_1 is equal to m_2. Therefore,
    $$\frac{-2-a}{3} = \frac{a-4}{-9}$$
    $$-9(-2-a) = 3(a-4)$$
    $$18+9a = 3a-12$$
    $$6a = -30 \qquad \text{and} \quad a = -5$$

25. An equation of a horizontal line is of the form $y = b$. In this case $b = -3$, so $y = -3$ is an equation of the line.

27. We use the point-slope form of an equation of a line with the point $(3,-4)$ and slope $m = 2$. Thus $\qquad y - y_1 = m(x - x_1)$,
    and $\qquad\qquad\qquad y - (-4) = 2(x - 3)$
    $$y + 4 = 2x - 6$$
    $$y = 2x - 10.$$

29. Since the slope $m = 0$, we know that the line is a horizontal line of the form $y = b$. Since the line passes through $(-3,2)$, we see that $b = 2$, and an equation of the line is $y = 2$.

31. We first compute the slope of the line joining the points $(2,4)$ and $(3,7)$. Thus,
    $$m = \frac{7-4}{3-2} = 3.$$
    Using the point-slope form of an equation of a line with the point $(2,4)$ and slope $m = 3$, we find

$$y - 4 = 3(x - 2)$$
$$y = 3x - 2.$$

33. We first compute the slope of the line joining the points (1,2) and (−3,−2). Thus,
$$m = \frac{-2-2}{-3-1} = \frac{-4}{-4} = 1.$$
Using the point-slope form of an equation of a line with the point (1,2) and slope $m = 1$, we find
$$y - 2 = x - 1$$
$$y = x + 1.$$

35. We use the slope-intercept form of an equation of a line: $y = mx + b$. Since $m = 3$, and $b = 4$, the equation is $y = 3x + 4$.

37. We use the slope-intercept form of an equation of a line: $y = mx + b$. Since $m = 0$, and $b = 5$, the equation is $y = 5$.

39. We first write the given equation in the slope-intercept form:
$$x - 2y = 0$$
$$-2y = -x$$
$$y = \tfrac{1}{2}x \ .$$
From this equation, we see that $m = 1/2$ and $b = 0$.

41. We write the equation in slope-intercept form:
$$2x - 3y - 9 = 0$$
$$-3y = -2x + 9$$
$$y = \tfrac{2}{3}x - 3.$$
From this equation, we see that $m = 2/3$ and $b = -3$.

43. We write the equation in slope-intercept form:
$$2x + 4y = 14$$
$$4y = -2x + 14$$
$$y = -\tfrac{2}{4}x + \tfrac{14}{4}$$
$$= -\tfrac{1}{2}x + \tfrac{7}{2}.$$
From this equation, we see that $m = -1/2$ and $b = 7/2$.

45. We first write the equation $2x - 4y - 8 = 0$ in slope- intercept form:
$$2x - 4y - 8 = 0$$
$$4y = 2x - 8$$
$$y = \tfrac{1}{2}x - 2$$
Now the required line is parallel to this line, and hence has the same slope. Using the point-slope equation of a line with $m = 1/2$ and the point $(-2,2)$, we have
$$y - 2 = \tfrac{1}{2}[x - (-2)]$$
$$y = \tfrac{1}{2}x + 3.$$

47. A line parallel to the x-axis has slope 0 and is of the form $y = b$. Since the line is 6 units below the axis, it passes through $(0,-6)$ and its equation is $y = -6$.

49. We use the point-slope form of an equation of a line to obtain
$$y - b = 0(x - a) \quad \text{or} \quad y = b.$$

51. Since the required line is parallel to the line joining $(-3,2)$ and $(6,8)$, it has slope
$$m = \frac{8 - 2}{6 - (-3)} = \frac{6}{9} = \frac{2}{3}.$$
We also know that the required line passes through $(-5,-4)$. Using the point-slope form of an equation of a line, we find
$$y - (-4) = \frac{2}{3}(x - (-5))$$
or $\qquad y = \tfrac{2}{3}x + \tfrac{10}{3} - 4$

that is $\qquad y = \tfrac{2}{3}x - \tfrac{2}{3}$.

53. Since the point $(-3,5)$ lies on the line $kx + 3y + 9 = 0$, it satisfies the equation. Substituting $x = -3$ and $y = 5$ into the equation gives
$$-3k + 15 + 9 = 0$$
or $\qquad k = 8.$

1 Preliminaries

55. $3x - 2y + 6 = 0$

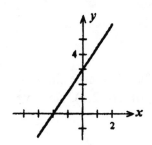

57. $x + 2y - 4 = 0$

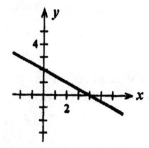

59. $y + 5 = 0$

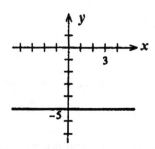

61. Since the line passes through the points $(a, 0)$ and $(0, b)$, its slope is

$m = \dfrac{b-0}{0-a} = -\dfrac{b}{a}$. Then, using the point-slope form of an equation of a line with the

point $(a, 0)$ we have

$$y - 0 = -\tfrac{b}{a}(x - a)$$
$$y = -\tfrac{b}{a}x + b$$

which may be written in the form

$$\tfrac{b}{a}x + y = b.$$

Multiplying this last equation by $1/b$, we have

$$\frac{x}{a} + \frac{y}{b} = 1.$$

63. Using the equation $\dfrac{x}{a} + \dfrac{y}{b} = 1$ with $a = -2$ and $b = -4$, we have $-\dfrac{x}{2} - \dfrac{y}{4} = 1$.

Then

$$-4x - 2y = 8$$
$$2y = -8 - 4x$$
$$y = -2x - 4.$$

65. Using the equation $\dfrac{x}{a} + \dfrac{y}{b} = 1$ with $a = 4$ and $b = -1/2$, we have

$$\dfrac{x}{4} + \dfrac{y}{-\frac{1}{2}} = 1$$
$$-\tfrac{1}{4}x + 2y = -1$$
$$2y = \tfrac{1}{4}x - 1$$
$$y = \tfrac{1}{8}x - \tfrac{1}{2}.$$

67. The slope of the line passing through A and B is $m = \dfrac{7-1}{1-(-2)} = \dfrac{6}{3} = 2$,

and the slope of the line passing through B and C is $m = \dfrac{13-7}{4-1} = \dfrac{6}{3} = 2$.

Since the slopes are equal, the points lie on the same line.

69. a. $y = 0.55x$

b. Solving the equation $1100 = 0.55x$ for x, we have $x = \dfrac{1100}{0.55} = 2000$.

71. Using the points $(0, 0.68)$ and $(10, 0.80)$, we see that the slope of the required line
is $\qquad\qquad m = \dfrac{0.80 - 0.68}{10 - 0} = \dfrac{0.12}{10} = .012.$

Next, using the point-slope form of the equation of a line, we have
$$y - 0.68 = 0.012(t - 0)$$
or $\qquad\qquad y = 0.012t + 0.68.$
Therefore, when $t = 12$, we have
$$y = 0.012(12) + 0.68 \ = 0.824 \text{ or } 82.4\%.$$
That is, in 2002 women's wages were, expected to be 82.4% of men's wages.

73. a. – b. (See the graph on the following page.)

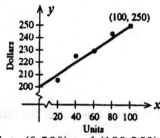

c. Using the points $(0,200)$ and $(100,250)$, we see that the slope of the required line

is $m = \dfrac{250-200}{100} = \dfrac{1}{2}$. Therefore, the required equation is

$$y - 200 = \tfrac{1}{2}x \quad \text{or} \quad y = \tfrac{1}{2}x + 200.$$

d. The approximate cost for producing 54 units of the commodity is
$\tfrac{1}{2}(54) + 200,$ or $227.

75. a. – b.

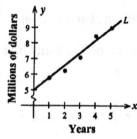

c. The slope of L is $m = \dfrac{9.0 - 5.8}{5-1} = \dfrac{3.2}{4} = 0.8$ .Using the point-slope form of an

equation of a line, we have $y - 5.8 = 0.8(x-1) = 0.8x - 0.8$, or $y = 0.8x + 5$.

d. Using the equation of part (c) with $x = 9$, we have
$$y = 0.8(9) + 5 = 12.2, \quad \text{or } 12.2 \text{ million.}$$

77. True. The slope of the line $Ax + By + C = 0$ is $-B/A$. (Write it in the slope-intercept form.) Similarly, the slope of the line $ax + by + c = 0$ is $-b/a$. They are parallel if and only if $-\dfrac{B}{A} = -\dfrac{b}{a}$, $Ab = aB$, or $Ab - aB = 0$.

79. True. the slope of the line $ax + by + c_1 = 0$ is $m_1 = -a/b$. The slope of the line $bx - ay + c_2 = 0$ is $m_2 = b/a$. since $m_1 m_2 = -1$, the straight lines are indeed

perpendicular.

81. Yes. A straight line with slope zero ($m = 0$) is a horizontal line, whereas a straight line whose slope does not exist is a vertical line (m cannot be computed).

83. The slope of L_1 is $m_1 = \dfrac{b-0}{1-0} = b$. The slope of L_2 is $m_2 = \dfrac{c-0}{1-0} = c$.

Applying the Pythagorean theorem to $\triangle OAC$ and $\triangle OCB$ gives
$$(OA)^2 = 1^2 + b^2 \quad \text{and} \quad (OB)^2 = 1^2 + c^2.$$
Adding these equations and applying the Pythagorean theorem to $\triangle OBA$ gives

$$(AB)^2 = (OA)^2 + (OB)^2 = 1^2 + b^2 + 1^2 + c^2 = 2 + b^2 + c^2.$$

Also $\qquad\qquad (AB)^2 = (b - c)^2.$

Therefore, $\qquad (b - c)^2 = 2 + b^2 + c^2$
$$b^2 - 2bc + c^2 = 2 + b^2 + c^2$$
$$-2bc = 2,\ 1 = -bc.$$

Next, $\qquad\qquad m_1 m_2 = b \cdot c = bc = -1,$ as was to be shown.

CHAPTER 1, REVIEW EXERCISES, page 56

1. Adding x to both sides yields $3 \le 3x + 9$ or $3x \ge -6$, and $x \ge -2$.
   We conclude that the solution set is $[-2, \infty)$.

3. The inequalities imply $x > 5$ or $x < -4$. So the solution set is
   $(-\infty, -4) \cup (5, \infty)$.

5. $|-5+7| + |-2| = |2| + |-2| = 2 + 2 = 4.$ 7. $|2\pi - 6| - \pi = 2\pi - 6 - \pi = \pi - 6.$

9. $\left(\dfrac{9}{4}\right)^{3/2} = \dfrac{9^{3/2}}{4^{3/2}} = \dfrac{27}{8}.$ 11. $(3 \cdot 4)^{-2} = 12^{-2} = \dfrac{1}{12^2} = \dfrac{1}{144}.$

13. $\dfrac{(3 \cdot 2^{-3})(4 \cdot 3^5)}{2 \cdot 9^3} = \dfrac{3 \cdot 2^{-3} \cdot 2^2 \cdot 3^5}{2 \cdot (3^2)^3} = \dfrac{2^{-1} \cdot 3^6}{2 \cdot 3^6} = \dfrac{1}{4}.$

15. $\dfrac{4(x^2 + y)^3}{x^2 + y} = 4(x^2 + y)^2.$

17. $\dfrac{\sqrt[4]{16x^5yz}}{\sqrt[4]{81xyz^5}} = \dfrac{(2^4x^5yz)^{1/4}}{(3^4xyz^5)^{1/4}} = \dfrac{2x^{5/4}y^{1/4}z^{1/4}}{3x^{1/4}y^{1/4}z^{5/4}} = \dfrac{2x}{3z}.$

19. $\left(\dfrac{3xy^2}{4x^3y}\right)^{-2}\left(\dfrac{3xy^3}{2x^2}\right)^3 = \left(\dfrac{3y}{4x^2}\right)^{-2}\left(\dfrac{3y^3}{2x}\right)^3 = \left(\dfrac{4x^2}{3y}\right)^2\left(\dfrac{3y^3}{2x}\right)^3 = \dfrac{(16x^4)(27y^9)}{(9y^2)(8x^3)} = 6xy^7.$

21. $2v^3w + 2vw^3 + 2u^2vw = 2vw(v^2 + w^2 + u^2).$

23. $12t^3 - 6t^2 - 18t = 6t(2t^2 - t - 3) = 6t(2t - 3)(t + 1).$

25. $-6x^2 - 10x + 4 = 0$, $3x^2 + 5x - 2 = (3x - 1)(x + 2) = 0$ and so $x = -2$ and $x = 1/3$.

27. $2x^4 + x^2 = 1$. Let $y = x^2$ and we can write the equation as
$$2y^2 + y - 1 = (2y - 1)(y + 1) = 0$$
giving $y = 1/2$ or $y = -1$. We reject the second root since $y = x^2$ must be nonnegative. Therefore, $x^2 = 1/2$ or $x = \pm 1/\sqrt{2} = \pm\sqrt{2}/2$.

29. Here we use the quadratic formula to solve the equation $2x^2 + 8x + 7 = 0$. Then $a = 2$, $b = 8$, and $c = 7$. So
$$x = \dfrac{-b \pm \sqrt{b^2 - 4ac}}{2a} = \dfrac{-(8) \pm \sqrt{(8)^2 - 4(2)(7)}}{4} = \dfrac{-8 \pm 2\sqrt{2}}{4} = -2 \pm \tfrac{1}{2}\sqrt{2}.$$

31. $\dfrac{6x}{2(3x^2 + 2)} + \dfrac{1}{4(x + 2)} = \dfrac{(6x)(2)(x + 2) + (3x^2 + 2)}{4(3x^2 + 2)(x + 2)} = \dfrac{12x^2 + 24x + 3x^2 + 2}{4(3x^2 + 2)(x + 2)}$
$$= \dfrac{15x^2 + 24x + 2}{4(3x^2 + 2)(x + 2)}.$$

33. $\dfrac{-2x}{\sqrt{x+1}} + 4\sqrt{x+1} = \dfrac{-2x + 4(x + 1)}{\sqrt{x+1}} = \dfrac{2(x + 2)}{\sqrt{x+1}}.$

35. $\dfrac{\sqrt{x} - 1}{2\sqrt{x}} = \dfrac{\sqrt{x} - 1}{2\sqrt{x}} \cdot \dfrac{\sqrt{x}}{\sqrt{x}} = \dfrac{x - \sqrt{x}}{2x}.$

37. The distance is
$$d = \sqrt{(-1/2-1/2)^2 + (2\sqrt{3}-\sqrt{3})^2} = \sqrt{1+3} = \sqrt{4} = 2$$

39. An equation is $y = 4$.

41. The line passes through the points $(-2, 4)$ and $(3, 0)$. So its slope is
$m = (4 - 0)/(-2 - 3)$ or $m = -4/5$. An equation is
$$y - 0 = -\tfrac{4}{5}(x-3) \quad \text{or} \quad y = -\tfrac{4}{5}x + \tfrac{12}{5}.$$

43. Writing the given equation in the form $y = -\tfrac{4}{3}x + 2$, we see that the slope of the given line is $-4/3$. Therefore, the slope of the required line is 3/4 and an equation of the line is
$$y - 4 = \tfrac{3}{4}(x+2) \qquad \text{or} \qquad y = \tfrac{3}{4}x + \tfrac{11}{2}.$$

45. The slope of the line joining the points $(-3,4)$ and $(2,1)$ is $m = \dfrac{1-4}{2-(-3)} = -\dfrac{3}{5}$.

Using the point-slope form of the equation of a line with the point $(-1,3)$ and slope $-3/5$, we have
$$y - 3 = -\tfrac{3}{5}[x-(-1)]$$
$$y = -\tfrac{3}{5}(x+1) + 3$$
$$= -\tfrac{3}{5}x + \tfrac{12}{5}.$$

47. Setting $x = 0$ gives $y = -6$ as the y-intercept. Setting $y = 0$ gives $x = 8$ as the x-intercept. The graph of the equation $3x - 4y = 24$ is follows.

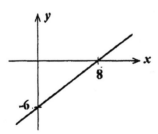

49. $2(1.5C + 80) \leq 2(2.5C - 20) \Rightarrow 1.5C + 80 \leq 2.5C - 20$, so $C \geq 100$ and the minimum cost is $100.

CHAPTER 2

EXERCISES 2.1, page 70

1. $f(x) = 5x + 6$. Therefore

$\quad f(3) = 5(3) + 6 = 21$

$\quad f(-3) = 5(-3) + 6 = -9$

$\quad f(a) = 5(a) + 6 = 5a + 6$

$\quad f(-a) = 5(-a) + 6 = -5a + 6$

$\quad f(a + 3) = 5(a + 3) + 6 = 5a + 15 + 6 = 5a + 21.$

3. $g(x) = 3x^2 - 6x - 3$

$\quad g(0) = 3(0) - 6(0) - 3 = -3$

$\quad g(-1) = 3(-1)^2 - 6(-1) - 3 = 3 + 6 - 3 = 6$

$\quad g(a) = 3(a)^2 - 6(a) - 3 = 3a^2 - 6a - 3$

$\quad g(-a) = 3(-a)^2 - 6(-a) - 3 = 3a^2 + 6a - 3$

$\quad g(x + 1) = 3(x + 1)^2 - 6(x + 1) - 3 = 3(x^2 + 2x + 1) - 6x - 6 - 3$

$\quad\quad\quad\quad\quad = 3x^2 + 6x + 3 - 6x - 9 = 3x^2 - 6.$

5. $s(t) = \dfrac{2t}{t^2 - 1}$. Therefore,

$\quad s(4) = \dfrac{2(4)}{(4)^2 - 1} = \dfrac{8}{15}.$

$\quad s(0) = \dfrac{2(0)}{0^2 - 1} = 0$

$\quad s(a) = \dfrac{2(a)}{a^2 - 1} = \dfrac{2a}{a^2 - 1}$

$\quad s(2 + a) = \dfrac{2(2 + a)}{(2 + a)^2 - 1} = \dfrac{2(2 + a)}{a^2 + 4a + 4 - 1} = \dfrac{2(2 + a)}{a^2 + 4a + 3}$

$$s(t+1) = \frac{2(t+1)}{(t+1)^2 - 1} = \frac{2(t+1)}{t^2 + 2t + 1 - 1} = \frac{2(t+1)}{t(t+2)}.$$

7. $f(t) = \dfrac{2t^2}{\sqrt{t-1}}$. Therefore, $f(2) = \dfrac{2(2^2)}{\sqrt{2-1}} = 8$

$$f(a) = \frac{2a^2}{\sqrt{a-1}}$$

$$f(x+1) = \frac{2(x+1)^2}{\sqrt{(x+1)-1}} = \frac{2(x+1)^2}{\sqrt{x}}$$

$$f(x-1) = \frac{2(x-1)^2}{\sqrt{(x-1)-1}} = \frac{2(x-1)^2}{\sqrt{x-2}}.$$

9. Since $x = -2 \le 0$, we see that $f(-2) = (-2)^2 + 1 = 4 + 1 = 5$

Since $x = 0 \le 0$, we see that $f(0) = (0)^2 + 1 = 1$

Since $x = 1 > 0$, we see that $f(1) = \sqrt{1} = 1$.

11. Since $x = -1 < 1$, $f(-1) = -\frac{1}{2}(-1)^2 + 3 = \frac{5}{2}$.

Since $x = 0 < 1$, $f(0) = -\frac{1}{2}(0)^2 + 3 = 3$.

Since $x = 1 \ge 1$, $f(1) = 2(1^2) + 1 = 3$.

Since $x = 2 \ge 1$, $f(2) = 2(2^2) + 1 = 9$.

13. a. $f(0) = -2$;
    b. (i) $f(x) = 3$ when $x \approx 2$ (ii) $f(x) = 0$ when $x = 1$
    c. $[0,6]$ d. $[-2, 6]$

15. $g(2) = \sqrt{2^2 - 1} = \sqrt{3}$ and the point $(2, \sqrt{3})$ lies on the graph of g.

17. $f(-2) = \dfrac{|-2-1|}{-2+1} = \dfrac{|-3|}{-1} = -3$ and the point $(-2,-3)$ does lie on the graph of f.

19. Since $f(x)$ is a real number for any value of x, the domain of f is $(-\infty, \infty)$.

21. $f(x)$ is not defined at $x = 0$ and so the domain of f is $(-\infty, 0) \cup (0, \infty)$.

23. $f(x)$ is a real number for all values of x. Note that $x^2 + 1 \geq 1$ for all x. Therefore, the domain of f is $(-\infty, \infty)$.

25. Since the square root of a number is defined for all real numbers greater than or equal to zero, we have
$$5 - x \geq 0, \text{ or } \qquad -x \geq -5$$
and $x \leq 5$. (Recall that multiplying by -1 reverses the sign of an inequality.) Therefore, the domain of g is $(-\infty, 5]$.

27. The denominator of f is zero when
$$x^2 - 1 = 0 \text{ or } x = \pm 1.$$
Therefore, the domain of f is $(-\infty, -1) \cup (-1, 1) \cup (1, \infty)$.

29. f is defined when $x + 3 \geq 0$, that is, when $x \geq -3$. Therefore, the domain of f is $[-3, \infty)$.

31. The numerator is defined when
$$1 - x \geq 0, \quad -x \geq -1 \quad \text{or} \quad x \leq 1.$$
Furthermore, the denominator is zero when $x = \pm 2$. Therefore, the domain is the set of all real numbers in $(-\infty, -2) \cup (-2, 1]$.

33. a. The domain of f is the set of all real numbers.
   b. $f(x) = x^2 - x - 6$. Therefore,
$$f(-3) = (-3)^2 - (-3) - 6 = 9 + 3 - 6 = 6;$$
$$f(-2) = (-2)^2 - (-2) - 6 = 4 + 2 - 6 = 0.$$
$$f(-1) = (-1)^2 - (-1) - 6 = 1 + 1 - 6 = -4; \quad f(0) = (0)^2 - (0) - 6 = -6.$$
$$f\left(\tfrac{1}{2}\right) = \left(\tfrac{1}{2}\right)^2 - \left(\tfrac{1}{2}\right) - 6 = \tfrac{1}{4} - \tfrac{2}{4} - \tfrac{24}{4} = -\tfrac{25}{4}; \quad f(1) = (1)^2 - 1 - 6 = -6.$$
$$f(2) = (2)^2 - 2 - 6 = 4 - 2 - 6 = -4; \quad f(3) = (3)^2 - 3 - 6 = 9 - 3 - 6 = 0.$$

c.

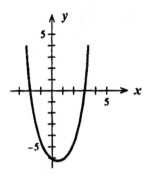

35.

x	-3	-2	-1	0	1	2	3
$f(x)$	19	9	3	1	3	9	19

$(-\infty, \infty)$; $[1, \infty)$

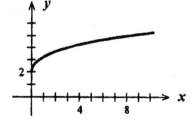

37.

x	0	1	2	4	9	16
$f(x)$	2	3	3.41	4	5	6

$[0, \infty)$; $[2, \infty)$

39.

x	0	-1	-3	-8	-15
$f(x)$	1	1.4	2	3	4

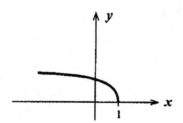

$(-\infty, 1]; [0, \infty)$

41.

x	-3	-2	-1	0	1	2	3
$f(x)$	2	1	0	-1	0	1	2

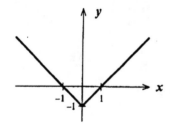

$(-\infty, \infty); [-1, \infty)$

43.

x	-3	-2	-1	0	1	2	3
$f(x)$	-3	-2	-1	1	3	5	7

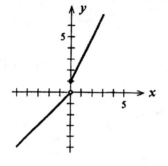

$(-\infty, \infty); (-\infty, 0) \cup [1, \infty)$

45. If $x \leq 1$, the graph of f is the half-line $y = -x + 1$. For $x > 1$, use the table

x	2	3	4
$f(x)$	3	8	15

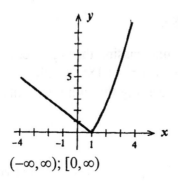

$(-\infty, \infty); [0, \infty)$

47. Each vertical line cuts the given graph at exactly one point, and so the graph represents y as a function of x.

49. Since there is a vertical line that intersects the graph at three points, the graph does not represent y as a function of x.

51. Each vertical line intersects the graph of f at exactly one point, and so the graph represents y as a function of x.

53. Each vertical line intersects the graph of f at exactly one point, and so the graph represents y as a function of x.

55. The circumference of a circle with a 5-inch radius is given by
$$C(5) = 2\pi(5) = 10\pi, \text{ or } 10\pi \text{ inches.}$$

57. $\frac{4}{3}(\pi)(2r)^3 = \frac{4}{3}\pi 8r^3 = 8(\frac{4}{3}\pi r^3)$. Therefore, the volume of the tumor is increased by a factor of 8.

59. a. From $t = 0$ to $t = 5$, the graph for cassettes lies above that for CDs so from 1985 to 1990, sales of prerecorded cassettes were greater than that of CDs.
b. Sales of prerecorded CDs were greater than that of prerecorded cassettes from 1990 on.

c. The graphs intersect at the point with coordinates $x = 5$ and $y \approx 3.5$, and this tells us that the sales of the two formats were the same in 1990 with the level of sales at approximately \$3.5 billion.

61. a. The slope of the straight line passing through the points $(0, 0.58)$ and $(20, 0.95)$ is

$$m = \frac{0.95 - 0.58}{20 - 0} = 0.0185,$$

and so an equation of the straight line passing through these two points is
$$y - 0.58 = 0.0185(t - 0) \quad \text{or} \quad y = 0.0185t + 0.58$$
Next, the slope of the straight line passing through the points $(20, 0.95)$ and $(30, 1.1)$ is

$$m = \frac{1.1 - 0.95}{30 - 20} = 0.015$$

and so an equation of the straight line passing through the two points is
$$y - 0.95 = 0.015(t - 20) \quad \text{or} \quad y = 0.015t + 0.65.$$
Therefore, the rule for f is

$$f(t) = \begin{cases} 0.0185t + 0.58 & 0 \le t \le 20 \\ 0.015t + 0.65 & 20 < t \le 30 \end{cases}$$

b. The ratios were changing at the rates of 0.0185/yr and 0.015/yr from 1960 through 1980, and from 1980 through 1990, respectively.
c. The ratio was 1 when $t \approx 20.3$. This shows that the number of bachelor's degrees earned by women equaled the number earned by men for the first time around 1983.

63. a. $T(x) = 0.06x$

b. $T(200) = 0.06(200) = 12$, or \$12.00.

$T(5.65) = 0.06(5.65) = 0.34$, or \$0.34.

65. The child should receive
$$D(4) = \tfrac{2}{25}(500)(4) = 160, \quad \text{or } 160 \text{ mg.}$$

67. a. The daily cost of leasing from Ace is
$$C_1(x) = 30 + 0.15x,$$
while the daily cost of leasing from Acme is
$$C_2(x) = 25 + 0.20x,$$
where x is the number of miles driven.

b.

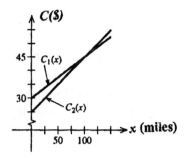

c. The costs will be the same when $C_1(x) = C_2(x)$, that is, when

$$30 + 0.15x = 25 + 0.20x$$
$$0.05x = 5, \text{ or } x = 100.$$

Since $\qquad C_1(70) = 30 + 0.15(70) = 40.5$

and $\qquad C_2(70) = 25 + 0.20(70) = 39,$

and the customer plans to drive less than 100 miles, she should rent from Acme.

69. Here $V = -20,000n + 1,000,000.$
    The book value in 1999 will be
    $$V = -20,000(15) + 1,000,000, \text{ or } \$700,000.$$
    The book value in 2003 will be
    $$V = -20,000(19) + 1,000,000, \text{ or } \$620,000.$$
    The book value in 2007 will be
    $$V = -20,000(23) + 1,000,000, \text{ or } \$540,000.$$

71. a. We require that $0.04 - r^2 \geq 0$ and $r \geq 0$. This is true if $0 \leq r \leq 0.2$. Therefore, the domain of v is $[0,0.2]$.
    b. Compute
    $$v(0) = 1000[0.04 - (0)^2] = 1000(0.04) = 40.$$
    $$v(0.1) = 1000[0.04 - (0.1)^2] = 1000(0.04 - .01)$$
    $$= 1000(0.03) = 30.$$
    $$v(0.2) = 1000[0.04 - (0.2)^2] = 1000(0.04 - 0.04) = 0.$$
    As the distance r increases, the velocity of the blood decreases.

73. Between 8 A.M. and 9 A.M., the average worker can be expected to assemble
    $$N(1) - N(0) = (-1 + 6 + 15) - 0 = 20,$$

or 20 walkie-talkies. Between 9 A.M. and 10 A.M., we can expect

$$N(2) - N(1) = [-2^3 + 6(2^2) + 15(2)] - (-1 + 6 + 15)$$
$$= 46 - 20,$$

or 26 walkie-talkies can be assembled by the average worker.

75. When the proportion of popular votes won by the Democratic presidential candidate is 0.60, the proportion of seats in the House of Representatives won by Democratic candidates is given by

$$s(0.6) = \frac{(0.6)^3}{(0.6)^3 + (1 - 0.6)^3} = \frac{0.216}{0.216 + 0.064} = \frac{0.216}{0.280} \approx 0.77.$$

77.

$$f(x) = \begin{cases} 76 & \text{if } 2 < x \le 3 \\ 97 & \text{if } 3 < x \le 4 \\ 118 & \text{if } 4 < x \le 5 \\ 139 & \text{if } 5 < x \le 6 \\ 160 & \text{if } 6 < x \le 7 \\ 181 & \text{if } 7 < x \le 8 \\ 202 & \text{if } 8 < x \le 9 \\ 223 & \text{if } 9 < x \le 10 \\ 244 & \text{if } 10 < x \le 11 \\ 265 & \text{if } 11 < x \le 12. \end{cases}$$

b.

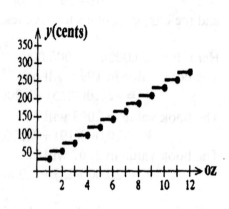

a. The domain of f is $(0, 12]$.

79. True, by definition of a function (page 60).

81. False. Let $f(x) = x^2$, then take $a = 1$, and $b = 2$. Then $f(a) = f(1) = 1$ and $f(b) = f(2) = 4$ and $f(a) + f(b) = 1 + 4 \neq f(a + b) = f(3) = 9$.

USING TECHNOLOGY EXERCISES 2.1, page 79

1.

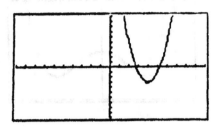

3.

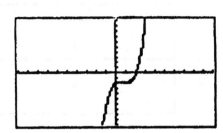

5.

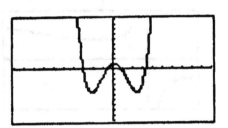

7.

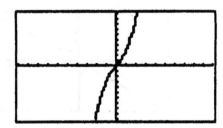

9. a.

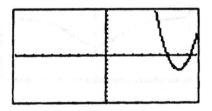

b.

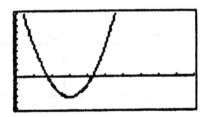

11. a.

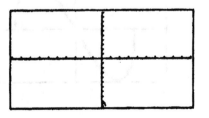

b.

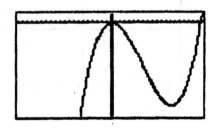

13. a.

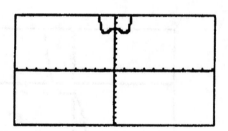

b.

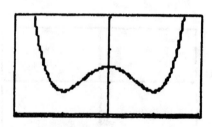

15. a.

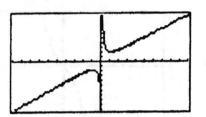

b.

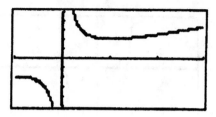

17. a

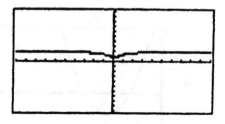

b.

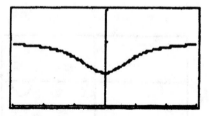

19. a.

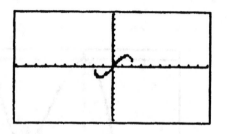

b.

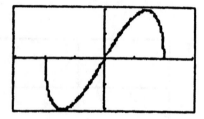

21.

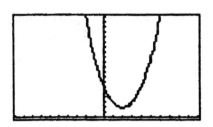

23.

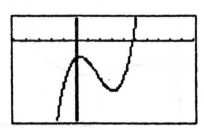

25.

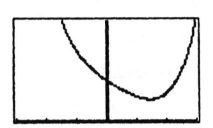

27.

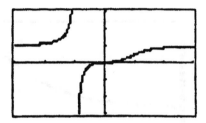

29.

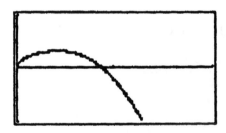

31. $18; f(-1) = -3(-1)^3 + 5(-1)^2 - 2(-1) + 8 = 3 + 5 + 2 + 8 = 18.$

33. $2; f(1) = \dfrac{(1)^4 - 3(1)^2}{1-2} = \dfrac{1-3}{-1} = 2.$

35. $f(2.145) \approx 18.5505$

37. $f(1.28) \approx 17.3850$

39. $f(2.41) \approx 4.1616$

41. $f(0.62) \approx 1.7214$

43. a.

b. $f(2) \approx 2.1762$, or approximately 2.2%/yr

$f(4) \approx 1.9095$, or approximately 1.91 %/yr.

45. a.

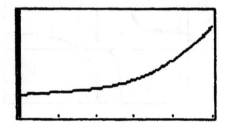

b. $f(6) = 44.7$;

$f(8) = 52.7$;

$f(11) = 129.2$.

EXERCISES 2.2, page 87

1. $(f+g)(x) = f(x) + g(x) = (x^3 + 5) + (x^2 - 2) = x^3 + x^2 + 3.$

3. $fg(x) = f(x)g(x) = (x^3 + 5)(x^2 - 2) = x^5 - 2x^3 + 5x^2 - 10.$

5. $\dfrac{f}{g}(x) = \dfrac{f(x)}{g(x)} = \dfrac{x^3 + 5}{x^2 - 2}.$

7. $\dfrac{fg}{h}(x) = \dfrac{f(x)g(x)}{h(x)} = \dfrac{(x^3 + 5)(x^2 - 2)}{2x + 4} = \dfrac{x^5 - 2x^3 + 5x^2 - 10}{2x + 4}$

9. $(f + g)(x) = x - 1 + \sqrt{x+1}.$

11. $(f\,g)(x) = (x - 1)\sqrt{x+1}$

13. $\dfrac{g}{h}(x) = \dfrac{g(x)}{h(x)} = \dfrac{\sqrt{x+1}}{2x^3-1}.$

15. $\dfrac{fg}{h}(x) = \dfrac{(x-1)(\sqrt{x+1})}{2x^3-1}.$

17. $\dfrac{f-h}{g}(x) = \dfrac{x-1-(2x^3-1)}{\sqrt{x+1}} = \dfrac{x-2x^3}{\sqrt{x+1}}.$

19. $(f+g)(x) = x^2+5+\sqrt{x}-2 = x^2+\sqrt{x}+3.$

$(f-g)(x) = x^2+5-(\sqrt{x}-2) = x^2-\sqrt{x}+7.$

$(fg)(x) = (x^2+5)(\sqrt{x}-2).$

$(\dfrac{f}{g})(x) = \dfrac{x^2+5}{\sqrt{x}-2}.$

21. $(f+g)(x) = \sqrt{x+3}+\dfrac{1}{x-1} = \dfrac{(x-1)\sqrt{x+3}+1}{x-1}.$

$(f-g)(x) = \sqrt{x+3}-\dfrac{1}{x-1} = \dfrac{(x-1)\sqrt{x+3}-1}{x-1}.$

$(fg)(x) = \sqrt{x+3}\left(\dfrac{1}{x-1}\right) = \dfrac{\sqrt{x+3}}{x-1}.$

$(\dfrac{f}{g}) = \sqrt{x+3}(x-1).$

23. $(f+g)(x) = \dfrac{x+1}{x-1}+\dfrac{x+2}{x-2} = \dfrac{(x+1)(x-2)+(x+2)(x-1)}{(x-1)(x-2)}$

$= \dfrac{x^2-x-2+x^2+x-2}{(x-1)(x-2)} = \dfrac{2x^2-4}{(x-1)(x-2)} = \dfrac{2(x^2-2)}{(x-1)(x-2)}.$

$(f-g)(x) = \dfrac{x+1}{x-1}-\dfrac{x+2}{x-2} = \dfrac{(x+1)(x-2)-(x+2)(x-1)}{(x-1)(x-2)}$

$= \dfrac{x^2-x-2-x^2-x+2}{(x-1)(x-2)} = \dfrac{-2x}{(x-1)(x-2)}.$

$(fg)(x) = \dfrac{(x+1)(x+2)}{(x-1)(x-2)}.$

$(\dfrac{f}{g}) = \dfrac{(x+1)(x-2)}{(x-1)(x+2)}.$

25. $(f \circ g)(x) = f(g(x)) = f(x^2) = (x^2)^2 + x^2 + 1 = x^4 + x^2 + 1.$
$(g \circ f)(x) = g(f(x)) = g(x^2 + x + 1) = (x^2 + x + 1)^2.$

27. $(f \circ g)(x) = f(g(x)) = f(x^2 - 1) = \sqrt{x^2 - 1} + 1.$
$(g \circ f)(x) = g(f(x)) = g(\sqrt{x} + 1) = (\sqrt{x} + 1)^2 - 1 = x + 2\sqrt{x} + 1 - 1 = x + 2\sqrt{x}.$

29. $(f \circ g)(x) = f(g(x)) = f\left(\dfrac{1}{x}\right) = \dfrac{1}{x} \div \left(\dfrac{1}{x^2} + 1\right) = \dfrac{1}{x} \cdot \dfrac{x^2}{x^2 + 1} = \dfrac{x}{x^2 + 1}.$
$(g \circ f)(x) = g(f(x)) = g\left(\dfrac{x}{x^2 + 1}\right) = \dfrac{x^2 + 1}{x}.$

31. $h(2) = g[f(2)]$. But $f(2) = 4 + 2 + 1 = 7$, so $h(2) = g(7) = 49$.

33. $h(2) = g[f(2)]$. But $f(2) = \dfrac{1}{2(2) + 1} = \dfrac{1}{5}$, so $h(2) = g(\dfrac{1}{5}) = \dfrac{1}{\sqrt{5}} = \dfrac{\sqrt{5}}{5}$.

35. $f(x) = 2x^3 + x^2 + 1, \ g(x) = x^5.$ 37. $f(x) = x^2 - 1, \ g(x) = \sqrt{x}.$

39. $f(x) = x^2 - 1, \ g(x) = \dfrac{1}{x}.$ 41. $f(x) = 3x^2 + 2, \ g(x) = \dfrac{1}{x^{3/2}}.$

43. $f(a + h) - f(a) = [3(a + h) + 4] - (3a + 4)$
$= 3a + 3h + 4 - 3a - 4 = 3h.$

45. $f(a + h) - f(a) = 4 - (a + h)^2 - (4 - a^2)$
$= 4 - a^2 - 2ah - h^2 - 4 + a^2$
$= -2ah - h^2 = -h(2a + h).$

47. $\dfrac{f(a + h) - f(a)}{h} = \dfrac{[(a + h)^2 + 1] - (a^2 + 1)}{h} = \dfrac{a^2 + 2ah + h^2 + 1 - a^2 - 1}{h}$
$= \dfrac{h(2a + h)}{h} = 2a + h.$

49. $C(x) = 0.6x + 12{,}100.$

51. a. $P(x) = R(x) - C(x)$
$$= -0.1x^2 + 500x - (0.000003x^3 - 0.03x^2 + 200x + 100{,}000)$$
$$= -0.000003x^3 - 0.07x^2 + 300x - 100{,}000.$$
   b. $P(1500) = -0.000003(1500)^3 - 0.07(1500)^2 + 300(1500) - 100{,}000$
$$= 182{,}375 \quad \text{or } \$182{,}375.$$

53. a.
$$N(r(t)) = \frac{7}{1 + 0.02\left(\dfrac{10t + 150}{t + 10}\right)^2}.$$

   b.
$$N(r(0)) = \frac{7}{1 + 0.02\left(\dfrac{10(0) + 150}{0 + 10}\right)^2}$$

$$= \frac{7}{1 + 0.02\left(\dfrac{150}{10}\right)^2} = \frac{7}{5.5} \approx 1.27, \text{ or } 1.27 \text{ million units.}$$

$$N(r(12)) = \frac{7}{1 + 0.02\left(\dfrac{120 + 150}{12 + 10}\right)^2} = \frac{7}{1 + 0.02\left(\dfrac{270}{22}\right)^2} = \frac{7}{4.01} \approx 1.74,$$

or 1.74 million units.

$$N(r(18)) = \frac{7}{1 + 0.02\left(\dfrac{180 + 150}{18 + 10}\right)^2} = \frac{7}{1 + 0.02\left(\dfrac{330}{28}\right)^2} = \frac{7}{3.78} \approx 1.85,$$

or 1.85 million units.

55. $N(t) = 1.42(x(t)) = \dfrac{(1.42)(7)(t + 10)^2}{(t + 10)^2 + 2(t + 15)^2} = \dfrac{9.94(t + 10)^2}{(t + 10)^2 + 2(t + 15)^2}.$
   The number of jobs created 6 months from now will be
$$N(6) = \frac{9.94(16)^2}{(16)^2 + 2(21)^2} = 2.24, \text{ or } 2.24 \text{ million jobs.}$$
   The number of jobs created 12 months from now will be
$$N(12) = \frac{9.94(22)^2}{(22)^2 + 2(27)^2} = 2.48, \text{ or } 2.48 \text{ million jobs.}$$

57. False. Let $(x) = x + 2$ and $g(x) = \sqrt{x}$. Then $(g \circ f)(x) = \sqrt{x+2}$ is defined at $x = -1$. But $(f \circ g)(x) = \sqrt{x} + 2$ is not defined at $x = -1$.

59. False. Take $f(x) = x + 1$. then $(f \circ f)(x) = f(f(x)) = x + 2$. But $f^2(x) = [f(x)]^2 = (x+1)^2 = x^2 + 2x + 1$.

EXERCISES 2.3, page 99

1. Yes. $2x + 3y = 6$ and so $y = -\frac{2}{3}x + 2$.

3. Yes. $2y = x + 4$ and so $y = \frac{1}{2}x + 2$.

5. Yes. $4y = 2x + 9$ and so $y = \frac{1}{2}x + \frac{9}{4}$.

7. No, because of the term x^2.

9. f is a polynomial function in x of degree 6.

11. Expanding $G(x) = 2(x^2 - 3)^3$, we have
$$G(x) = 2x^6 - 18x^4 + 54x^2 - 54,$$
and we conclude that G is a polynomial function in x of degree 6.

13. f is neither a polynomial nor a rational function.

15. $f(0) = 2$ gives $g(0) = m(0) + b = b = 2$. Next, $f(3) = -1$ gives
$f(3) = m(3) + b = -1$.
Substituting $b = 2$ in this last equation, we have $3m + 2 = -1$, and $3m = -3$, or
$m = -1$. So $m = -1$ and $b = 2$.

17. a. $C(x) = 8x + 40,000$
    b. $R(x) = 12x$
    c. $P(x) = R(x) - C(x) = 12x - (8x + 40,000) = 4x - 40,000.$
    d. $P(8000) = 4(8000) - 40,000 = -8000$, or a loss of $8000.
       $P(12,000) = 4(12,000) - 40,000 = 8000$, or a profit of $8000.

19. The individual's disposable income is
$$D = (1 - 0.28)40{,}000 = 28{,}800, \quad \text{or } \$28{,}800.$$

21. The child should receive
$$D(4) = \left(\frac{4+1}{24}\right)(500) = 104.17, \quad \text{or } 104 \text{ mg.}$$

23. When 1000 units are produced,
$$R(1000) = -0.1(1000)^2 + 500(1000) = 400{,}000, \quad \text{or } \$400{,}000.$$

25. a.

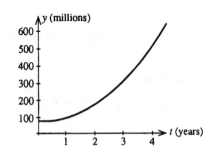

  b.
$$f(2) = 38.57(4) - 24.29(2) + 79.14$$
$$= 184.84, \quad \text{or} \quad 184{,}840{,}000.$$

27. a. The given data implies that $R(40) = 50$, that is,
$$\frac{100(40)}{b+40} = 50$$
$$50(b+40) = 4000,$$
  or
$$b = 40.$$

Therefore, the required response function is $R(x) = \dfrac{100x}{40+x}$.

  b. The response will be $R(60) = \dfrac{100(60)}{40+60} = 60$, or approximately 60 percent.

29. a. We are given that $T = aN + b$ where a and b are constants to be determined. The given conditions imply
$$70 = 120a + b$$
  and
$$80 = 160a + b$$

Subtracting the first equation from the second gives
$$10 = 40a, \quad \text{or} \quad a = \tfrac{1}{4}.$$

Substituting this value of a into the first equation gives
$$70 = 120(\tfrac{1}{4}) + b, \quad \text{or} \quad b = 40.$$

Therefore, $T = \tfrac{1}{4}N + 40$.

b. Solving the equation in (a) for N, we find
$$\tfrac{1}{4}N = T - 40,$$
or $\quad N = f(t) = 4T - 160.$

When $T = 102$, we find $N = 4(102) - 160 = 248$, or 248 times per minute.

31. Using the formula given in Problem 30, we have
$$V(2) = 100{,}000 - \frac{(100{,}000 - 30{,}000)}{5}(2) = 100{,}000 - \frac{70{,}000}{5}(2)$$
$$= 72{,}000, \text{ or } \$72{,}000.$$

33. $f(t) = 0.1714t^2 + 0.6657t + 0.7143$
  a. $f(0) = 0.7143$ or $714{,}300$.
  b. $f(5) = 0.1714(25) + 0.6657(5) + 0.7143 = 8.3278$, or 8.33 million.

35. $h(t) = f(t) - g(t) = \dfrac{110}{\frac{1}{2}t + 1} - 26(\tfrac{1}{4}t^2 - 1)^2 - 52.$

$$h(0) = f(0) - g(0) = \frac{110}{\frac{1}{2}(0) + 1} - 26\left[\tfrac{1}{4}(0)^2 - 1\right]^2 - 52 = 110 - 26 - 52 = 32, \text{ or } \$32.$$

$$h(1) = f(1) - g(1) = \frac{110}{\frac{1}{2}(1) + 1} - 26\left[\tfrac{1}{4}(1)^2 - 1\right]^2 - 52 = 6.71, \text{ or } \$6.71.$$

$$h(2) = f(2) - g(2) = \frac{110}{\frac{1}{2}(2) + 1} - 26\left[\tfrac{1}{4}(2)^2 - 1\right]^2 - 52 = 3, \text{ or } \$3.$$

We conclude that the price gap was narrowing.

37. a.

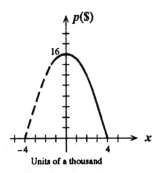

Units of a thousand

b. If $p = 7$, we have $7 = -x^2 + 16$, or $x^2 = 9$, so that $x = \pm 3$. Therefore, the quantity demanded when the unit price is \$7 is 3000 units.

39. a.

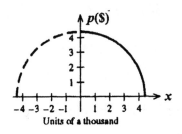

Units of a thousand

b. If $p = 3$, then $3 = \sqrt{18 - x^2}$, and $9 = 18 - x^2$, so that $x^2 = 9$ and $x = \pm 3$. Therefore, the quantity demanded when the unit price is \$3 is 3000 units.

41. a.

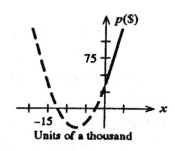

Units of a thousand

b. If $x = 2$, then $p = 2^2 + 16(2) + 40 = 76$, or \$76.

43. a.

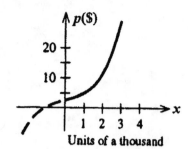

b. $p = 2^3 + 2(2) + 3 = 15$, or $15.

45. The slope of L_2 is greater than that of L_1. This tells us that for each increase of a dollar in the price of a clock radio, more model A clock radios will be made available in the market place than model B clock radios.

47. Substituting $x = 6$ and $p = 8$ into the given equation gives
$$8 = \sqrt{-36a + b}, \quad \text{or } -36a + b = 64.$$
Next, substituting $x = 8$ and $p = 6$ into the equation gives
$$6 = \sqrt{-64a + b}, \quad \text{or } -64a + b = 36.$$

Solving the system $\quad \begin{aligned} -36a + b &= 64 \\ -64a + b &= 36 \end{aligned}$

for a and b, we find $a = 1$ and $b = 100$. Therefore the demand equation is
$$p = \sqrt{-x^2 + 100}.$$
When the unit price is set at $7.50, we have
$$7.5 = \sqrt{-x^2 + 100}, \quad \text{or } 56.25 = -x^2 + 100$$
from which we deduce that $x = \pm 6.614$. So, the quantity demanded is 6614 units.

49. Substituting $x = 10,000$ and $p = 20$ into the given equation yields
$$20 = a\sqrt{10,000} + b = 100a + b.$$
Next, substituting $x = 62,500$ and $p = 35$ into the equation yields
$$35 = a\sqrt{62,500} + b = 250a + b.$$
Subtracting the first equation from the second yields $15 = 150a$, or $a = \frac{1}{10}$.
Substituting this value of a into the first equation gives $b = 10$. Therefore, the required equation is $p = \frac{1}{10}\sqrt{x} + 10$. The graph of the supply function follows.

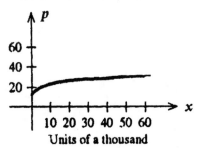

Units of a thousand

Substituting $x = 40,000$ into the supply equation yields

$$p = \tfrac{1}{10}\sqrt{40,000} + 10 = 30, \text{ or } \$30.$$

51. We solve the equation $-2x^2 + 80 = 15x + 30$, or $2x^2 + 15x - 50 = 0$ for x. Thus,
$$(2x - 5)(x + 10) = 0,$$
or $x = 5/2$ or $x = -10$. Rejecting the negative root, we have $x = 5/2$. The corresponding value of p is $p = -2(\tfrac{5}{2})^2 + 80 = 67.5$.

We conclude that the equilibrium quantity is 2500 and the equilibrium price is $67.50.

53. Solving both equations for x, we have $x = -(11/3)p + 22$ and $x = 2p^2 + p - 10$. Equating these two equations, we have
$$-\tfrac{11}{3}p + 22 = 2p^2 + p - 10,$$
or $-11p + 66 = 6p^2 + 3p - 30$
and $6p^2 + 14p - 96 = 0.$
Dividing this last equation by 2 and then factoring, we have
$$(3p + 16)(p - 3) = 0,$$
or $p = 3$. The corresponding value of x is $2(3)^2 + 3 - 10 = 11$. We conclude that the equilibrium quantity is 11,000 and the equilibrium price is $3.

55. Equating the two equations, we have
$$0.1x^2 + 2x + 20 = -0.1x^2 - x + 40$$
$$0.2x^2 + 3x - 20 = 0$$
$$2x^2 + 30x - 200 = 0$$
$$x^2 + 15x - 100 = 0$$
$$(x + 20)(x - 5) = 0,$$
and $x = -20$ or 5.
Substituting $x = 5$ into the first equation gives

$$p = -0.1(25) - 5 + 40 = 32.5.$$

Therefore, the equilibrium quantity is 500 tents (x is measured in hundreds) and the equilibrium price is \$32.50.

57. a.

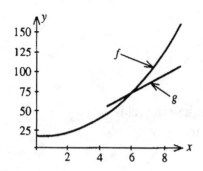

b.
$$5x^2 + 5x + 30 = 33x + 30$$
$$5x^2 - 28x = 0$$
$$x(5x - 28) = 0$$
$$x = 0 \text{ or } x = \frac{28}{5} = 5.6, \text{ or } 5.6 \text{ mph.}$$
$$g(x) = 11(5.6) + 10 = 71.6, \text{ or } 71.6 \text{ mL/lb/min}$$

c. The oxygen consumption of the walker is greater than that of the runner.

59. The area of Juanita's garden is 250 sq ft. Therefore $xy = 250$ and $y = \dfrac{250}{x}$.

The amount of fencing needed is given by $2x + 2y$.

Therefore, $f = 2x + 2\left(\dfrac{250}{x}\right) = 2x + \dfrac{500}{x}$. The domain of f is $x > 0$.

61. Since the volume of the box is given by
$$V = (\text{area of the base}) \times \text{the height of the box}$$
$$= x^2 y = 20,$$

we have $y = \dfrac{20}{x^2}$.

Next, the amount of material used in constructing the box is given by the area of the base of the box, plus the area of the 4 sides, plus the area of the top of the box, or $x^2 + 4xy + x^2$. Then, the cost of constructing the box is given by

$$f(x) = 0.30x^2 + 0.40x \cdot \frac{20}{x^2} + .20x^2 = 0.5x^2 + \frac{8}{x}.$$

63. The average yield of the apple orchard is 36 bushels/tree when the density is 22 trees/acre. Let $x =$ the unit increase in tree density beyond 22. Then the yield of the apple orchard in bushels/acre is given by $(22 + x)(36 - 2x)$.

65. True. If $P(x)$ is a polynomial function, then $P(x) = \frac{P(x)}{1}$ and so it is a rational function. The converse if false. For example, $R(x) = \frac{x+1}{x-1}$ is a rational function that is not a polynomial.

67. False. A function has the form x^r, where r is a real number.

USING TECHNOLOGY EXERCISES 2.3, page 106

1. (-3.0414, 0.1503); (3.0414, 7.4497)   3. (-2.3371, 2.4117); (6.0514, -2.5015)

5. (-1.0219, -6.3461); (1.2414, -1.5931), and (5.7805, 7.9391)

7. a. 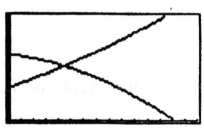   b. 438 wall clocks; $40.92

9. a. $y = 0.1375t^2 + 0.675t + 3.1$
   b.
   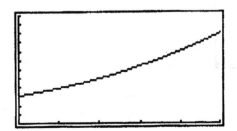

c. 3.1; 3.9; 5; 6.4; 8; 9.9

11. a. $y = -0.02028t^3 + 0.31393t^2 + 0.40873t + 0.66024$

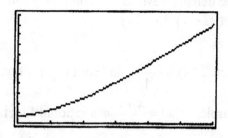

c. 0.66; 1.36; 2.57; 4.16; 6.02; 8.02; 10.03

13. a. $y = 0.05833t^3 - 0.325t^2 + 1.8881t + 5.07143$
    b.

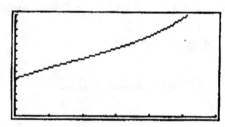

c. 6.7; 8.0; 9.4; 11.2; 13.7

15. a. $y = 0.0125t^4 - 0.01389t^3 + 0.55417t^2 + 0.53294t + 4.95238$ $(0 \le t \le 5)$

b.

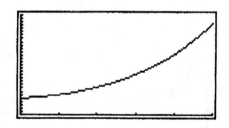

c. 5.0; 6.0; 8.3; 12.2; 18.3; 27.5

EXERCISES 2.4, page 127

1. $\lim\limits_{x\to-2} f(x) = 3.$ 3. $\lim\limits_{x\to3} f(x) = 3.$ 5. $\lim\limits_{x\to-2} f(x) = 3.$

7. The limit does not exist. If we consider any value of x to the right of $x = -2$, $f(x) \le 2$. If we consider values of x to the left of $x = -2$, $f(x) \ge -2$. Since $f(x)$ does not approach any one number as x approaches $x = -2$, we conclude that the limit does not exist.

9.

x	1.9	1.99	1.999	2.001	2.01	2.1
$f(x)$	4.61	4.9601	4.9960	5.004	5.0401	5.41

$\lim\limits_{x\to2} (x^2 + 1) = 5.$

11.

x	-0.1	-0.01	-0.001	0.001	0.01	0.1
$f(x)$	-1	-1	-1	1	1	1

The limit does not exist.

13.

x	0.9	0.99	0.999	1.001	1.01	1.1
$f(x)$	100	10,000	1,000,000	1,000,000	10,000	100

The limit does not exist.

15.

x	0.9	0.99	0.999	1.001	1.01	1.1
$f(x)$	2.9	2.99	2.999	3.001	3.01	3.1

$$\lim_{x \to 1} \frac{x^2 + x - 2}{x - 1} = 3.$$

17.

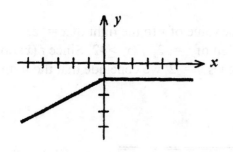

$$\lim_{x \to 0} f(x) = -1$$

19.

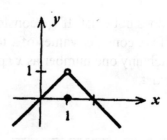

$$\lim_{x \to 1} f(x) = 1$$

21.

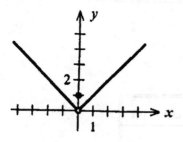

$$\lim_{x \to 0} f(x) = 0$$

23. $\lim\limits_{x \to 2} 3 = 3$

25. $\lim\limits_{x \to 3} x = 3$

27. $\lim\limits_{x \to 1}(1 - 2x^2) = 1 - 2(1)^2 = -1$

29. $\lim\limits_{x \to 1}(2x^3 - 3x^2 + x + 2) = 2(1)^3 - 3(1)^2 + 1 + 2 = 2.$

31. $\lim\limits_{s \to 0}(2s^2 - 1)(2s + 4) = (-1)(4) = -4.$

33. $\lim\limits_{x \to 2} \dfrac{2x + 1}{x + 2} = \dfrac{2(2) + 1}{2 + 2} = \dfrac{5}{4}.$

35. $\lim\limits_{x \to 2} \sqrt{x + 2} = \sqrt{2 + 2} = 2.$

37. $\lim_{x \to -3} \sqrt{2x^4 + x^2} = \sqrt{2(-3)^4 + (-3)^2} = \sqrt{162 + 9} = \sqrt{171} = 3\sqrt{19}.$

39. $\lim_{x \to -1} \dfrac{\sqrt{x^2 + 8}}{2x + 4} = \dfrac{\sqrt{(-1)^2 + 8}}{2(-1) + 4} = \dfrac{\sqrt{9}}{2} = \dfrac{3}{2}.$

41. $\lim_{x \to a}[f(x) - g(x)] = \lim_{x \to a} f(x) - \lim_{x \to a} g(x) = 3 - 4 = -1.$

43. $\lim_{x \to a}[2f(x) - 3g(x)] = \lim_{x \to a} 2f(x) - \lim_{x \to a} 3g(x) = 2(3) - 3(4) = -6.$

45. $\lim_{x \to a} \sqrt{g(x)} = \lim_{x \to a} \sqrt{4} = 2.$

47. $\lim_{x \to a} \dfrac{2f(x) - g(x)}{f(x)g(x)} = \dfrac{2(3) - (4)}{(3)(4)} = \dfrac{2}{12} = \dfrac{1}{6}.$

49. $\lim_{x \to 1} \dfrac{x^2 - 1}{x - 1} = \lim_{x \to 1} \dfrac{(x-1)(x+1)}{x-1} = \lim_{x \to 1}(x + 1) = 1 + 1 = 2.$

51. $\lim_{x \to 0} \dfrac{x^2 - x}{x} = \lim_{x \to 0} \dfrac{x(x-1)}{x} = \lim_{x \to 0}(x - 1) = 0 - 1 = -1.$

53. $\lim_{x \to -5} \dfrac{x^2 - 25}{x + 5} = \lim_{x \to -5} \dfrac{(x+5)(x-5)}{x+5} = \lim_{x \to -5}(x - 5) = -10.$

55. $\lim_{x \to 1} \dfrac{x}{x - 1}$ does not exist.

57. $\lim_{x \to -2} \dfrac{x^2 - x - 6}{x^2 + x - 2} = \lim_{x \to -2} \dfrac{(x-3)(x+2)}{(x+2)(x-1)} = \lim_{x \to -2} \dfrac{x-3}{x-1} = \dfrac{-2-3}{-2-1} = \dfrac{5}{3}.$

59. $\lim_{x \to 1} \dfrac{\sqrt{x} - 1}{x - 1} = \lim_{x \to 1} \dfrac{\sqrt{x} - 1}{x - 1} \cdot \dfrac{\sqrt{x} + 1}{\sqrt{x} + 1} = \lim_{x \to 1} \dfrac{x - 1}{(x-1)(\sqrt{x} - 1)} = \lim_{x \to 1} \dfrac{1}{(\sqrt{x} - 1)} = \dfrac{1}{2}.$

61. $\displaystyle\lim_{x\to 1}\frac{x-1}{x^3+x^2-2x}=\lim_{x\to 1}\frac{x-1}{x(x-1)(x+2)}=\lim_{x\to 1}\frac{1}{x(x+2)}=\frac{1}{3}.$

63. $\displaystyle\lim_{x\to\infty}f(x)=\infty$ (does not exist) and $\displaystyle\lim_{x\to-\infty}f(x)=\infty$ (does not exist).

65. $\displaystyle\lim_{x\to\infty}f(x)=0$ and $\displaystyle\lim_{x\to-\infty}f(x)=0.$

67. $\displaystyle\lim_{x\to\infty}f(x)=-\infty$ (does not exist) and $\displaystyle\lim_{x\to-\infty}f(x)=-\infty$ (does not exist).

69.

x	1	10	100	1000
$f(x)$	0.5	0.009901	0.0001	0.000001

x	-1	-10	-100	-1000
$f(x)$	0.5	0.009901	0.0001	0.000001

$\displaystyle\lim_{x\to\infty}f(x)=0$ and $\displaystyle\lim_{x\to-\infty}f(x)=0$

71.

x	1	5	10	100	1000
$f(x)$	12	360	2910	2.99×10^6	2.999×10^9

x	-1	-5	-10	-100	-1000
$f(x)$	6	-390	-3090	-3.01×10^6	-3.0×10^9

$\displaystyle\lim_{x\to\infty}f(x)=\infty$ (does not exist) and $\displaystyle\lim_{x\to-\infty}f(x)=-\infty$ (does not exist).

73. $\displaystyle\lim_{x\to\infty}\frac{3x+2}{x-5}=\lim_{x\to\infty}\frac{3+\dfrac{2}{x}}{1-\dfrac{5}{x}}=\frac{3}{1}=3.$

75. $\displaystyle\lim_{x\to-\infty}\frac{3x^3+x^2+1}{x^3+1}=\lim_{x\to-\infty}\frac{3+\dfrac{1}{x}+\dfrac{1}{x^3}}{1+\dfrac{1}{x^3}}=3.$

77. $\displaystyle\lim_{x\to-\infty}\frac{x^4+1}{x^3-1}=\lim_{x\to-\infty}\frac{x+\dfrac{1}{x^3}}{1-\dfrac{1}{x^3}}=-\infty$; that is, the limit does not exist.

79. $\displaystyle\lim_{x\to\infty}\frac{x^5-x^3+x-1}{x^6+2x^2+1}=\lim_{x\to\infty}\frac{\dfrac{1}{x}-\dfrac{1}{x^3}+\dfrac{1}{x^5}-\dfrac{1}{x^6}}{1+\dfrac{2}{x^4}+\dfrac{1}{x^6}}=0.$

81. a. The cost of removing 50 percent of the pollutant is

$$C(50)=\frac{0.5(50)}{100-50}=0.5\text{, or \$500,000.}$$

Similarly, we find that the cost of removing 60, 70, 80, 90, and 95 percent of the pollutants is \$750,000; \$1,166,667; \$2,000,000, \$4,500,000, and \$9,500,000, respectively.

b. $\displaystyle\lim_{x\to100}\frac{0.5x}{100-x}=\infty,$

which means that the cost of removing the pollutant increases astronomically if we wish to remove almost all of the pollutant.

83. $\displaystyle\lim_{x\to\infty}\overline{C}(x)=\lim_{x\to\infty}2.2+\frac{2500}{x}=2.2,\text{ or \$2.20 per video disc.}$

In the long-run, the average cost of producing x video discs will approach \$2.20/disc.

85. a. $\displaystyle T(1)=\frac{120}{1+4}=24,\text{ or \$24 million.}$ $\qquad$ $\displaystyle T(2)=\frac{120(4)}{8}=60,\text{ \$60 million.}$

$\displaystyle T(3)=\frac{120(9)}{13}=83.1,\text{ or \$83.1 million.}$

b. In the long run, the movie will gross

$$\lim_{x \to \infty} \frac{120x^2}{x^2+4} = \lim_{x \to \infty} \frac{120}{1 + \dfrac{4}{x^2}} = 120, \text{ or } \$120 \text{ million.}$$

87. a. The average cost of driving 5000 miles per year is

$$C(5) = \frac{2010}{5^{2.2}} + 17.80 = 76.07,$$

or 76.1 cents per mile. Similarly, we see that the average cost of driving 10,000 miles per year; 15,000 miles per year; 20,000 miles per year; and 25,000 miles per year is 30.5, 23; 20.6, and 19.5 cents per mile, respectively.

b.

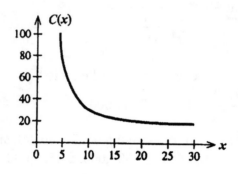

c. It approaches 17.80 cents per mile.

89. False. Let $f(x) = \begin{cases} -1 & \text{if } x < 0 \\ 1 & \text{if } x > 0 \end{cases}$. Then $\lim_{x \to 0} f(x) = 1$, but $f(1)$ is not defined.

91. True. Division by zero is not permitted.

93. True. Each limit in the sum exists. Therefore,

$$\lim_{x \to 2}\left(\frac{x}{x+1} + \frac{3}{x+1}\right) = \lim_{x \to 2}\frac{x}{x+1} + \lim_{x \to 2}\frac{3}{x-1}$$

$$= \frac{2}{3} + \frac{3}{1} = \frac{11}{3}.$$

95. $\lim_{x \to \infty}\dfrac{ax}{x+b} = \lim_{x \to \infty}\dfrac{a}{1 + \dfrac{b}{x}} = a.$ As the amount of substrate becomes very large, the initial

speed approaches the constant a moles per liter per second.

97. Consider the functions
$$f(x) = \begin{cases} -1 & \text{if } x < 0 \\ 1 & \text{if } x \geq 0 \end{cases} \quad \text{and} \quad g(x) = \begin{cases} 1 & \text{if } x < 0 \\ -1 & \text{if } x \geq 0 \end{cases}.$$
Then $\lim_{x \to 0} f(x)$ and $\lim_{x \to 0} g(x)$ do not exist, but $\lim_{x \to 0} [f(x)g(x)] = \lim_{x \to 0}(-1) = -1$.
This example does not contradict Theorem 1 for a reason similar to that given in Example 96 (replace "sum" by "product.")

USING TECHNOLOGY EXERCISES 2.4, page 132

1. 5   3. 3   5. $\dfrac{2}{3}$ 7. $\dfrac{1}{2}$

9. e^2, or 7.38906

11. From the graph we see that $f(x)$ does not approach any finite number as x approaches 3.

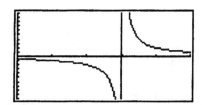

13. a.

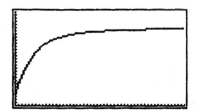

     b. $\quad \lim_{t \to \infty} \dfrac{25t^2 + 125t + 200}{t^2 + 5t + 40} = 25$, so in the long run the population will approach 25,000.

EXERCISES 2.5, page 145

1. $\lim_{x \to 2^-} f(x) = 3, \ \lim_{x \to 2^+} f(x) = 2, \ \lim_{x \to 2} f(x)$ does not exist.

3. $\lim\limits_{x\to-1^-} f(x) = \infty$, $\lim\limits_{x\to-1^+} f(x) = 2$. Therefore $\lim\limits_{x\to-1} f(x)$ does not exist.

5. $\lim\limits_{x\to1^-} f(x) = 0$, $\lim\limits_{x\to1^+} f(x) = 2$, $\lim\limits_{x\to1} f(x)$ does not exist.

7. $\lim\limits_{x\to0^-} f(x) = -2$, $\lim\limits_{x\to0^+} f(x) = 2$, $\lim\limits_{x\to0} f(x)$ does not exist.

9. True   11. True   13. False   15. True   17. False   19. True

21. $\lim\limits_{x\to1^+}(2x+4) = 6$.

23. $\lim\limits_{x\to2^-} \dfrac{x-3}{x+2} = \dfrac{2-3}{2+2} = -\dfrac{1}{4}$.

25. $\lim\limits_{x\to0^+} \dfrac{1}{x}$ does not exist because $1/x \to \infty$ as $x \to 0$ from the right..

27. $\lim\limits_{x\to0^+} \dfrac{x-1}{x^2+1} = \dfrac{-1}{1} = -1$..

29. $\lim\limits_{x\to0^+} \sqrt{x} = \sqrt{\lim\limits_{x\to0^+} x} = 0$.

31. $\lim\limits_{x\to-2^+}(2x+\sqrt{2+x}) = \lim\limits_{x\to-2^+} 2x + \lim\limits_{x\to-2^+} \sqrt{2+x} = -4+0 = -4$.

33. $\lim\limits_{x\to1^-} \dfrac{1+x}{1-x} = \infty$, that is, the limit does not exist.

35. $\lim\limits_{x\to2^-} \dfrac{x^2-4}{x-2} = \lim\limits_{x\to2^-} \dfrac{(x+2)(x-2)}{x-2} = \lim\limits_{x\to2^-}(x+2) = 4$.

37. $\lim\limits_{x\to3^+} \dfrac{x^2-9}{x+3} = \dfrac{9-9}{3+3} = 0$.

39. $\lim\limits_{x\to0^+} f(x) = \lim\limits_{x\to0^+} x^2 = 0$, $\lim\limits_{x\to0^-} f(x) = \lim\limits_{x\to0^-} 2x = 0$

41. $\lim\limits_{x\to1^+} f(x) = \lim\limits_{x\to1^+} \sqrt{x+3} = \sqrt{4} = 2$.

$\lim\limits_{x\to1^-} f(x) = \lim\limits_{x\to1^-}(2+\sqrt{x}) = 2+\sqrt{x} = 2+\sqrt{1} = 3$.

43. The function is discontinuous at $x = 0$. Conditions 2 and 3 are violated.

45. The function is continuous everywhere.

47. The function is discontinuous at $x = 0$. Condition 3 is violated.

49. The function is discontinuous at $x = 0$. Condition 3 is violated.

51. f is continuous for all values of x.

53. f is continuous for all values of x. Note that $x^2 + 1 \geq 1 > 0$.

55. f is discontinuous at $x = 1/2$, where the denominator is 0.

57. Observe that $x^2 + x - 2 = (x + 2)(x - 1) = 0$ if $x = -2$ or $x = 1$. So, f is discontinuous at these values of x.

59. f is continuous everywhere since all three conditions are satisfied.

61. f is continuous everywhere because all three conditions are satisfied.

63. f is continuous everywhere since all three conditions are satisfied. Observe that

$$\lim_{x \to 1} f(x) = \lim_{x \to 1} \frac{x^2 - 1}{x - 1} = \lim_{x \to 1} \frac{(x-1)(x+1)}{x-1} = \lim_{x \to 1}(x + 1) = 2 = f(1).$$

65. f is continuous everywhere since all three conditions are satisfied.

67. Since the denominator $x^2 - 1 = (x - 1)(x + 1) = 0$ if $x = -1$ or 1, we see that f is discontinuous at these points.

69. Since $x^2 - 3x + 2 = (x - 2)(x - 1) = 0$ if $x = 1$ or 2, we see that the denominator is zero at these points and so f is discontinuous at these points.

71. The function f is discontinuous at $x = 1, 2, 3, ..., 11$ because the limit of f does not exist at these points.

73. Having made steady progress up to $x = x_1$, Michael's progress came to a standstill.

Then at $x = x_2$ a sudden break-through occurs and he then continues to successfully complete the solution to the problem.

75. Conditions 2 and 3 are not satisfied at each of these points.

77. The graph of f follows.

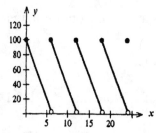

f is discontinuous at $x = 6, 12, 18, 24$.

79.

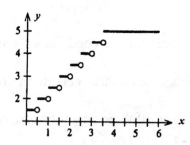

f is discontinuous at $x = \frac{1}{2}, 1, 1\frac{1}{2}, \ldots, 4$.

81. a. $\lim\limits_{v \to u^+} \dfrac{aLv^3}{v - u} = \infty$ and this shows that, when the speed of the fish is very close to that of the current, the energy expended by the fish will be enormous.

b. $\lim\limits_{v \to \infty} \dfrac{aLv^3}{v - u} = \infty$ and this says that if the speed of the fish increases greatly, so does the amount of energy required to swim a distance of L ft.

83. Since $\lim\limits_{x \to -2} \dfrac{x^2 - 4}{x + 2} = \lim\limits_{x \to -2} \dfrac{(x - 2)(x + 2)}{x + 2} = \lim\limits_{x \to -2} (x - 2) = -4,$

we define $f(-2) = k = -4$, that is, take $k = -4$.

85. a. No. Consider the function $f(x) = 0$ for all x in $(-\infty, \infty)$ and

$$g(x) = \begin{cases} 1 & \text{if } x < 0 \\ -1 & \text{if } x \geq 0 \end{cases}.$$

b. No. Consider the functions f and g of Exercise 84b.

87. f is a polynomial and is therefore continuous on $[-1,1]$.
$$f(-1) = (-1)^3 - 2(-1)^2 + 3(-1) + 2 = -1 - 2 - 3 + 2 = -4.$$
$$f(1) = 1 - 2 + 3 + 2 = 4.$$
Since $f(-1)$ and $f(1)$ have opposite signs, we see that f has at least one zero in $(-1,1)$.

89. f is continuous on $[14,16]$ and
$$f(14) = 2(14)^{5/3} - 5(14)^{4/3} \approx -6.06$$
$$f(16) = 2(16)^{5/3} - 5(16)^{4/3} \approx 1.60,$$
and so f has at least one zero in $(14,16)$.

91. $f(0) = 6$ and $f(3) = 3$ and f is continuous on $[0,3]$. So the Intermediate Value Theorem guarantees that there is at least one value of x for which $f(x) = 2$. Solving
$$f(x) = x^2 - 4x + 6 = 2$$
we find, $x^2 - 4x + 4 = (x - 2)^2 = 0$, or $x = \pm 2$. Since -2 does not lie in $[0,3]$, we see that $x = 2$.

93.

Step	Root of f(x) = 0 lies in
1	(1,2)
2	(1,1.5)
3	(1.25,1.5)
4	(1.25,1.375)
5	(1.3125,1.375)
6	(1.3125,1.34375)
7	(1.328125,1.34375)
8	(1.3359375,1.34375)
9	(1.33984375,1.34375)

We see that the required root is approximately 1.34.

95. False. Consider the function $f(x) = x^2 - 1$ on the interval $[-2, 2]$. Here, $f(-2) = f(2) = 3$, but f has zeros at $x = -1$ and $x = 1$.

97. False. Let $f(x) = \begin{cases} x & \text{if } x \neq 0 \\ 1 & \text{if } x = 0 \end{cases}$. Then $\lim\limits_{x \to 0^+} f(x) = \lim\limits_{x \to 0^-} f(x)$, but $f(0) = 1$.

99. True. Such a number is guaranteed by the Intermediate Value Theorem.

101.a. f is a rational function whose denominator is never zero, and so it is continuous for all values of x.

b. Since the numerator, x^2, is nonnegative and the denominator is $x^2 + 1 \geq 1$ for all values of x, we see that $f(x)$ is nonnegative for all values of x.

c. $f(0) = \dfrac{0}{0+1} = \dfrac{0}{1} = 0$ and so f has a zero at $x = 0$. This does not contradict Theorem 4.

103. Consider the function f defined by
$$f(x) = \begin{cases} -1 & \text{if } -1 \leq x < 0 \\ 1 & \text{if } 0 \leq x < 1 \end{cases}.$$
Then $f(-1) = -1$ and $f(1) = 1$. But, if we take the number 1/2 which lies between $y = -1$ and $y = 1$, there is no value of x such that $f(x) = 1/2$.

USING TECHNOLOGY EXERCISES 2.5, page 152

1. $x = 0, 1$ 3. $x = 2$ 5. $x = 0, \frac{1}{2}$ 7. $x = -\frac{1}{2}, 2$ 9. $x = -2, 1$

11.

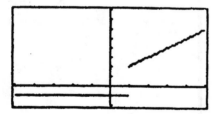

13

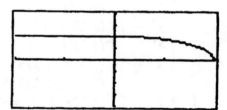

15.

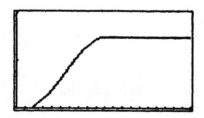

EXERCISES 2.6, page 169

1. The rate of change of the average infant's weight when $t = 3$ is (7.5)/5, or 1.5 lb/month. The rate of change of the average infant's weight when $t = 18$ is (3.5)/6, or approximately 0.6 lb/month. The average rate of change over the infant's first year of life is $(22.5 - 7.5)/(12)$, or 1.25 lb/month.

3. The rate of change of the percentage of households watching television at 4 P.M. is (12.3)/4, or approximately 3.1 percent per hour. The rate at 11 P.M. is $(-42.3)/2 = -21.15$; that is, it is dropping off at the rate of 21.15 percent per hour.

5. a. Car A is travelling faster than Car B at t_1 because the slope of the tangent line to the graph of f is greater than the slope of the tangent line to the graph of g at t_1.

    b. Their speed is the same because the slope of the tangent lines are the same at t_2.

    c. Car B is travelling faster than Car A.

    d. They have both covered the same distance and are once again side by side at t_3.

7. a. P_2 is decreasing faster at t_1 because the slope of the tangent line to the graph of g at t_1 is greater than the slope of the tangent line to the graph of f at t_1.
    b. P_1 is decreasing faster than P_2 at t_2.
    c. Bactericide B is more effective in the short run, but bactericide A is more effective in the long run.

9. **Step 1** $f(x + h) = 13$

    Step 2 $f(x + h) - f(x) = 13 - 13 = 0$

    Step 3 $\dfrac{f(x+h) - f(x)}{h} = \dfrac{0}{h} = 0$

    Step 4 $f'(x) = \lim\limits_{h \to 0} \dfrac{f(x+h) - f(x)}{h} = \lim\limits_{h \to 0} 0 = 0$

11. **Step 1** $f(x + h) = 2(x + h) + 7$
    Step 2 $f(x + h) - f(x) = 2(x + h) + 7 - (2x + 7) = 2h$
    Step 3 $\dfrac{f(x+h) - f(x)}{h} = \dfrac{2h}{h} = 2$
    Step 4 $f'(x) = \lim\limits_{h \to 0} \dfrac{f(x+h) - f(x)}{h} = \lim\limits_{h \to 0} 2 = 2$

13. **Step 1** $f(x+h) = 3(x+h)^2 = 3x^2 + 6xh + 3h^2$

Step 2 $f(x+h) - f(x) = (3x^2 + 6xh + 3h^2) - 3x^2 = 6xh + 3h^2 = h(6x + 3h)$

Step 3 $\dfrac{f(x+h) - f(x)}{h} = \dfrac{h(6x + 3h)}{h} = 6x + 3h$

Step 4 $f'(x) = \lim\limits_{h \to 0} \dfrac{f(x+h) - f(x)}{h} = \lim\limits_{h \to 0} (6x + 3h) = 6x.$

15. **Step 1** $f(x+h) = -(x+h)^2 + 3(x+h) = -x^2 - 2xh - h^2 + 3x + 3h$

Step 2 $f(x+h) - f(x) = (-x^2 - 2xh - h^2 + 3x + 3h) - (-x^2 + 3x)$

$$= -2xh - h^2 + 3h = h(-2x - h + 3)$$

Step 3 $\dfrac{f(x+h) - f(x)}{h} = \dfrac{h(-2x - h + 3)}{h} = -2x - h + 3$

Step 4 $f'(x) = \lim\limits_{h \to 0} \dfrac{f(x+h) - f(x)}{h} = \lim\limits_{h \to 0} (-2x - h + 3) = -2x + 3.$

17. Using the four-step process

Step 1 $f(x+h) = 2(x+h) + 7 = 2x + 2h + 7$

Step 2 $f(x+h) - f(x) = 2x + 2h + 7 - 2x - 7 = 2h$

Step 3 $\dfrac{f(x+h) - f(x)}{h} = \dfrac{2h}{h} = 2$

Step 4 $f'(x) = \lim\limits_{h \to 0} \dfrac{f(x+h) - f(x)}{h} = \lim\limits_{h \to 0} 2 = 2$

we find that $f'(x) = 2$. In particular, the slope at $x = 2$ is also 2. Therefore, a required equation is

$$y - 11 = 2(x - 2) \qquad \text{or} \qquad y = 2x + 7.$$

19. We first compute $f'(x) = 6x$ (see Problem 13). Since the slope of the tangent line is $f'(1) = 6$, we use the point-slope form of the equation of a line and find that a required equation is

$$y - 3 = 6(x - 1), \text{ or } y = 6x - 3.$$

21. We first compute $f'(x)$ using the four-step process.

Step 1 $f(x+h) = -\dfrac{1}{x+h}$

Step 2 $f(x+h) - f(x) = -\dfrac{1}{x+h} + \dfrac{1}{x} = \dfrac{-x+(x+h)}{x(x+h)} = \dfrac{h}{x(x+h)}$

Step 3 $\dfrac{f(x+h) - f(x)}{h} = \dfrac{\frac{h}{x(x+h)}}{h} = \dfrac{1}{x(x+h)}$

Step 4 $f'(x) = \lim\limits_{h \to 0} \dfrac{f(x+h) - f(x)}{h} = \lim\limits_{h \to 0} \dfrac{1}{x(x+h)} = \dfrac{1}{x^2}.$

The slope of the tangent line is $f'(3) = 1/9$. Therefore, a required equation is
$$y - (-\tfrac{1}{3}) = \tfrac{1}{9}(x-3) \quad \text{or} \quad y = \tfrac{1}{9}x - \tfrac{2}{3}.$$

23. a. We use the four-step process.

Step 1 $f(x+h) = 2(x+h)^2 + 1 = 2x^2 + 4xh + 2h^2 + 1$

Step 2 $f(x+h) - f(x) = (2x^2 + 4xh + 2h^2 + 1) - (2x^2 + 1) = 4xh + 2h^2$
$\qquad\qquad\qquad\quad = h(4x + 2h)$

Step 3 $\dfrac{f(x+h) - f(x)}{h} = \dfrac{h(4x+2h)}{h} = 4x + 2h$

Step 4 $f'(x) = \lim\limits_{h \to 0} \dfrac{f(x+h) - f(x)}{h} = \lim\limits_{h \to 0}(4x + 2h) = 4x$

b. The slope of the tangent line is $f'(1) = 4(1) = 4$. Therefore, an equation is
$y - 3 = 4(x - 1)$ or $y = 4x - 1$.

c.

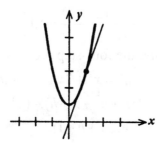

25. a. We use the four-step process:
Step 1 $f(x+h) = (x+h)^2 - 2(x+h) + 1 = x^2 + 2xh + h^2 - 2x - 2h + 1$

Step 2 $f(x+h)-f(x) = (x^2 +2xh+h^2 -2x-2h+1)-(x^2 -2x+1)]$
$$= 2xh+h^2 -2h = h(2x+h-2)$$

Step 3 $\dfrac{f(x+h)-f(x)}{h} = \dfrac{h(2x+h-2)}{h} = 2x+h-2$

Step 4 $f'(x)=\lim\limits_{h\to 0}\dfrac{f(x+h)-f(x)}{h} = \lim\limits_{h\to 0}(2x+h-2)=2x-2.$

b. At a point on the graph of f where the tangent line to the curve is horizontal, $f'(x)=0$. Then $2x-2=0$, or $x=1$. Since $f(1)=1-2+1=0$, we see that the required point is $(1,0)$.

c.

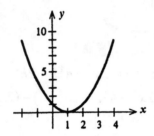

d. It is changing at the rate of 0 units per unit change in x.

27. a. $\dfrac{f(3)-f(2)}{3-2} = \dfrac{(3^2 +3)-(2^2 +2)}{1}=6$

$\dfrac{f(2.5)-f(2)}{2.5-2} = \dfrac{(2.5^2 +2.5)-(2^2 +2)}{0.5}=5.5$

$\dfrac{f(2.1)-f(2)}{2.1-2} = \dfrac{(2.1^2 +2.1)-(2^2 +2)}{0.1}=5.1$

b. We first compute $f'(x)$ using the four-step process.

Step 1 $f(x+h) = (x+h)^2 +(x+h) = x^2 +2xh+h^2 +x+h$

Step 2 $f(x+h)-f(x) = (x^2 +2xh+h^2 +x+h)-(x^2 +x)]$
$$= 2xh+h^2 +h = h(2x+h+1)$$

Step 3 $\dfrac{f(x+h)-f(x)}{h} = \dfrac{h(2x+h+1)}{h} = 2x+h+1$

Step 4 $f'(x)=\lim\limits_{h\to 0}\dfrac{f(x+h)-f(x)}{h} = \lim\limits_{h\to 0}(2x+h+1)=2x+1.$

The instantaneous rate of change of y at $x = 2$ is $f'(2) = 5$ or 5 units per unit change in x.

c. The results in (a) suggest that the average rates of change of f at $x = 2$ approach 5 as the interval $[2, 2+h]$ gets smaller and smaller ($h = 1, 0.5$, and 0.1). This number is the instantaneous rate of change of f at $x = 2$ as computed in (b).

29. a. The average velocity of the car over $[20,21]$ is
$$\frac{f(21) - f(20)}{21 - 20} = \frac{[2(21)^2 + 48(21)] - [2(20)^2 + 48(20)]}{1} = 130 \text{ ft / sec}$$
Its average velocity over $[20,20.1]$ is
$$\frac{f(20.1) - f(20)}{20.1 - 20} = \frac{[2(20.1)^2 + 48(20.1)] - [2(20)^2 + 48(20)]}{0.1} = 128.2 \text{ ft / sec}$$
Its average velocity over $[20,20.01]$
$$\frac{f(20.01) - f(20)}{20.01 - 20} = \frac{[2(20.01)^2 + 48(20.01)] - [2(20)^2 + 48(20)]}{0.01} = 128.02 \text{ ft / sec}$$
b. We first compute $f'(t)$ using the four-step process.

Step 1 $f(t + h) = 2(t + h)^2 + 48(t + h) = 2t^2 + 4th + 2h^2 + 48t + 48h$
Step 2 $f(t + h) - f(t) = (2t^2 + 4th + 2h^2 + 48t + 48h) - (2t^2 + 48t)]$
$$= 4th + 2h^2 + 48h = h(4t + 2h + 48).$$
Step 3 $\dfrac{f(t + h) - f(t)}{h}$

The instantaneous velocity of the car at $t = 20$ is
$$f'(20) = 4(20) + 48, \text{ or } 128 \text{ ft/sec.}$$
c. Our results shows that the average velocities do approach the instantaneous velocity as the intervals over which they are computed decreases.

31. a. We solve the equation $16t^2 = 400$ obtaining $t = 5$ which is the time it takes the screw driver to reach the ground.
b. The average velocity over the time $[0,5]$ is
$$\frac{f(5) - f(0)}{5 - 0} = \frac{16(25) - 0}{5} = 80, \text{ or } 80 \text{ ft/sec.} \quad [\text{Let } s = f(t) = 16t^2.]$$
c. The velocity of the screwdriver at time t is
$$v(t) = \lim_{h \to 0} \frac{f(t + h) - f(t)}{h} = \lim_{h \to 0} \frac{16(t + h)^2 - 16t^2}{h}$$

$$= \lim_{h\to 0}\frac{16t^2 + 32th + 16h^2 - 16t^2}{h} = \lim_{h\to 0}\frac{(32t + 16h)h}{h}.$$

In particular, the velocity of the screwdriver when it hits the ground (at $t = 5$) is
$v(5) = 32(5) = 160$, or 160 ft/sec.

33. a. The average rate of change of V is
$$\frac{f(3) - f(2)}{3 - 2} = \frac{\frac{1}{3} - \frac{1}{2}}{1} = -\frac{1}{6}, \qquad \left[\text{Write } V = f(p) = \frac{1}{p}.\right]$$
or a decrease of $\frac{1}{6}$ liter/atmosphere.

b. $\qquad V'(t) = \lim_{h\to 0}\frac{f(p+h) - f(p)}{h} = \lim_{h\to 0}\frac{\frac{1}{p+h} - \frac{1}{p}}{h}$

$$= \lim_{h\to 0}\frac{p - (p+h)}{hp(p+h)} = \lim_{h\to 0}-\frac{1}{p(p+h)} = -\frac{1}{p^2}.$$

In particular, the rate of change of V when $p = 2$ is
$$V'(2) = -\frac{1}{2^2}, \text{ or a decrease of } \frac{1}{4} \text{ liter/atmosphere}$$

35. a. Using the four-step process, we find that
$$P'(x) = \lim_{h\to 0}(-\tfrac{2}{3}x - \tfrac{1}{3}h + 7) = -\tfrac{2}{3}x + 7.$$

b. $\quad P'(10) = -\tfrac{2}{3}(10) + 7 \approx 0.333$, or \$333 per quarter.

$P'(30) = -\tfrac{2}{3}(30) + 7 \approx -13$, or a decrease of \$13,000 per quarter.

37. We first compute $N'(t)$ using the four–step process.
Step 1 $\quad N(t + h) = (t + h)^2 + 2(t + h) + 50$
$\qquad\qquad = t^2 + 2th + h^2 + 2t + 2h + 50$
Step 2 $\quad N(t + h) - N(t)$
$\qquad\qquad = (t^2 + 2th + h^2 + 2t + 2h + 50) - (t^2 + 2t + 50)$
$\qquad\qquad = 2th + h^2 + 2h = h(2t + h + 2).$
Step 3 $\quad \dfrac{N(t+h) - N(t)}{h} = 2t + h + 2.$
Step 4 $\quad N'(t) = \lim_{h\to 0}(2t + h + 2) = 2t + 2.$

The rate of change of the country's GNP two years from now will be $N'(2) = 6$, or
\$6 billion/yr. The rate of change four years from now will be $N'(4) = 10$, or
\$10 billion/yr.

39. $\dfrac{f(a+h)-f(a)}{h}$ gives the average rate of change of the seal population over the time interval $[a, a+h]$. $\displaystyle\lim_{h\to 0}\dfrac{f(a+h)-f(a)}{h}$ gives the instantaneous rate of change of the seal population at $x=a$.

41. $\dfrac{f(a+h)-f(a)}{h}$ gives the average rate of change of the country's industrial production over the time interval $[a, a+h]$. $\displaystyle\lim_{h\to 0}\dfrac{f(a+h)-f(a)}{h}$ gives the instantaneous rate of change of the country's industrial production at $x=a$.

43. $\dfrac{f(a+h)-f(a)}{h}$ gives the average rate of change of the atmospheric pressure over the altitudes $[a, a+h]$. $\displaystyle\lim_{h\to 0}\dfrac{f(a+h)-f(a)}{h}$ gives the instantaneous rate of change of the atmospheric pressure at $x=a$.

45. a. f has a limit at $x=a$.
    b. f is not continuous at $x=a$ because $f(a)$ is not defined.
    c. f is not differentiable at $x=a$ because it is not continuous there.

47. a. f has a limit at $x=a$.   b. f is continuous at $x=a$.
    c. f is not differentiable at $x=a$ because f has a kink at the point $x=a$.

49. a. f does not have a limit at $x=a$ because it is unbounded in the neighborhood of a.
    b. f is not continuous at $x=a$.
    c. f is not differentiable at $x=a$ because it is not continuous there.

51. Our computations yield the following results:
    5.06060, 5.06006, 5.060006, 5.0600006, 5.06000006;
    The rate of change of the total cost function when the level of production is 100 cases a day is approximately $5.06.

53. True. If g is differentiable at $x=a$, then it is continuous there. Therefore, the product fg is continuous. Therefore,

$$\lim_{x \to a} f(x)g(x) = \left[\lim_{x \to a} f(x)\right]\left[\lim_{x \to 1} g(x)\right] = f(a)g(a).$$

55. f does not have a derivative at $x = 1$ because it is not continuous there.

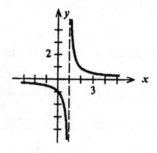

57. f is continuous at $x = 0$, but $f'(0)$ does not exist because the graph of f has a vertical tangent line at $x = 0$. The graph of f follows.

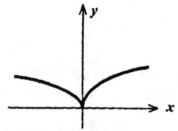

59. From $\qquad f(x) - f(a) = \left[\dfrac{f(x) - f(a)}{x - a}\right](x - a) \qquad$ we see that

$$\lim_{x \to a}[f(x) - f(a)] = \lim_{x \to a}\left[\frac{f(x) - f(a)}{x - a}\right]\lim_{x \to a}(x - a)$$
$$= f'(a) \cdot 0 = 0$$

and so $\lim_{x \to a} f(x) = f(a)$. This shows that f is continuous at $x = a$.

1. a. $y = 4x - 3$
    b.

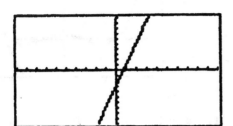

3. a. $y = -7x - 8$
    b.

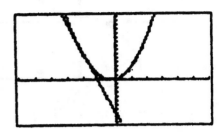

5. a. $y = 9x - 11$

    b.

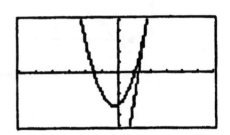

7. a. $y = 2$
    b.

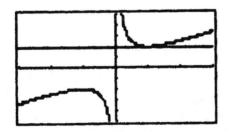

9. a. $y = \frac{1}{4}x + 1$
    b.

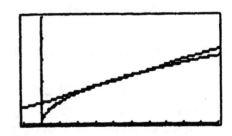

11. a. 4 b. $y = 4x - 1$
    c.

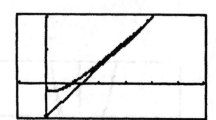

13. a. 20 b. $y = 20x - 35$
    c.

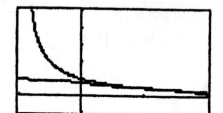

15. a. $\frac{3}{4}$ b. $y = \frac{3}{4}x - 1$
    c.

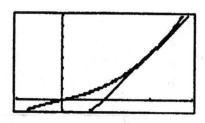

17. a. $-\frac{1}{4}$ b. $y = -\frac{1}{4}x + \frac{3}{4}$
    c.

19. a. 4.02 b. $y = 4.02x - 3.57$
    c.

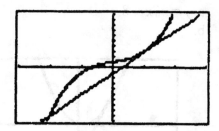

CHAPTER 2 REVIEW, page 179

1. a. $9 - x \geq 0$ gives $x \leq 9$ and the domain is $(-\infty, 9]$.
    b. $2x^2 - x - 3 = (2x - 3)(x + 1)$, and $x = 3/2$ or $x = -1$.
    Since the denominator of the given expression is zero at these points, we see that the domain of f cannot include these points and so the domain of f is
    $(-\infty, -1) \cup (-1, \frac{3}{2}) \cup (\frac{3}{2}, \infty)$.

3. a.

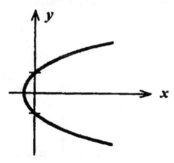

b. For each value of $x > 0$, there are two values of y. We conclude that y is not a function of x. Equivalently, the function fails the vertical line test.

c. Yes. For each value of y, there is only 1 value of x.

5. a. $f(x)g(x) = \dfrac{2x+3}{x}$

b. $\dfrac{f(x)}{g(x)} = \dfrac{1}{x(2x+3)}$

c. $f(g(x)) = \dfrac{1}{2x+3}$.

d. $g(f(x)) = 2\left(\dfrac{1}{x}\right) + 3 = \dfrac{2}{x} + 3$.

7. $\lim\limits_{x\to 1}(x^2 + 1) = \lim\limits_{x\to 1}[(1)^2 + 1] = 1 + 1 = 2.$

9. $\lim\limits_{x\to 3}\dfrac{x-3}{x+4} = \dfrac{3-3}{3+4} = 0.$

11. $\lim\limits_{x\to -2}\dfrac{x^2 - 2x - 3}{x^2 + 5x + 6}$ does not exist. (The denominator is 0 at $x = -2$.)

13. $\lim\limits_{x\to 3}\dfrac{4x-3}{\sqrt{x+1}} = \dfrac{12-3}{\sqrt{4}} = \dfrac{9}{2}.$

15. $\lim\limits_{x\to 1^-}\dfrac{\sqrt{x}-1}{x-1} = \lim\limits_{x\to 1^-}\dfrac{(\sqrt{x}-1)(\sqrt{x}+1)}{(x-1)(\sqrt{x}+1)} = \lim\limits_{x\to 1^-}\dfrac{x-1}{(x-1)(\sqrt{x}+1)}$

$= \lim\limits_{x\to 1^-}\dfrac{1}{\sqrt{x}+1} = \dfrac{1}{2}.$

17. $\lim\limits_{x\to -\infty}\dfrac{x+1}{x} = \lim\limits_{x\to -\infty}\left(1 + \dfrac{1}{x}\right) = 1.$

19. $\lim\limits_{x\to-\infty}\dfrac{x^2}{x+1} = \lim\limits_{x\to-\infty} x\cdot\dfrac{1}{1+\dfrac{1}{x}} = -\infty$, so the limit does not exist.

21. $\lim\limits_{x\to 2^+} f(x) = \lim\limits_{x\to 2^+}(x+2) = 4$;
$\lim\limits_{x\to 2^-} f(x) = \lim\limits_{x\to 2^-}(4-x) = 2.$
Therefore, $\lim\limits_{x\to 2} f(x)$ does not exist.

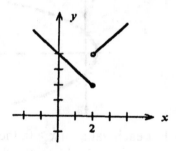

23. Since the denominator
$$4x^2 - 2x - 2 = 2(2x^2 - x - 1) = 2(2x+1)(x-1) = 0$$
if $x = -1/2$ or 1, we see that f is discontinuous at these points.

25. The function is discontinuous at $x = 0$.

27. Using the four-step process, we find
Step 1 $f(x+h) = 3(x+h) + 5 = 3x + 3h + 5$
Step 2 $f(x+h) - f(x) = 3x + 3h + 5 - 3x - 5 = 3h$
Step 3 $\dfrac{f(x+h)-f(x)}{h} = \dfrac{3h}{h} = 3.$
Step 4 $f'(x) = \lim\limits_{h\to 0}\dfrac{f(x+h)-f(x)}{h} = \lim\limits_{h\to 0}(3) = 3.$

29. We use the four-step process to obtain
Step 1 $f(x+h) = \frac{3}{2}(x+h) + 5 = \frac{3}{2}x + \frac{3}{2}h + 5.$

Step 2 $f(x+h) - f(x) = \frac{3}{2}x + \frac{3}{2}h + 5 - \frac{3}{2}x - 5 = \frac{3}{2}h.$

Step 3 $\dfrac{f(x+h)-f(x)}{h} = \dfrac{3}{2}.$

Step 4 $f'(x) = \lim\limits_{h\to 0}\dfrac{f(x+h)-f(x)}{h} = \lim\limits_{h\to 0}\dfrac{3}{2} = \dfrac{3}{2}.$

Therefore, the slope of the tangent line to the graph of the function f at the point
(-2,2) is 3/2. To find the equation of the tangent line to the curve at the point
(-2,2), we use the point–slope form of the equation of a line obtaining
$$y - 2 = \tfrac{3}{2}[x-(-2)] \quad \text{or} \quad y = \tfrac{3}{2}x + 5.$$

31. a. f is continuous at $x = a$ because the three conditions for continuity are satisfied at $x = a$; that is,
      i. $f(x)$ is defined
      ii. $\lim\limits_{x \to a} f(x)$ exists
      iii. $\lim\limits_{x \to a} f(x) = f(a)$
  b. f is not differentiable at $x = a$ because the graph of f has a kink at $x = a$.

33. a. The line passes through (0, 2.4) and (5, 7.4) and has slope $m = \dfrac{7.4 - 2.4}{5 - 0} = 1.$

    Letting y denote the sales, we see that an equation of the line is
$$y - 2.4 = 1(t - 0), \text{ or } y = t + 2.4.$$
    We can also write this in the form $S(t) = t + 2.4$.
  b. The sales in 1999 were $S(3) = 3 + 2.4 = 5.4$, or \$5.4 million.

35. Substituting the first equation into the second yields
$$3x - 2(\tfrac{3}{4}x + 6) + 3 = 0 \quad \text{or} \quad \tfrac{3}{2}x - 12 + 3 = 0$$

    or $x = 6$. Substituting this value of x into the first equation then gives $y = 21/2$, so the point of intersection is $(6, \tfrac{21}{2})$.

37. We solve the system
$$3x + p - 40 = 0$$
$$2x - p + 10 = 0.$$
    Adding, we obtain $5x - 30 = 0$, or $x = 6$. So,
$$p = 2x + 10 = 12 + 10 = 22.$$
    Therefore, the equilibrium quantity is 6000 and the equilibrium price is \$22.

39. $R(30) = -\tfrac{1}{2}(30)^2 + 30(30) = 450$, or \$45,000.

41. $T = f(n) = 4n\sqrt{n - 4}$.
    $f(4) = 0$, $f(5) = 20\sqrt{1} = 20$, $f(6) = 24\sqrt{2} \approx 33.9$, $f(7) = 28\sqrt{3} \approx 48.5$,
    $f(8) = 32\sqrt{4} = 64$, $f(9) = 36\sqrt{5} \approx 80.5$, $f(10) = 40\sqrt{6} \approx 98$,
    $f(11) = 44\sqrt{7} \approx 116$ and $f(12) = 48\sqrt{8} \approx 135.8$.

The graph of f follows:

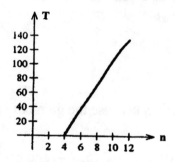

43.

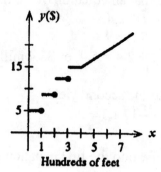

The function is discontinuous at $x = 100$, 200, and 300.

CHAPTER 3

EXERCISES 3.1, page 191

1. $f'(x) = \dfrac{d}{dx}(-3) = 0.$

3. $f'(x) = \dfrac{d}{dx}(x^5) = 5x^4.$

5. $f'(x) = \dfrac{d}{dx}(x^{2.1}) = 2.1x^{1.1}.$

7. $f'(x) = \dfrac{d}{dx}(3x^2) = 6x.$

9. $f'(r) = \dfrac{d}{dr}(\pi r^2) = 2\pi r.$

11. $f'(x) = \dfrac{d}{dx}(9x^{1/3}) = \dfrac{1}{3}(9)x^{(1/3-1)} = 3x^{-2/3}.$

13. $f'(x) = \dfrac{d}{dx}(3\sqrt{x}) = \dfrac{d}{dx}(3x^{1/2}) = \dfrac{1}{2}(3)x^{-1/2} = \dfrac{3}{2}x^{-1/2} = \dfrac{3}{2\sqrt{x}}.$

15. $f'(x) = \dfrac{d}{dx}(7x^{-12}) = (-12)(7)x^{(-12-1)} = -84x^{-13}.$

17. $f'(x) = \dfrac{d}{dx}(5x^2 - 3x + 7) = 10x - 3.$

19. $f'(x) = \dfrac{d}{dx}(-x^3 + 2x^2 - 6) = -3x^2 + 4x.$

21. $f'(x) = \dfrac{d}{dx}(0.03x^2 - 0.4x + 10) = 0.06x - 0.4.$

23. If $f(x) = \dfrac{x^3 - 4x^2 + 3}{x} = x^2 - 4x + \dfrac{3}{x},$

    then $f'(x) = \dfrac{d}{dx}(x^2 - 4x + 3x^{-1}) = 2x - 4 - \dfrac{3}{x^2}.$

25. $f'(x) = \dfrac{d}{dx}(4x^4 - 3x^{5/2} + 2) = 16x^3 - \tfrac{15}{2}x^{3/2}.$

3 Differentiation

27. $f'(x) = \dfrac{d}{dx}\left(3x^{-1} + 4x^{-2}\right) = -3x^{-2} - 8x^{-3}$.

29. $f'(t) = \dfrac{d}{dt}\left(4t^{-4} - 3t^{-3} + 2t^{-1}\right) = -16t^{-5} + 9t^{-4} - 2t^{-2}$.

31. $f'(x) = \dfrac{d}{dx}\left(2x - 5x^{1/2}\right) = 2 - \dfrac{5}{2}x^{-1/2} = 2 - \dfrac{5}{2\sqrt{x}}$.

33. $f'(x) = \dfrac{d}{dx}\left(2x^{-2} - 3x^{-1/3}\right) = -4x^{-3} + x^{-4/3} = -\dfrac{4}{x^3} + \dfrac{1}{x^{4/3}}$.

35. $f'(x) = \dfrac{d}{dx}\left(2x^3 - 4x\right) = 6x^2 - 4$.

    a. $f'(-2) = 6(-2)^2 - 4 = 20$.

    b. $f'(0) = 6(0) - 4 = -4$.

    c. $f'(2) = 6(2)^2 - 4 = 20$.

37. The given limit is $f'(1)$ where $f(x) = x^3$. Since $f'(x) = 3x^2$, we have

$$\lim_{h \to 0} \dfrac{(1+h)^3 - 1}{h} = f'(1) = 3$$

39. Let $f(x) = 3x^2 - x$. Then

$$\lim_{h \to 0} \dfrac{3(2+h)^2 - (2+h) - 10}{h} = \lim_{h \to 0} \dfrac{f(2+h) - f(2)}{h}$$

because $f(2 + h) - f(2) = 3(2 + h)^2 - (2 + h) - [3(4) - 2]$
$\qquad\qquad\qquad\qquad = 3(2 + h)^2 - (2 + h) - 10$.

But the last limit is $f'(2)$. Since $f'(x) = 6x - 1$, we have $f'(2) = 11$.

Therefore, $\lim\limits_{h \to 0} \dfrac{3(2+h)^2 - (2+h) - 10}{h} = 11$.

41.   The slope of the tangent line at any point $(x, f(x))$ on the graph of f is
$$f'(x) = 4x - 3.$$
In particular, the slope of the tangent line at the point $(2,6)$ is
$$f'(2) = 4(2) - 3 = 5.$$

An equation of the required tangent line is
$$y - 6 = 5(x - 2) \qquad \text{or} \qquad y = 5x - 4.$$

43. $f'(x) = 4x^3 - 9x^2 + 4x - 1$. The slope is $f'(1) = 4 - 9 + 4 - 1 = -2$. An equation of the tangent line is $y - 0 = -2(x - 1)$ or $y = -2x + 2$.

45. a. $f'(x) = 3x^2$. At a point where the tangent line is horizontal, $f'(x) = 0$, or $3x^2 = 0$ giving $x = 0$. Therefore, the point is $(0,0)$.

b.
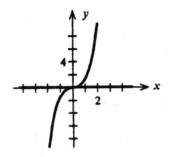

47. a. The slope of the tangent line at any point $(x, f(x))$ on the graph of f is
$$f'(x) = 3x^2.$$
At the point(s) where the slope is 12, we have
$$3x^2 = 12, \text{ or } x = \pm 2.$$
The required points are $(-2,-7)$ and $(2,9)$.
b. The tangent line at $(-2,-7)$ has equation
$$y - (-7) = 12[x - (-2)], \qquad \text{or} \qquad y = 12x + 17,$$
and the tangent line at $(2,9)$ has equation
$$y - 9 = 12(x - 2), \qquad \text{or} \qquad y = 12x - 15.$$
c.
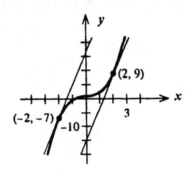

49. If $f(x) = \frac{1}{4}x^4 - \frac{1}{3}x^3 - x^2$, then $f'(x) = x^3 - x^2 - 2x$.
   a. $f'(x) = x^3 - x^2 - 2x = -2x$
   $$x^3 - x^2 = 0$$
   $$x^2(x - 1) = 0$$
   and $\quad x = 0$ or $x = 1$.

   $$f(1) = \frac{1}{4}(1)^4 - \frac{1}{3}(1)^3 - (1)^2 = -\frac{13}{12}.$$
   $$f(0) = \frac{1}{4}(0)^4 - \frac{1}{3}(0)^3 - (0)^2 = 0.$$
   We conclude that the corresponding points on the graph are $(1, -\frac{13}{12})$ and $(0,0)$.

   b. $\quad f'(x) = x^3 - x^2 - 2x = 0$
   $$x(x^2 - x - 2) = 0$$
   $$x(x - 2)(x + 1) = 0$$
   and $\quad x = 0, 2,$ or -1.
   $$f(0) = 0$$
   $$f(2) = \frac{1}{4}(2)^4 - \frac{1}{3}(2)^3 - (2)^2 = 4 - \frac{8}{3} - 4 = -\frac{8}{3}.$$
   $$f(-1) = \frac{1}{4}(-1)^4 - \frac{1}{3}(-1)^3 - (-1)^2 = \frac{1}{4} + \frac{1}{3} - 1 = -\frac{5}{12}.$$
   We conclude that the corresponding points are $(0,0)$, $(2, -\frac{8}{3})$ and $(-1, -\frac{5}{12})$.

   c. $\quad f'(x) = x^3 - x^2 - 2x = 10x$
   $$x^3 - x^2 - 12x = 0$$
   $$x(x^2 - x - 12) = 0$$
   $$x(x - 4)(x + 3) = 0$$
   and $x = 0, 4,$ or -3.
   $$f(0) = 0$$
   $$f(4) = \frac{1}{4}(4)^4 - \frac{1}{3}(4)^3 - (4)^2 = 48 - \frac{64}{3} = \frac{80}{3}.$$
   $$f(-3) = \frac{1}{4}(-3)^4 - \frac{1}{3}(-3)^3 - (-3)^2 = \frac{81}{4} + 9 - 9 = \frac{81}{4}.$$
   We conclude that the corresponding points are $(0,0)$, $(4, \frac{80}{3})$ and $(-3, \frac{81}{4})$.

51. $V'(r) = 4\pi r^2$.
   a. $V'(\frac{2}{3}) = 4\pi(\frac{4}{9}) = \frac{16}{9}\pi$ cm³/cm.
   b. $V'(\frac{5}{4}) = 4\pi(\frac{25}{16}) = \frac{25}{4}\pi$ cm³/cm.

53. $\dfrac{dA}{dx} = 26.5\dfrac{d}{dx}(x^{-0.45}) = 26.5(-0.45)x^{-1.45} = -\dfrac{11.925}{x^{1.45}}.$

Therefore, $\left.\dfrac{dA}{dx}\right|_{x=0.25} = -\dfrac{11.925}{(0.25)^{1.45}} \approx -89.01$ and $\left.\dfrac{dA}{dx}\right|_{x=2} = -\dfrac{11.925}{(2)^{1.45}} \approx -4.36$

Our computations reveal that if you make 0.25 stops per mile, your average speed will decrease at the rate of approximately 89.01 mph per stop per mile. If you make 2 stops per mile, your average speed will decrease at the rate of approximately 4.36 mph per stop per mile.

55. $I'(t) = -0.6t^2 + 6t$.
  a. In 1995, it was changing at a rate of $I'(5) = -0.6(25) + 6(5)$, or 15 points/yr. In 1997, it was $I'(7) = -0.6(49) + 6(7)$, or 12.6 pts/yr. In 2000, it was $I'(10) = -0.6(100) + 6(10)$, or 0 pts/yr.
  b. The average rate of increase of the CPI over the period from 1995 to 2000 was
$$\dfrac{I(10) - I(5)}{5} = \dfrac{[-0.2(1000) + 3(100) + 100] - [-0.2(125) + 3(25) + 100]}{5}$$
$$= \dfrac{200 - 150}{5} = 10, \text{ or } 10 \text{ pts/yr.}$$

57. The rate at which the population will be increasing at any time t is
$$P'(t) = 45t^{1/2} + 20.$$
  Nine months from now the population will be increasing at
$$P'(9) = 45(3) + 20, \text{ or } 155 \text{ people/month.}$$
  Sixteen months from now the population will be increasing at
$$P'(16) = 45(4) + 20, \text{ or } 200 \text{ people/month.}$$

59. $N'(t) = 6t^2 + 6t - 4$.
  $N'(2) = 6(4) + 6(2) - 4 = 32$, or 32 turtles/yr.
  $N'(8) = 6(64) + 6(8) - 4 = 428$, or 428 turtles/yr.
  The population ten years after implementation of the conservation measures will be $N(10) = 2(1000) + 3(100) - 4(10) + 1000$, or 3260 turtles.

61. a. $v = f'(t) = 120 - 30t$ b. $v(0) = 120$ ft/sec
  c. Setting $v = 0$ gives $120 - 30t = 0$, or $t = 4$. Therefore, the stopping distance is
$$f(4) = 120(4) - 15(16) \text{ or } 240 \text{ ft.}$$

63. a. The number of temporary workers at the beginning of 1994 ($t = 3$) was
$$N(3) = 0.025(9) + 0.255(3) + 1.505 = 2.495 \text{ million.}$$
  b. $N'(t) = 0.05t + 0.255$.

So, at the beginning of 1994 ($t = 3$), the number of temporary workers was growing at the rate of

$$N'(3) = 0.05(3) + 0.255 = 0.405, \text{ or } 405{,}000 \text{ per year.}$$

65. a. $f'(x) = \dfrac{d}{dx}\left[0.0001x^{5/4} + 10\right] = \dfrac{5}{4}(0.0001x^{1/4}) = 0.000125x^{1/4}$

b. $f'(10{,}000) = 0.000125(10{,}000)^{1/4} = 0.00125$, or $0.00125/radio.

67. a. $f'(t) = 20 - 40\left(\dfrac{1}{2}\right)t^{-1/2} = 20\left(1 - \dfrac{1}{\sqrt{t}}\right)$.

b. $f(0) = 20(0) - 40\sqrt{0} + 50 = 50$
    $f(1) = 20(1) - 40\sqrt{1} + 50 = 30$
    $f(2) = 20(2) - 40\sqrt{2} + 50 \approx 33.43$.
The average velocity at 6, 7, and 8 A.M. is 50 mph, 30 mph, and 33.43 mph, respectively.

c. $f'(\tfrac{1}{2}) = 20 - 20(\tfrac{1}{2})^{-1/2} \approx -8.28$.
    $f'(1) = 20 - 20(1)^{-1/2} \approx 0$.
    $f'(2) = 20 - 20(2)^{-1/2} \approx 5.86$.
At 6:30 A.M. the average velocity is decreasing at the rate of 8.28 mph/hr; at 7 A.M., it is unchanged, and at 8 A.M., it is increasing at the rate of 5.86 mph.

69. a. $\dfrac{d}{dx}[0.075t^3 + 0.025t^2 + 2.45t + 2.4] = 0.225t^2 + 0.05t + 2.45$

b. $f'(3) = 0.225(3)^2 + 0.05(3) + 2.45 = 4.625$, or $4.625 billion/yr.

c. $f(3) = 0.075(3)^3 + 0.025(3)^2 + 2.45(3) + 2.4 = 12$, or $12 billion/yr.

71. False. f is *not* a power function.

USING TECHNOL0GY EXERCISES 3.1, page 193

1. 1   3. 0.4226   5.   0.1613

7.   a.

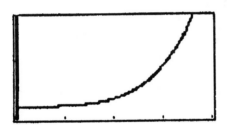

b. 3.4295 parts/million;
    105.4332 parts/million

9.   a.

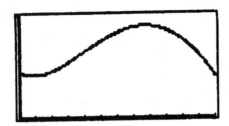

b. decreasing at the rate of 9 days/yr
   increasing at the rate of 13 days/yr

11.   a.

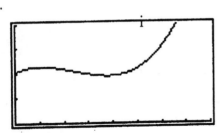

b. decreasing at the rate of 0.188887%/yr

   increasing at the rate of 0.0777812%/yr

EXERCISES 3.2, page 205

1. $f(x) = 2x(x^2 + 1)$.
$$f'(x) = 2x\frac{d}{dx}\left(x^2 + 1\right) + (x^2 + 1)\frac{d}{dx}(2x)$$
$$= 2x(2x) + (x^2 + 1)(2) = 6x^2 + 2.$$

3. $f(t) = (t - 1)(2t + 1)$
$$f'(t) = (t - 1)\frac{d}{dt}(2t + 1) + (2t + 1)\frac{d}{dt}(t - 1) = (t - 1)(2) + (2t + 1)(1)$$
$$= 4t - 1.$$

3 Differentiation

5. $f(x) = (3x+1)(x^2 - 2)$

$$f'(x) = (3x+1)\frac{d}{dx}\left(x^2 - 2\right) + (x^2 - 2)\frac{d}{dx}(3x+1)$$

$$= (3x+1)(2x) + (x^2 - 2)(3) = 9x^2 + 2x - 6.$$

7. $f(x) = (x^3 - 1)(x+1).$

$$f'(x) = (x^3 - 1)\frac{d}{dx}(x+1) + (x+1)\frac{d}{dx}(x^3 - 1)$$

$$= (x^3 - 1)(1) + (x+1)(3x^2) = 4x^3 + 3x^2 - 1.$$

9. $f(w) = (w^3 - w^2 + w - 1)(w^2 + 2).$

$$f'(w) = (w^3 - w^2 + w - 1)\frac{d}{dw}\left(w^2 + 2\right) + (w^2 + 2)\frac{d}{dw}(w^3 - w^2 + w - 1)$$

$$= (w^3 - w^2 + w - 1)(2w) + (w^2 + 2)(3w^2 - 2w + 1)$$

$$= 2w^4 - 2w^3 + 2w^2 - 2w + 3w^4 - 2w^3 + w^2 + 6w^2 - 4w + 2$$

$$= 5w^4 - 4w^3 + 9w^2 - 6w + 2.$$

11. $f(x) = (5x^2 + 1)(2\sqrt{x} - 1)$

$$f'(x) = (5x^2 + 1)\frac{d}{dx}(2x^{1/2} - 1) + (2x^{1/2} - 1)\frac{d}{dx}(5x^2 + 1)$$

$$= (5x^2 + 1)(x^{-1/2}) + (2x^{1/2} - 1)(10x)$$

$$= 5x^{3/2} + x^{-1/2} + 20x^{3/2} - 10x$$

$$= \frac{25x^2 - 10x\sqrt{x} + 1}{\sqrt{x}}.$$

13. $f(x) = (x^2 - 5x + 2)(x - \frac{2}{x})$

$$f'(x) = (x^2 - 5x + 2)\frac{d}{dx}(x - \frac{2}{x}) + (x - \frac{2}{x})\frac{d}{dx}(x^2 - 5x + 2)$$

$$= \frac{(x^2 - 5x + 2)(x^2 + 2)}{x^2} + \frac{(x^2 - 2)(2x - 5)}{x}$$

$$= \frac{(x^2 - 5x + 2)(x^2 + 2) + x(x^2 - 2)(2x - 5)}{x^2}$$

$$= \frac{x^4 + 2x^2 - 5x^3 - 10x + 2x^2 + 4 + 2x^4 - 5x^3 - 4x^2 + 10x}{x^2}$$

$$= \frac{3x^4 - 10x^3 + 4}{x^2}.$$

15. $f(x) = \dfrac{1}{x-2}.$ $f'(x) = \dfrac{(x-2)\dfrac{d}{dx}(1) - (1)\dfrac{d}{dx}(x-2)}{(x-2)^2} = \dfrac{0 - 1(1)}{(x-2)^2} = -\dfrac{1}{(x-2)^2}.$

17. $f(x) = \dfrac{x-1}{2x+1}.$

$$f'(x) = \frac{(2x+1)\dfrac{d}{dx}(x-1) - (x-1)\dfrac{d}{dx}(2x+1)}{(2x+1)^2}$$

$$= \frac{2x+1 - (x-1)(2)}{(2x+1)^2} = \frac{3}{(2x+1)^2}.$$

19. $f(x) = \dfrac{1}{x^2+1}.$

$$f'(x) = \frac{(x^2+1)\dfrac{d}{dx}(1) - (1)\dfrac{d}{dx}(x^2+1)}{(x^2+1)^2}$$

$$= \frac{(x^2+1)(0) - 1(2x)}{(x^2+1)^2} = -\frac{2x}{(x^2+1)^2}.$$

21. $f(s) = \dfrac{s^2 - 4}{s+1}.$

$$f'(s) = \frac{(s+1)\dfrac{d}{ds}(s^2-4) - (s^2-4)\dfrac{d}{ds}(s+1)}{(s+1)^2} = \frac{s^2 + 2s + 4}{(s+1)^2}.$$

23. $f(x) = \dfrac{\sqrt{x}}{x^2+1}.$

$$f'(x) = \frac{(x^2+1)\frac{d}{dx}(x^{1/2}) - (x^{1/2})\frac{d}{dx}(x^2+1)}{(x^2+1)^2}$$

$$= \frac{(\frac{1}{2}x^{-1/2})[(x^2+1) - 4x^2]}{(x^2+1)^2} = \frac{1-3x^2}{2\sqrt{x}(x^2+1)^2}.$$

25. $f(x) = \dfrac{x^2+2}{x^2+x+1}.$

$$f'(x) = \frac{(x^2+x+1)\frac{d}{dx}(x^2+2) - (x^2+2)\frac{d}{dx}(x^2+x+1)}{(x^2+x+1)^2}$$

$$= \frac{(x^2+x+1)(2x) - (x^2+2)(2x+1)}{(x^2+x+1)^2}$$

$$= \frac{2x^3+2x^2+2x-2x^3-x^2-4x-2}{(x^2+x+1)^2} = \frac{x^2-2x-2}{(x^2+x+1)^2}.$$

27. $f(x) = \dfrac{(x+1)(x^2+1)}{x-2} = \dfrac{(x^3+x^2+x+1)}{x-2}.$

$$f'(x) = \frac{(x-2)\frac{d}{dx}(x^3+x^2+x+1) - (x^3+x^2+x+1)\frac{d}{dx}(x-2)}{(x-2)^2}$$

$$= \frac{(x-2)(3x^2+2x+1) - (x^3+x^2+x+1)}{(x-2)^2}$$

$$= \frac{3x^3+2x^2+x-6x^2-4x-2-x^3-x^2-x-1}{(x-2)^2}$$

$$= \frac{2x^3-5x^2-4x-3}{(x-2)^2}.$$

29. $f(x) = \dfrac{x}{x^2-4} - \dfrac{x-1}{x^2+4} = \dfrac{x(x^2+4)-(x-1)(x^2-4)}{(x^2-4)(x^2+4)} = \dfrac{x^2+8x-4}{(x^2-4)(x^2+4)}.$

$$f'(x) = \frac{(x^2-4)(x^2+4)\frac{d}{dx}(x^2+8x-4)-(x^2+8x-4)\frac{d}{dx}(x^4-16)}{(x^2-4)^2(x^2+4)^2}$$

$$= \frac{(x^2-4)(x^2+4)(2x+8)-(x^2+8x-4)(4x^3)}{(x^2-4)^2(x^2+4)^2}$$

$$= \frac{2x^5+8x^4-32x-128-4x^5-32x^4+16x^3}{(x^2-4)^2(x^2+4)^2}$$

$$= \frac{-2x^5-24x^4+16x^3-32x-128}{(x^2-4)^2(x^2+4)^2}.$$

31. $h'(x) = f(x)g'(x) + f'(x)g(x)$, by the Product Rule. Therefore,
$h'(1) = f(1)g'(1) + f'(1)g(1) = (2)(3) + (-1)(-2) = 8.$

33. Using the Quotient Rule followed by the Product Rule, we have

$$h'(x) = \frac{[x+g(x)]\frac{d}{dx}[xf(x)]-xf(x)\frac{d}{dx}[x+g(x)]}{[x+g(x)]^2}$$

$$= \frac{[x+g(x)][xf'(x)+f(x)]-xf(x)[1+g'(x)]}{[x+g(x)]^2}$$

Therefore,

$$h'(1) = \frac{[1+g(1)][f'(1)+f(1)]-f(1)[1+g'(1)]}{[1+g(1)]^2}$$

$$= \frac{(1-2)(-1+2)-2(1+3)}{(1-2)^2} = \frac{-1-8}{1} = -9.$$

35. $f(x) = (2x-1)(x^2+3)$

$$f'(x) = (2x-1)\frac{d}{dx}(x^2+3) + (x^2+3)\frac{d}{dx}(2x-1)$$

$$= (2x-1)(2x) + (x^2+3)(2) = 6x^2 - 2x + 6 = 2(3x^2 - x + 3).$$

At $x = 1, f'(1) = 2[3(1)^2 - (1) + 3] = 2(5) = 10.$

37. $f(x) = \dfrac{x}{x^4 - 2x^2 - 1}.$

$$f'(x) = \frac{(x^4 - 2x^2 - 1)\dfrac{d}{dx}(x) - x\dfrac{d}{dx}(x^4 - 2x^2 - 1)}{(x^4 - 2x^2 - 1)^2}$$

$$= \frac{(x^4 - 2x^2 - 1)(1) - x(4x^3 - 4x)}{(x^4 - 2x^2 - 1)^2} = \frac{-3x^4 + 2x^2 - 1}{(x^4 - 2x^2 - 1)^2}.$$

Therefore, $f'(-1) = \dfrac{-3 + 2 - 1}{(1 - 2 - 1)^2} = -\dfrac{2}{4} = -\dfrac{1}{2}.$

39. $f(x) = (x^3 + 1)(x^2 - 2).$

$$f'(x) = (x^3 + 1)\frac{d}{dx}(x^2 - 2) + (x^2 - 2)\frac{d}{dx}(x^3 + 1)$$

$$= (x^3 + 1)(2x) + (x^2 - 2)(3x^2).$$

The slope of the tangent line at (2,18) is
$$f'(2) = (8 + 1)(4) + (4 - 2)(12) = 60.$$
An equation of the tangent line is
$$y - 18 = 60(x - 2), \quad \text{or} \quad y = 60x - 102.$$

41. $f(x) = \dfrac{x + 1}{x^2 + 1}.$

$$f'(x) = \frac{(x^2 + 1)\dfrac{d}{dx}(x + 1) - (x + 1)\dfrac{d}{dx}(x^2 + 1)}{(x^2 + 1)^2}$$

$$= \frac{(x^2 + 1)(1) - (x + 1)(2x)}{(x^2 + 1)^2} = \frac{-x^2 - 2x + 1}{(x^2 + 1)^2}.$$

At $x = 1$, $f'(1) = \dfrac{-1 - 2 + 1}{4} = -\dfrac{1}{2}.$

Therefore, the slope of the tangent line at $x = 1$ is -1/2. Then an equation of the tangent line is
$$y - 1 = -\tfrac{1}{2}(x - 1) \quad \text{or} \quad y = -\tfrac{1}{2}x + \tfrac{3}{2}.$$

43. $f(x) = (x^3 + 1)(3x^2 - 4x + 2)$

$$f'(x) = (x^3+1)\frac{d}{dx}(3x^2-4x+2) + (3x^2-4x+2)\frac{d}{dx}(x^3+1)$$

$$= (x^3+1)(6x-4) + (3x^2-4x+2)(3x^2)$$

$$= 6x^4 + 6x - 4x^3 - 4 + 9x^4 - 12x^3 + 6x^2$$

$$= 15x^4 - 16x^3 + 6x^2 + 6x - 4.$$

At $x = 1$, $f'(1) = 15(1)^4 - 16(1)^3 + 6(1) + 6(1) - 4 = 7$. The slope of the tangent line at the point $x = 1$ is 7. The equation of the tangent line is

$$y - 2 = 7(x - 1), \quad \text{or} \quad y = 7x - 5.$$

45. $f(x) = (x^2+1)(2-x)$

$$f'(x) = (x^2+1)\frac{d}{dx}(2-x) + (2-x)\frac{d}{dx}(x^2+1)$$

$$= (x^2+1)(-1) + (2-x)(2x) = -3x^2 + 4x - 1.$$

At a point where the tangent line is horizontal, we have

$$f'(x) = -3x^2 + 4x - 1 = 0$$

or $\quad 3x^2 - 4x + 1 = (3x-1)(x-1) = 0$,

giving $x = 1/3$ or $x = 1$.

Since $\quad f(\frac{1}{3}) = (\frac{1}{9}+1)(2-\frac{1}{3}) = \frac{50}{27}$, and $f(1) = 2(2-1) = 2$, we see that the required points are $(\frac{1}{3}, \frac{50}{27})$ and $(1, 2)$.

47. $f(x) = (x^2+6)(x-5)$

$$f'(x) = (x^2+6)\frac{d}{dx}(x-5) + (x-5)\frac{d}{dx}(x^2+6)$$

$$= (x^2+6)(1) + (x-5)(2x) = x^2 + 6 + 2x^2 - 10x = 3x^2 - 10x + 6.$$

At a point where the slope of the tangent line is -2, we have

$$f'(x) = 3x^2 - 10x + 6 = -2.$$

This gives

$$3x^2 - 10x + 8 = (3x-4)(x-2) = 0.$$

So $x = \frac{4}{3}$ or $x = 2$.

Since $\quad f(\frac{4}{3}) = (\frac{16}{9}+6)(\frac{4}{3}-5) = -\frac{770}{27}$ and $f(2) = (4+6)(2-5) = -30$,

the required points are $(\frac{4}{3}, -\frac{770}{27})$ and $(2, -30)$.

49. $$y' = \frac{(1+x^2)\frac{d}{dx}(1)-(1)\frac{d}{dx}(1+x^2)}{(1+x^2)^2} = \frac{-2x}{(1+x^2)^2}.$$

So, the slope of the tangent line at $(1,\frac{1}{2})$ is

$$y'|_{x=1} = \frac{-2x}{(1+x^2)^2}\bigg|_{x=1} = \frac{-2}{4} = -\frac{1}{2}$$

and the equation of the tangent line is

$$y-\tfrac{1}{2}=-\tfrac{1}{2}(x-1), \quad \text{or} \quad y=-\tfrac{1}{2}x+1.$$

Next, the slope of the required normal line is 2 and its equation is

$$y-\tfrac{1}{2}=2(x-1), \quad \text{or} \quad y=2x-\tfrac{3}{2}.$$

51. $$C(x) = \frac{0.5x}{100-x}.$$

$$C'(x) = \frac{(100-x)(0.5)-0.5x(-1)}{(100-x)^2} = \frac{50}{(100-x)^2}.$$

$$C'(80) = \frac{50}{20^2} = 0.125, \ C'(90) = \frac{50}{10^2} = 0.5,$$

$$C'(95) = \frac{50}{5^2} = 2; \ C'(99) = \frac{50}{1} = 50.$$

The rates of change of the cost in removing 80%, 90%, and 99% of the toxic waste are 0.125, 0.5, 2, and 50 million dollars per 1% more of the waste to be removed, respectively.

53. $$N(t) = \frac{10,000}{1+t^2} + 2000$$

$$N'(t) = \frac{d}{dt}[10,000(1+t^2)^{-1}+2000] = -\frac{10,000}{(1+t^2)^2}(2t) = -\frac{20,000t}{(1+t^2)^2}.$$

The rate of change after 1 minute and after 2 minutes is

$$N'(1) = -\frac{20,000}{(1+1^2)^2} = -5000; \ N'(2) = -\frac{20,000(2)}{(1+2^2)^2} = -1600.$$

The population of bacteria after one minute is

$$N(1) = \frac{10,000}{1+1} + 2000 = 7000.$$

The population after two minutes is

$$N(2) = \frac{10,000}{1+4} + 2000 = 4000.$$

55. a. $N(t) = \dfrac{60t + 180}{t + 6}.$

$$N'(t) = \frac{(t+6)\dfrac{d}{dt}(60t+180) - (60t+180)\dfrac{d}{dt}(t+6)}{(t+6)^2}$$

$$= \frac{(t+6)(60) - (60t+180)(1)}{(t+6)^2}$$

$$= \frac{180}{(t+6)^2}.$$

b. $N'(1) = \dfrac{180}{(1+6)^2} = 3.7, \ N'(3) = \dfrac{180}{(3+6)^2} = 2.2, \ N'(4) = \dfrac{180}{(4+6)^2} = 1.8,$

$N'(7) = \dfrac{180}{(7+6)^2} = 1.1$

We conclude that the rate at which the average student is increasing his or her speed one week, three weeks, four weeks, and seven weeks into the course is 3.7, 2.2, 1.8, and 1.1 words per minute, respectively.

c.

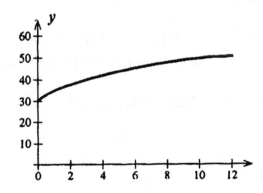

d. $N(12) = \dfrac{60(12) + 180}{12 + 6} = 50,$ or 50 words/minute.

57. $f'(t) = \dfrac{(t+2)(0.055) - (0.055t + 0.26)(1)}{(t+2)^2} = -\dfrac{0.15}{(t+2)^2}$. At the beginning, the

formaldehyde level is changing at the rate of

$$f'(0) = -\frac{0.15}{4} = -0.0375;$$

that is, it is dropping at the rate of 0.0375 parts per million per year. Next,

$$f'(3) = -\frac{0.15}{5^2} = -0.006,$$

and so the level is dropping at the rate of 0.006 parts per million per year at the beginning of the fourth year ($t = 3$).

59. False. Take $f(x) = x$ and $g(x) = x$. Then $f(x)g(x) = x^2$. So

$$\frac{d}{dx}[f(x)g(x)] = \frac{d}{dx}(x^2) = 2x \neq f'(x)g'(x) = 1.$$

61. False. Let $f(x) = x^3$. Then

$$\frac{d}{dx}\left[\frac{f(x)}{x^2}\right] = \frac{d}{dx}\left(\frac{x^3}{x^2}\right) = \frac{d}{dx}(x) = 1 \neq \frac{f'(x)}{2x} = \frac{3x^2}{2x} = \frac{3}{2}x.$$

63. Let $f(x) = u(x)v(x)$ and $g(x) = w(x)$. Then $h(x) = f(x)g(x)$. Therefore,
$$h'(x) = f'(x)g(x) + f(x)g'(x).$$
But $\quad f'(x) = u(x)v'(x) + u'(x)v(x).$
Therefore,
$$h'(x) = [u(x)v'(x) + u'(x)v(x)]g(x) + u(x)v(x)w'(x)$$
$$= u(x)v(x)w'(x) + u(x)v'(x)w(x) + u'(x)v(x)w(x).$$

USING TECHNOLOGY EXERCISES 3.2, page 209

1. 0.8750   3. 0.0774   5. -0.5000   7. 87,322 per year

EXERCISES 3.3, page 220

1. $f(x) = (2x-1)^4$. $f'(x) = 4(2x-1)^3 \dfrac{d}{dx}(2x-1) = 4(2x-1)^3(2) = 8(2x-1)^3$.

3. $f(x) = (x^2+2)^5$. $f'(x) = 5(x^2+2)^4(2x) = 10x(x^2+2)^4$.

5. $f(x) = (2x - x^2)^3.$

$f'(x) = 3(2x - x^2)^2 \dfrac{d}{dx}(2x - x^2) = 3(2x - x^2)^2(2 - 2x)$

$\qquad = 6x^2(1 - x)(2 - x)^2.$

7. $f(x) = (2x + 1)^{-2}.$

$f'(x) = -2(2x + 1)^{-3} \dfrac{d}{dx}(2x + 1) = -2(2x + 1)^{-3}(2) = -4(2x + 1)^{-3}.$

9. $f(x) = (x^2 - 4)^{3/2}.$

$f'(x) = \tfrac{3}{2}(x^2 - 4)^{1/2} \dfrac{d}{dx}(x^2 - 4) = \tfrac{3}{2}(x^2 - 4)^{1/2}(2x) = 3x(x^2 - 4)^{1/2}.$

11. $f(x) = \sqrt{3x - 2} = (3x - 2)^{1/2}.$

$f'(x) = \dfrac{1}{2}(3x - 2)^{-1/2}(3) = \dfrac{3}{2}(3x - 2)^{-1/2} = \dfrac{3}{2\sqrt{3x - 2}}.$

13. $f(x) = \sqrt[3]{1 - x^2}.$

$f'(x) = \dfrac{d}{dx}(1 - x^2)^{1/3} = \dfrac{1}{3}(1 - x^2)^{-2/3} \dfrac{d}{dx}(1 - x^2)$

$\qquad = \dfrac{1}{3}(1 - x^2)^{-2/3}(-2x) = -\dfrac{2}{3}x(1 - x^2)^{-2/3}.$

15. $f(x) = \dfrac{1}{(2x + 3)^3} = (2x + 3)^{-3}.$

$f'(x) = -3(2x + 3)^{-4}(2) = -6(2x + 3)^{-4} = -\dfrac{6}{(2x + 3)^4}.$

17. $f(t) = \dfrac{1}{\sqrt{2t - 3}}.$

$f'(t) = \dfrac{d}{dt}(2t - 3)^{-1/2} = -\dfrac{1}{2}(2t - 3)^{-3/2}(2) = -(2t - 3)^{-3/2}.$

19. $y = \dfrac{1}{(4x^4 + x)^{3/2}}.$

$$\frac{dy}{dx} = \frac{d}{dx}(4x^4 + x)^{-3/2} = -\frac{3}{2}(4x^4 + x)^{-5/2}(16x^3 + 1) = -\frac{3}{2}(16x^3 + 1)(4x^4 + x)^{-5/2}.$$

21. $f(x) = (3x^2 + 2x + 1)^{-2}$.

$$f'(x) = -2(3x^2 + 2x + 1)^{-3}\frac{d}{dx}(3x^2 + 2x + 1)$$
$$= -2(3x^2 + 2x + 1)^{-3}(6x + 2) = -4(3x + 1)(3x^2 + 2x + 1)^{-3}.$$

23. $f(x) = (x^2 + 1)^3 - (x^3 + 1)^2$.

$$f'(x) = 3(x^2 + 1)^2\frac{d}{dx}(x^2 + 1) - 2(x^3 + 1)\frac{d}{dx}(x^3 + 1)$$
$$= 3(x^2 + 1)^2(2x) - 2(x^3 + 1)(3x^2)$$
$$= 6x[(x^2 + 1)^2 - x(x^3 + 1)] = 6x(2x^2 - x + 1).$$

25. $f(t) = (t^{-1} - t^{-2})^3$.

$$f'(t) = 3(t^{-1} - t^{-2})^2\frac{d}{dt}(t^{-1} - t^{-2}) = 3(t^{-1} - t^{-2})^2(-t^{-2} + 2t^{-3}).$$

27. $f(x) = \sqrt{x+1} + \sqrt{x-1} = (x+1)^{1/2} + (x-1)^{1/2}$.
$$f'(x) = \tfrac{1}{2}(x+1)^{-1/2}(1) + \tfrac{1}{2}(x-1)^{-1/2}(1) = \tfrac{1}{2}[(x+1)^{-1/2} + (x-1)^{-1/2}].$$

29. $f(x) = 2x^2(3 - 4x)^4$.
$$f'(x) = 2x^2(4)(3 - 4x)^3(-4) + (3 - 4x)^4(4x) = 4x(3 - 4x)^3(-8x + 3 - 4x)$$
$$= 4x(3 - 4x)^3(-12x + 3) = (-12x)(4x - 1)(3 - 4x)^3.$$

31. $f(x) = (x - 1)^2(2x + 1)^4$.
$$f'(x) = (x - 1)^2\frac{d}{dx}(2x + 1)^4 + (2x + 1)^4\frac{d}{dx}(x - 1)^2 \quad \text{[Product Rule]}$$
$$= (x - 1)^2(4)(2x + 1)^3\frac{d}{dx}(2x + 1) + (2x + 1)^4(2)(x - 1)\frac{d}{dx}(x - 1)$$
$$= 8(x - 1)^2(2x + 1)^3 + 2(x - 1)(2x + 1)^4$$
$$= 2(x - 1)(2x + 1)^3(4x - 4 + 2x + 1) = 6(x - 1)(2x - 1)(2x + 1)^3.$$

33. $f(x) = \left(\dfrac{x+3}{x-2}\right)^3$.

$$f'(x) = 3\left(\frac{x+3}{x-2}\right)^2 \frac{d}{dx}\left(\frac{x-3}{x-2}\right) = 3\left(\frac{x+3}{x-2}\right)^2 \left[\frac{(x-2)(1)-(x+3)(1)}{(x-2)^2}\right]$$

$$= 3\left(\frac{x+3}{x-2}\right)^2 \left[-\frac{5}{(x-2)^2}\right] = -\frac{15(x+3)^2}{(x-2)^4}.$$

35. $s(t) = \left(\dfrac{t}{2t+1}\right)^{3/2}$.

$$s'(t) = \frac{3}{2}\left(\frac{t}{2t+1}\right)^{1/2} \frac{d}{dt}\left(\frac{t}{2t+1}\right) = \frac{3}{2}\left(\frac{t}{2t+1}\right)^{1/2}\left[\frac{(2t+1)(1)-t(2)}{(2t+1)^2}\right]$$

$$= \frac{3}{2}\left(\frac{t}{2t+1}\right)^{1/2}\left[\frac{1}{(2t+1)^2}\right] = \frac{3t^{1/2}}{2(2t+1)^{5/2}}.$$

37. $g(u) = \left(\dfrac{u+1}{3u+2}\right)^{1/2}$.

$$g'(u) = \frac{1}{2}\left(\frac{u+1}{3u+2}\right)^{-1/2} \frac{d}{du}\left(\frac{u+1}{3u+2}\right)$$

$$= \frac{1}{2}\left(\frac{u+1}{3u+2}\right)^{-1/2}\left[\frac{(3u+2)(1)-(u+1)(3)}{(3u+2)^2}\right] = -\frac{1}{2\sqrt{u+1})(3u+2)^{3/2}}.$$

39. $f(x) = \dfrac{x^2}{(x^2-1)^4}$.

$$f'(x) = \frac{(x^2-1)\dfrac{d}{dx}(x^2) - (x^2)\dfrac{d}{dx}(x^2-1)^4}{\left[(x^2-1)^4\right]^2}$$

$$= \frac{(x^2-1)^4(2x) - x^2(4)(x^2-1)^3(2x)}{(x^2-1)^8}$$

$$= \frac{(x^2-1)^3(2x)(x^2-1-4x^2)}{(x^2-1)^8} = \frac{(-2x)(3x^2+1)}{(x^2-1)^5}.$$

41. $h(x) = \dfrac{(3x^2+1)^3}{(x^2-1)^4}$.

$h'(x) = \dfrac{(x^2-1)^4(3)(3x^2+1)^2(6x)-(3x^2+1)^3(4)(x^2-1)^3(2x)}{(x^2-1)^8}$

$= \dfrac{2x(x^2-1)^3(3x^2+1)^2[9(x^2-1)-4(3x^2+1)]}{(x^2-1)^8}$

$= -\dfrac{2x(3x^2+13)(3x^2+1)^2}{(x^2-1)^5}$.

43. $f(x) = \dfrac{\sqrt{2x+1}}{x^2-1}$.

$f'(x) = \dfrac{(x^2-1)(\frac{1}{2})(2x+1)^{-1/2}(2)-(2x+1)^{1/2}(2x)}{(x^2-1)^2}$

$= \dfrac{(2x+1)^{-1/2}[(x^2-1)-(2x+1)(2x)]}{(x^2-1)^2} = -\dfrac{3x^2+2x+1}{\sqrt{2x+1}(x^2-1)^2}$.

45. $g(t) = \dfrac{(t+1)^{1/2}}{(t^2+1)^{1/2}}$.

$g'(t) = \dfrac{(t^2+1)^{1/2}\dfrac{d}{dt}(t+1)^{1/2}-(t+1)^{1/2}\dfrac{d}{dt}(t^2+1)^{1/2}}{t^2+1}$

$= \dfrac{(t^2+1)^{1/2}(\frac{1}{2})(t+1)^{-1/2}(1)-(t+1)^{1/2}(\frac{1}{2})(t^2+1)^{-1/2}(2t)}{t^2+1}$

$= \dfrac{\frac{1}{2}(t+1)^{-1/2}(t^2+1)^{-1/2}[(t^2+1)-2t(t+1)]}{t^2+1}$

$= -\dfrac{t^2+2t-1}{2\sqrt{t+1}(t^2+1)^{3/2}}$.

47. $y = g(u) = u^{4/3}$ and $\dfrac{dy}{du} = \dfrac{4}{3}u^{1/3}$, $u = f(x) = 3x^2-1$, and $\dfrac{du}{dx} = 6x$.

$\dfrac{dy}{dx} = \dfrac{dy}{du}\cdot\dfrac{du}{dx} = \frac{4}{3}u^{1/3}(6x)$

So $\qquad = \frac{4}{3}(3x^2-1)^{1/3}6x$

$\qquad = 8x(3x^2-1)^{1/3}$.

49. $\dfrac{dy}{du} = -\dfrac{2}{3}u^{-5/3} = -\dfrac{2}{3u^{5/3}}, \quad \dfrac{du}{dx} = 6x^2 - 1.$

$\dfrac{dy}{dx} = \dfrac{dy}{du}\cdot\dfrac{du}{dx} = -\dfrac{2(6x^2-1)}{3u^{5/3}} = -\dfrac{2(6x^2-1)}{3(2x^3-x+1)^{5/3}}.$

51. $\dfrac{dy}{du} = \tfrac{1}{2}u^{-1/2} - \tfrac{1}{2}u^{-3/2}, \quad \dfrac{du}{dx} = 3x^2 - 1.$

$\dfrac{dy}{dx} = \dfrac{dy}{du}\cdot\dfrac{du}{dx} = \left[\dfrac{1}{2\sqrt{x^3-x}} - \dfrac{1}{2(x^3-x)^{3/2}}\right](3x^2-1)$

$= \dfrac{(3x^2-1)(x^3-x-1)}{2(x^3-x)^{3/2}}.$

53. $F(x) = g(f(x)); \; F'(x) = g'(f(x))f'(x) \;$ and $\; F'(2) = g'(3)(-3) = (4)(-3) = -12$

55. Let $g(x) = x^2 + 1$, then $F(x) = f(g(x))$. Next, $F'(x) = f'(g(x))g'(x)$
and $F'(1) = f'(2)(2x) = (3)(2) = 6.$

57. No. Suppose $h = g(f(x))$. Let $f(x) = x$ and $g(x) = x^2$. Then
$h = g(f(x)) = g(x) = x^2$ and $h'(x) = 2x \neq g'(f'(x)) = g'(1) = 2(1) = 2.$

59. $f(x) = (1-x)(x^2-1)^2.$

$f'(x) = (1-x)2(x^2-1)(2x) + (-1)(x^2-1)^2$

$= (x^2-1)(4x-4x^2-x^2+1)$

$= (x^2-1)(-5x^2+4x+1).$

Therefore, the slope of the tangent line at $(2,-9)$ is
$f'(2) = [(2)^2-1][-5(2)^2+4(2)+1] = -33.$

Then the required equation is
$y + 9 = -33(x-2)$, or $\quad y = -33x + 57.$

61. $f'(x) = \sqrt{2x^2+7} + x(\tfrac{1}{2})(2x^2+7)^{-1/2}(4x).$

The slope of the tangent line is $\; f'(3) = \sqrt{25} + (\tfrac{3}{2})(25)^{-1/2}(12) = \tfrac{43}{5}.$

An equation of the tangent line is $y - 15 = \frac{43}{5}(x - 3)$ or $y = \frac{43}{5}x - \frac{54}{5}$.

63. $N(x) = (60 + 2x)^{2/3}$.

$N'(x) = \frac{2}{3}(60 + 2x)^{-1/3} \frac{d}{dx}(60 + 2x) = \frac{4}{3}(60 + 2x)^{-1/3}$.

The rate of increase at the end of the second week is
$$N'(2) = \frac{4}{3}(64)^{-1/3} = \frac{1}{3}, \text{ or } \frac{1}{3} \text{ million/week}$$

At the end of the 12th week, $N'(12) = \frac{4}{3}(84)^{-1/3} \approx 0.3$ million/wk. The number of viewers in the 2nd and 24th week are $N(2) = (60 + 4)^{2/3} = 16$ million and $N(24) = (60 + 48)^{2/3} = 22.7$ million, respectively.

65. $C(t) = 0.01(0.2t^2 + 4t + 64)^{2/3}$.

a. $C'(t) = 0.01(\frac{2}{3})(0.2t^2 + 4t + 64)^{-1/3} \frac{d}{dt}(0.2t^2 + 4t + 64)$

$= (0.01)(0.667)(0.4t + 4)(0.2t^2 + 4t + 4)^{-1/3}$

$= 0.027(0.1t + 1)(0.2t^2 + 4t + 64)^{-1/3}$.

b. $C'(5) = 0.007[0.4(5) + 4][0.2(25) + 4(5) + 64]^{-1/3} \approx 0.009$,
or 0.009 parts per million per year.

67. a. $A(t) = 0.03t^3(t - 7)^4 + 60.2$
$A'(t) = 0.03[3t^2(t - 7)^4 + t^3(4)(t - 7)^3)] = 0.03t^2(t - 7)^3[3(t - 7) + 4t]$
$= 0.21t^2(t - 3)(t - 7)^3$.

b. $A'(1) = 0.21(-2)(-6)^3 = 90.72$; $A'(3) = 0$. $A'(4) = 0.21(16)(1)(-3)^3 = -90.72$.

The amount of pollutant is increasing at the rate of 90.72 units/hr at 8 A.M. Its rate of change is 0 units/hr at 10 A.M.;and its rate of change is -90.72 units/hr at 11 A.M.

69. $P(t) = \dfrac{300\sqrt{\frac{1}{2}t^2 + 2t + 25}}{t + 25} = \dfrac{300(\frac{1}{2}t^2 + 2t + 25)^{1/2}}{t + 25}$.

$P'(t) = 300\left[\dfrac{(t + 25)\frac{1}{2}(\frac{1}{2}t^2 + 2t + 25)^{-1/2}(t + 2) - (\frac{1}{2}t^2 + 2t + 25)^{1/2}(1)}{(t + 25)^2}\right]$

$= 300\left[\dfrac{(\frac{1}{2}t^2 + 2t + 25)^{-1/2}[(t + 25)(t + 2) - 2(\frac{1}{2}t^2 + 2t + 25)^{1/2}]}{(t + 25)^2}\right]$

$$= \frac{3450t}{(t+25)^2 \sqrt{\frac{1}{2}t^2 + 2t + 25}}.$$

Ten seconds into the run, the athlete's pulse rate is increasing at

$$P'(10) = \frac{3450(10)}{(35)^2 \sqrt{50+20+25}} \approx 2.9,$$

or approximately 2.9 beats per minute per minute. Sixty seconds into the run, it is increasing at

$$P'(60) = \frac{3450(60)}{(85)^2 \sqrt{1800+120+25}} \approx 0.65,$$

or approximately 0.7 beats per minute per minute. Two minutes into the run, it is increasing at

$$P'(120) = \frac{3450(120)}{(145)^2 \sqrt{7200+240+25}} \approx 0.23,$$

or approximately 0.2 beats per minute per minute. The pulse rate two minutes into the run is given by

$$P(120) = \frac{300\sqrt{7200+240+25}}{120+25} \approx 178.8,$$

or approximately 179 beats per minute.

71. The area is given by $A = \pi r^2$. The rate at which the area is increasing is given by dA/dt, that is, $\dfrac{dA}{dt} = \dfrac{d}{dt}(\pi r^2) = \dfrac{d}{dt}(\pi r^2)\dfrac{dr}{dt} = 2\pi r \dfrac{dr}{dt}.$

If $r = 40$ and $dr/dt = 2$, then $\dfrac{dA}{dt} = 2\pi(40)(2) = 160\pi$, that is, it is increasing at the rate of 160π, or approximately 503, sq ft/sec.

73. $\dfrac{dS}{dt} = g'(x)f'(t) = (-0.0015x)(12.5t + 19.75).$

When $t = 4$, we have $x = f(4) = 6.25(16) + 19.75(4) + 74.75 = 253.75$

and $\left.\dfrac{dS}{dt}\right|_{t=4} = (-0.0015)(253.75)[12.5(4) + 19.75] \approx -26.55;$

that is, the average speed will be dropping at the rate of approximately 27 mph per decade. The average speed of traffic flow at that time will be

$$S = g(f(4)) = -0.00075(253.75^2) + 67.5 = 19.2,$$

or approximately 19 mph.

75. $N(x) = 1.42x$ and $x(t) = \dfrac{7t^2 + 140t + 700}{3t^2 + 80t + 550}$. The number of construction jobs as a function of time is $n(t) = N[x(t)]$. Using the Chain Rule,

$$n'(t) = \frac{dN}{dx} \cdot \frac{dx}{dt} = 1.42 \frac{dx}{dt}$$

$$= (1.42) \left[\frac{(3t^2 + 80t + 550)(14t + 140) - (7t^2 + 140t + 700)(6t + 80)}{(3t^2 + 80t + 550)^2} \right]$$

$$= \frac{1.42(140t^2 + 3500t + 21000)}{(3t^2 + 80t + 550)^2}.$$

$$n'(1) = \frac{1.42(140 + 3500 + 21000)}{(3 + 80 + 550)^2} \approx 0.0873216,$$

or approximately 87,322 jobs per year.

77. We want $\dfrac{dR}{dt} = \dfrac{dR}{dp} \cdot \dfrac{dp}{dt}$. Now $\dfrac{dR}{dp} = 1000 \left[\dfrac{(p+2)(1) - (p+4)(1)}{(p+2)^2} \right] = -\dfrac{2400}{(p+2)^2}$

$$\frac{dp}{dt} = 50 \left[\frac{(t^2 + 4t + 8)(2t + 2) - (t^2 + 2t + 4)(2t + 4)}{(t^2 + 4t + 8)^2} \right]$$

$$= \frac{100t(t+4)}{(t^2 + 4t + 8)^2}.$$

When $t = 2$,

$$\frac{dR}{dt} = -\frac{2000}{(p+2)^2} \cdot \frac{100t(t+4)}{(t^2 + 4t + 8)^2} \bigg|_{t=2} = -\frac{2000}{(32)^2} \cdot \frac{100(2)(6)}{(4+8+8)^2}$$

$$\approx -5.86,$$

that is, the passage will decrease at the rate of approximately $5.86 per passenger per year.

79. True. $\dfrac{d}{dx}[f(cx)] = f'(cx)\dfrac{d}{dx}(cx) = f'(cx) \cdot c$.

81. False. Let $f(x) = x$. Then $f\left(\dfrac{1}{x}\right) = \dfrac{1}{x}$ and so $f'(x) = -\dfrac{1}{x^2}$. But $f'(x) = 1$ and so and so $f'\left(\dfrac{1}{x}\right) = 1$.

83. Let $f(x) = x^r = x^{m/n} = (x^m)^{1/n}$. Then $[f(x)]^n = x^m$.

$$n[f(x)]^{n-1} f'(x) = \frac{m}{n}[f(x)]^{-n+1} x^{m-1} = \frac{m}{n}(x^{m/n})^{-n+1} x^{m-1}$$

$$= \frac{m}{n} x^{[m(-n+1)/n]+m-1} = \frac{m}{n} x^{(m-n)/n} = \frac{m}{n} x^{(m/n)-1}$$

$$= rx^{r-1}.$$

USING TECHNOLOGY EXERCISES 3.3 page 219

1. 0.5774   3. 0.9390   5. –4.9498

7. a. 10,146,200/decade   b. 7,810,520/decade

EXERCISES 3.4, page 236

1. a. $C(x)$ is always increasing because as x, the number of units produced, increases, the greater the amount of money that must be spent on production.
   b. This occurs at $x = 4$, or a production level of 4000. You can see this by looking at the slopes of the tangent lines for x less than, equal to, and a little larger then $x = 4$.

3. a. $C(101) - C(100) = [0.0002(101)^3 - 0.06(101)^2 + 120(101) + 5000]$
   $$- [0.0002(100)^3 - 0.06(100)^2 + 120(100) + 5000]$$
   $$\approx 114, \text{ or approximately } \$114.$$
   Similarly, we find $C(201) - C(200) \approx \$120.16$; $C(301) - C(300) \approx \$138.12$.
   b. We compute $C'(x) = 0.0006x^2 - 0.12x + 120$. So the required quantities are
   $$C'(100) = 0.0006(100)^2 - 0.12(100) + 120 = 114, \text{ or } \$114,$$
   $$C'(200) = 0.0006(200)^2 - 0.12(200) + 120 = 120, \text{ or } \$120,$$
   and $C'(300) = 0.0006(300)^2 - 0.12(300) + 120 = 138$, or $138.

5. a. $\overline{C}(x) = \dfrac{C(x)}{x} = \dfrac{5000}{x} + 2.$ b. $\overline{C}'(x) = -\dfrac{5000}{x^2}.$
   c. Since the marginal average cost function is negative for $x > 0$, the rate of change of the average cost function is negative for all $x > 0$.

7. $\overline{C}(x) = \dfrac{C(x)}{x} = \dfrac{0.0002x^3 - 0.06x^2 + 120x + 5000}{x}$

$$= 0.0002x^2 - 0.06x + 120 + \frac{5000}{x}.$$

$$\overline{C}'(x) = 0.0004x - 0.06 - \frac{5000}{x^2}.$$

9. a. $R(x) = px = x(-0.04x + 800) = -0.04x^2 + 800x$
   b. $R'(x) = -0.08x + 800$
   c. $R'(5000) = -0.08(5000) + 800 = 400.$
   This says that when the level of production is 5000 units the production of the next speaker system will bring an additional revenue of $400.

11. a. $P(x) = R(x) - C(x) = (-0.04x^2 + 800x) - (200x + 300,000)$
       $= -0.04x^2 + 600x - 300,000.$
    b. $P'(x) = -0.08x + 600$
    c. $P'(5000) = -0.08(5000) + 600 = 200$
       $P'(8000) = -0.08(8000) + 600 = -40.$
    d.

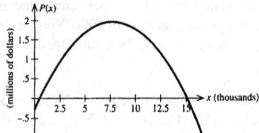

The profit realized by the company increases as production increases, peaking at a level of production of 7500 units. Beyond this level of production, the profit begins to fall.

13. a. $R(x) = xp(x) = -0.006x^2 + 180x$
       $P(x) = R(x) - C(x)$
       $= -0.006x^2 + 180x - (0.000002x^3 - 0.02x^2 + 120x + 60,000)$
       $= -0.000002x^3 + 0.014x^2 + 60x - 60,000.$
    b. $C'(x) = 0.000006x^2 - 0.04x + 120$
       $R'(x) = -0.012x + 180$
       $P'(x) = -0.000006x^2 + 0.028x + 60.$
    c. $C'(2000) = 0.000006(2000)^2 - 0.04(2000) + 120 = 64$
       $R'(2000) = -0.012(2000) + 180 = 156$
       $P'(2000) = -0.000006(2000)^2 + 0.028(2000) + 60 = 92.$

d.

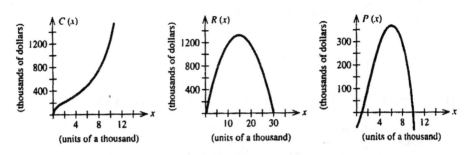

(units of a thousand)   (units of a thousand)   (units of a thousand)

15. $C(x) = 0.000002x^3 - 0.02x^2 + 120x + 60{,}000$

$\overline{C}(x) = 0.000002x^2 - 0.02x + 120 + \dfrac{60{,}000}{x}.$

a. The marginal average cost function is given by

$$\overline{C}'(x) = 0.000004x - 0.02 - \frac{60{,}000}{x^2}$$

b. $\overline{C}'(5000) = 0.000004(5000) - 0.02 - \dfrac{60{,}000}{(5000)^2}$

$= 0.02 - 0.02 - 0.0024 = -0.0024.$

$\overline{C}'(10000) = 0.000004(10000) - 0.02 - \dfrac{60{,}000}{(10000)^2}$

$= 0.04 - 0.02 - 0.0006 = 0.0194.$

We conclude that the average cost is decreasing when 5000 TV sets are produced and increasing when 10,000 units are produced.

17. $\dfrac{dC}{dx} = \dfrac{d}{dx}(0.712x + 95.05) = 0.712.$

19. $\dfrac{dS}{dx} = \dfrac{d}{dx}[x - C(x)] = 1 - \dfrac{dC}{dx}.$

21. Here $C(x) = 0.873x^{1.1} + 20.34$. So $C'(x) = 0.9603x^{0.1}$ and

$$\dfrac{dS}{dx} = 1 - \dfrac{dC}{dx} = 1 - 0.9603x^{0.1}.$$

When $x = 10$, we have

$$\frac{dS}{dx} = 1 - 0.9603(10)^{0.1} = -0.209,$$

−$0.209 billion per billion dollars.

23. Here $x = f(p) = -\frac{5}{4}p + 20$ and so $f'(p) = -\frac{5}{4}$. Therefore,

$$E(p) = -\frac{pf'(p)}{f(p)} = -\frac{p(-\frac{5}{4})}{-\frac{5}{4}p + 20} = \frac{5p}{80 - 5p}.$$

$$E(10) = \frac{5(10)}{80 - 5(10)} = \frac{50}{30} = \frac{5}{3} > 1,$$

and so the demand is elastic.

25. Solving the demand equation for x, we have
$$0.4x = -p + 20 \quad \text{or} \quad x = f(p) = -\frac{5}{2}p + 50.$$
Then $f'(p) = -\frac{5}{2}$, and so

$$E(p) = -\frac{pf'(p)}{f(p)} = -\frac{p(-\frac{5}{2})}{-\frac{5}{2}p + 50} = \frac{5p}{100 - 5p}.$$

$$E(10) = \frac{50}{50} = 1, \text{ and so the demand is unitary.}$$

27. $x^2 = 169 - p$ and $f(p) = (169 - p)^{1/2}$.
Next, $f'(p) = \frac{1}{2}(169 - p)^{-1/2}(-1) = -\frac{1}{2}(169 - p)^{-1/2}$.
Then the elasticity of demand is given by

$$E(p) = -\frac{pf'(p)}{f(p)} = -\frac{p(-\frac{1}{2})(169 - p)^{-1/2}}{(169 - p)^{1/2}} = \frac{\frac{1}{2}p}{169 - p}.$$

Therefore, when $p = 29$,

$$E(p) = \frac{\frac{1}{2}(29)}{169 - 29} = \frac{14.5}{140} = 0.104.$$

Since $E(p) < 1$, we conclude that demand is inelastic at this price.

29. $f(p) = \frac{1}{5}(225 - p^2); \quad f'(p) = \frac{1}{5}(-2p) = -\frac{2}{5}p.$
Then the elasticity of demand is given by

$$E(p) = -\frac{pf'(p)}{f(p)} = -\frac{p(-\frac{2}{5}p)}{\frac{1}{3}(225-p^2)} = \frac{2p^2}{225-p^2}.$$

a. When $p = 8$,

$$E(8) = \frac{2(64)}{225-64} = 0.8 < 1$$

and the demand is inelastic. When $p = 10$,

$$E(10) = \frac{2(100)}{225-100} = 1.6 > 1$$

and the demand is elastic when $p = 10$.

b. The demand is unitary when $E = 1$. Solving $\dfrac{2p^2}{225-p^2} = 1$

we find $2p^2 = 225 - p^2$, $3p^2 = 225$, and $p = 8.66$. So the demand is unitary when $p = 8.66$.

c. Since demand is elastic when $p = 10$, lowering the unit price will cause the revenue to increase.

d. Since the demand is inelastic at $p = 8$, a slight increase in the unit price will cause the revenue to increase.

31. $f(p) = \frac{2}{3}(36-p^2)^{1/2}$

$f'(p) = \frac{2}{3}(\frac{1}{2})(36-p^2)^{-1/2}(-2p) = -\frac{2}{3}p(36-p^2)^{-1/2}$.

Then the elasticity of demand is given by

$$E(p) = -\frac{pf'(p)}{f(p)} = -\frac{-\frac{2}{3}p(36-p^2)^{-1/2}\,p}{\frac{2}{3}(36-p^2)^{1/2}} = \frac{p^2}{36-p^2}.$$

When $p = 2$,

$$E(2) = \frac{4}{36-4} = \frac{1}{8} < 1,$$

and we conclude that the demand is inelastic.

b. Since the demand is inelastic, the revenue will increase when the rental price is increased.

33. $f(p) = 10\left(\dfrac{50-p}{p}\right)^{1/2} = 10\left(\dfrac{50}{p}-1\right)^{1/2}$.

$f'(p) = 10(\frac{1}{2})\left(\dfrac{50}{p}-1\right)^{-1/2}\left(-\dfrac{50}{p^2}\right) = -\dfrac{250}{p^2}\left(\dfrac{50}{p}-1\right)^{-1/2}$.

Then the elasticity of demand is given by

$$E(p) = -\frac{pf'(p)}{f(p)} = -\frac{p\left(-\dfrac{250}{p^2}\right)\left(\dfrac{50}{p}-1\right)^{-1/2}}{10\left(\dfrac{50}{p}-1\right)^{1/2}}$$

$$= -\frac{\dfrac{250}{p}}{10\left(\dfrac{50}{p}-1\right)} = \frac{25}{p\left(\dfrac{50-p}{p}\right)} = \frac{25}{50-p}.$$

Setting $E = 1$, gives

$$1 = \frac{25}{50-p} \quad \text{and so } 25 = 50 - p, \text{ and } p = 25.$$

Thus, if $p > 25$, then $E > 1$, and the demand is elastic; if $p = 25$, then $E = 1$ and the demand is unitary; and if $p < 25$, then $E < 1$ and the demand is inelastic.

35. False. In fact, it makes good sense to *increase* the level of production since, in this instance, the profit increases by $f'(a)$ units/unit increase in x.

EXERCISES 3.5, page 245

1. $f(x) = 4x^2 - 2x + 1;\ f'(x) = 8x - 2;\ f''(x) = 8.$

3. $f(x) = 2x^3 - 3x^2 + 1;\ f'(x) = 6x^2 - 6x;\ f''(x) = 12x - 6 = 6(2x-1).$

5. $h(t) = t^4 - 2t^3 + 6t^2 - 3t + 10;\ h'(t) = 4t^3 - 6t^2 + 12t - 3$
    $h''(t) = 12t^2 - 12t + 12 = 12(t^2 - t + 1).$

7. $f(x) = (x^2 + 2)^5;\ f'(x) = 5(x^2 + 2)^4(2x) = 10x(x^2 + 2)^4$ and
    $f''(x) = 10(x^2 + 2)^4 + 10x(x^2 + 2)^3(2x)$
    $\qquad = 10(x^2 + 2)^3[(x^2 + 2) + 8x^2] = 10(9x^2 + 2)(x^2 + 2)^3.$

9. $g(t) = (2t^2 - 1)^2(3t^2);$

$$g'(t) = 2(2t^2 - 1)(4t)(3t^2) - (2t^2 - 1)^2(6t)$$
$$= 6t(2t^2 - 1)[4t^2 + (2t^2 - 1)] = 6t(2t^2 - 1)(6t^2 - 1)$$
$$= 6t(12t^4 - 8t^2 + 1) = 72t^5 - 48t^3 + 6t.$$
$$g''(t) = 360t^4 - 144t^2 + 6 = 6(60t^4 - 24t^2 + 1)$$

11. $f(x) = (2x^2 + 2)^{7/2}$; $f'(x) = \frac{7}{2}(2x^2 + 2)^{5/2}(4x) = 14x(2x^2 + 2)^{5/2}$;
$$f''(x) = 14(2x^2 + 2)^{5/2} + 14x(\tfrac{5}{2})(2x^2 + 2)^{3/2}(4x)$$
$$= 14(2x^2 + 2)^{3/2}[(2x^2 + 2) + 10x^2] = 28(6x^2 + 1)(2x^2 + 2)^{3/2}.$$

13. $f(x) = x(x^2 + 1)^2$;
$$f'(x) = (x^2 + 1)^2 + x(2)(x^2 + 1)(2x)$$
$$= (x^2 + 1)[(x^2 + 1) + 4x^2] = (x^2 + 1)(5x^2 + 1);$$
$$f''(x) = 2x(5x^2 + 1) + (x^2 + 1)(10x) = 2x(5x^2 + 1 + 5x^2 + 5)$$
$$= 4x(5x^2 + 3).$$

15. $f(x) = \dfrac{x}{2x+1}$; $f'(x) = \dfrac{(2x+1)(1) - x(2)}{(2x+1)^2} = \dfrac{1}{(2x+1)^2}$;

$$f''(x) = \frac{d}{dx}(2x+1)^{-2} = -2(2x+1)^{-3}(2) = -\frac{4}{(2x+1)^3}.$$

17. $f(s) = \dfrac{s-1}{s+1}$; $f'(s) = \dfrac{(s+1)(1) - (s-1)(1)}{(s+1)^2} = \dfrac{2}{(s+1)^2}$.

$$f''(s) = 2\frac{d}{ds}(s+1)^{-2} = -4(s+1)^{-3} = -\frac{4}{(s+1)^3}.$$

19. $f(u) = \sqrt{4 - 3u} = (4 - 3u)^{1/2}$.

$$f'(u) = \tfrac{1}{2}(4 - 3u)^{-1/2}(-3) = -\frac{3}{2\sqrt{4 - 3u}}.$$

$$f''(u) = -\frac{3}{2} \cdot \frac{d}{du}(4 - 3u)^{-1/2} = -\frac{3}{2}\left(-\frac{1}{2}\right)(4 - 3u)^{-3/2}(-3) = -\frac{9}{4(4 - 3u)^{3/2}}.$$

21. $f(x) = 3x^4 - 4x^3;\ f'(x) = 12x^3 - 12x^2;\ f''(x) = 36x^2 - 24x;\ f'''(x) = 72x - 24.$

23. $f(x) = \dfrac{1}{x};\ f'(x) = \dfrac{d}{dx}(x^{-1}) = -x^{-2};\ f''(x) = 2x^{-3};\ f'''(x) = -6x^{-4} = -\dfrac{6}{x^4}.$

25. $g(s) = (3s-2)^{1/2};\ g'(s) = \dfrac{1}{2}(3s-2)^{-1/2}(3) = \dfrac{3}{2(3s-2)^{1/2}};$

$g''(s) = \dfrac{3}{2}\left(-\dfrac{1}{2}\right)(3s-2)^{-3/2}(3) = -\dfrac{9}{4}(3s-2)^{-3/2} = -\dfrac{9}{4(3s-2)^{3/2}};$

$g'''(s) = \dfrac{27}{8}(3s-2)^{-5/2}(3) = \dfrac{81}{8}(3s-2)^{-5/2} = \dfrac{81}{8(3s-2)^{5/2}}.$

27. $f(x) = (2x-3)^4;\ f'(x) = 4(2x-3)^3(2) = 8(2x-3)^3$

$f''(x) = 24(2x-3)^2(2) = 48(2x-3)^2$

$f'''(x) = 96(2x-3)(2) = 192(2x-3).$

29. Its velocity at any time t is $v(t) = \dfrac{d}{dt}(16t^2) = 32t.$ The hammer strikes the ground when $16t^2 = 256$ or $t = 4$ (we reject the negative root). Therefore, its velocity at the instant it strikes the ground is $v(4) = 32(4) = 128$ ft/sec. Its acceleration at time t is $a(t) = \dfrac{d}{dt}(32t) = 32.$ In particular, its acceleration at $t = 4$ is 32 ft/sec².

31. $N(t) = -0.1t^3 + 1.5t^2 + 100.$
a. $N'(t) = -0.3t^2 + 3t = 0.3t(10 - t).$ Since $N'(t) > 0$ for $t = 0, 1, 2, ..., 7$, it is evident that $N(t)$ (and therefore the crime rate) was increasing from 1988 through 1995.
b. $N''(t) = -0.6t + 3 = 0.6(5 - t).$ Now $N''(4) = 0.6 > 0$, $N''(5) = 0$, $N''(6) = -0.6 < 0$ and $N''(7) = -1.2 < 0.$ This shows that the rate of the rate of change is decreasing beyond $t = 5$ (1990) and this shows that the program was working.

33. $h(t) = \tfrac{1}{16}t^4 - t^3 + 4t^2.$
a. $h'(t) = \tfrac{1}{4}t^3 - 3t^2 + 8t$
b. $h'(0) = 0$ or zero feet per second.
$h'(4) = \tfrac{1}{4}(64) - 3(16) + 8(4) = 0,$ or zero feet per second.

$h'(8) = \frac{1}{4}(8)^3 - 3(64) + 8(8) = 0$, or zero feet per second.

c. $h''(t) = \frac{3}{4}t^2 - 6t + 8$

d. $h''(0) = 8$ ft/sec^2

$h''(4) = \frac{3}{4}(16) - 6(4) + 8 = -4$ ft/sec^2.

$h''(8) = \frac{3}{4}(64) - 6(8) + 8 = 8$ ft/sec^2.

e. $h(0) = 0$ feet

$h(4) = \frac{1}{16}(4)^4 - (4)^3 + 4(4)^2 = 16$ feet.

$h(8) = \frac{1}{16}(8)^4 - (8)^3 + 4(8)^2 = 0$ feet.

35. $A(t) = 100 - 17.63t + 1.915t^2 - 0.1316t^3 + 0.00468t^4 - 0.00006t^5$

$A'(t) = -17.63 + 3.83t - 0.3948t^2 + 0.01872t^3 - 0.0003t^4$

$A''(t) = 3.83 - 0.7896t + 0.05616t^2 - 0.0012t^3$.

So, $A'(10) = -3.09$ and $A''(10) = 0.35$.

Our computations show that 10 minutes after the start of the test, the smoke remaining is decreasing at a rate of 3 percent per minute but the rate at which the rate of smoke is decreasing is increasing at the rate of 0.35 percent per minute per minute.

37. True. If $h = fg$ where f and g have derivatives of order 2. Then

$h''(x) = f''(x)g(x) = 2f'(x)g'(x) + f(x)g''(x)$.

39. True. Suppose $P(t)$ represents the population of bacteria at time t and suppose $P'(t) > 0$ and $P''(t) < 0$, then the population is increasing at time t but at a decreasing rate.

41. Consider the function $f(x) = x^{(2n+1)/2} = x^{n+(1/2)}$.

Then $f'(x) = (n + \frac{1}{2})x^{n-(1/2)}$

$f''(x) = (n + \frac{1}{2})(n - \frac{1}{2})x^{n-(3/2)}$

....

$f^{(n)}(x) = (n + \frac{1}{2})(n - \frac{1}{2}) \cdots \frac{3}{2}x^{1/2}$

$f^{(n+1)}(x) = (n + \frac{1}{2})(n - \frac{1}{2}) \cdots \frac{1}{2}x^{-1/2}$.

The first n derivatives exist at $x = 0$, but the $(n + 1)$st derivative fails to be defined there.

1. −18   3. 15.2762   5. −0.6255   7. 0.1973

9. $f''(6) = -68.46214$ and it tells us that at the beginning of 1988, the rate of the rate of the rate at which banks were failing was 68 banks per year per year per year.

EXERCISES 3.6, page 257

1.   a. Solving for y in terms of x, we have $y = -\frac{1}{2}x + \frac{5}{2}$. Therefore, $y' = -\frac{1}{2}$.

     b. Next, differentiating $x + 2y = 5$ implicitly, we have $1 + 2y' = 0$, or $y' = -\frac{1}{2}$.

3.   a. $xy = 1, y = \dfrac{1}{x}$, and $\dfrac{dy}{dx} = -\dfrac{1}{x^2}$.

     b. $x\dfrac{dy}{dx} + y = 0$

$$x\frac{dy}{dx} = -y$$

$$\frac{dy}{dx} = -\frac{y}{x} = \frac{-\frac{1}{x}}{x} = -\frac{1}{x^2}.$$

5. $x^3 - x^2 - xy = 4$.

     a. $-xy = 4 - x^3 + x^2$

$$y = -\frac{4}{x} + x^2 - x$$

$$y' = \frac{4}{x^2} + 2x - 1.$$

     b. $x^3 - x^2 - xy = 4$

$$3x^2 - 2x - x\frac{dy}{dx} - y = 0$$

$$-x\frac{dy}{dx} = -3x^2 + 2x + y$$

$$\frac{dy}{dx} = 3x - 2 - \frac{y}{x}$$

$$= 3x - 2 - \frac{1}{x}(-\frac{4}{x} + x^2 - x)$$

$$= 3x - 2 + \frac{4}{x^2} - x + 1$$

$$= \frac{4}{x^2} + 2x - 1.$$

7. a. $\frac{x}{y} - x^2 = 1$ is equivalent to $\frac{x}{y} = x^2 + 1$, or $y = \frac{x}{x^2 + 1}$. Therefore,

$$y' = \frac{(x^2 + 1) - x(2x)}{(x^2 + 1)^2} = \frac{1 - x^2}{(x^2 + 1)^2}.$$

b. Next, differentiating the equation $x - x^2 y = y$ implicitly, we obtain

$$1 - 2xy - x^2 y' = y', \ y'(1 + x^2) = 1 - 2xy, \text{ or } \ y' = \frac{1 - 2xy}{(1 + x^2)}.$$

(This may also be written in the form $-2y^2 + \frac{y}{x}$.)

To show that this is equivalent to the results obtained earlier, use the value of y obtained before, to get

$$y' = \frac{1 - 2x\left(\dfrac{x}{x^2 + 1}\right)}{1 + x^2} = \frac{x^2 + 1 - 2x^2}{(1 + x^2)^2} = \frac{1 - x^2}{(1 + x^2)^2}.$$

9. $x^2 + y^2 = 16$. Differentiating both sides of the equation implicitly, we obtain

$$2x + 2yy' = 0 \text{ and so } y' = -\frac{x}{y}.$$

11. $x^2 - 2y^2 = 16$. Differentiating implicitly with respect to x, we have

$$2x - 4y\frac{dy}{dx} = 0 \text{ and } \frac{dy}{dx} = \frac{x}{2y}.$$

13. $x^2 - 2xy = 6$. Differentiating both sides of the equation implicitly, we obtain

$2x - 2y - 2xy' = 0$ and so $y' = \dfrac{x-y}{x} = 1 - \dfrac{y}{x}$.

15. $x^2y^2 - xy = 8$. Differentiating both sides of the equation implicitly, we obtain
$$2xy^2 + 2x^2yy' - y - xy' = 0, \; 2xy^2 - y + y'(2x^2y - x) = 0$$

and so $y' = \dfrac{y(1 - 2xy)}{x(2xy - 1)} = -\dfrac{y}{x}$.

17. $x^{1/2} + y^{1/2} = 1$. Differentiating implicitly with respect to x, we have
$$\tfrac{1}{2}x^{-1/2} + \tfrac{1}{2}y^{-1/2}\dfrac{dy}{dx} = 0.$$
Therefore, $\dfrac{dy}{dx} = -\dfrac{x^{-1/2}}{y^{-1/2}} = -\dfrac{\sqrt{y}}{\sqrt{x}}$.

19. $\sqrt{x+y} = x$. Differentiating both sides of the equation implicitly, we obtain
$$\tfrac{1}{2}(x+y)^{-1/2}(1+y') = 1, \; 1+y' = 2(x+y)^{1/2},$$
or $\;\; y' = 2\sqrt{x+y} - 1$.

21. $\dfrac{1}{x^2} + \dfrac{1}{y^2} = 1$. Differentiating both sides of the equation implicitly, we obtain
$$-\dfrac{2}{x^3} - \dfrac{2}{y^3}y' = 0, \; \text{ or } \; y' = -\dfrac{y^3}{x^3}.$$

23. $\sqrt{xy} = x + y$. Differentiating both sides of the equation implicitly, we obtain
$$\tfrac{1}{2}(xy)^{-1/2}(xy'+y) = 1 + y'$$
$$xy' + y = 2\sqrt{xy}(1 + y')$$
$$y'(x - 2\sqrt{xy}) = 2\sqrt{xy} - y$$
or $\qquad\qquad y' = -\dfrac{(2\sqrt{xy} - y)}{(2\sqrt{xy} - x)} = \dfrac{2\sqrt{xy} - y}{x - 2\sqrt{xy}}.$

25. $\dfrac{x+y}{x-y} = 3x$, or $x + y = 3x^2 - 3xy$. Differentiating both sides of the equation

implicitly, we obtain $1 + y' = 6x - 3xy' - 3y$ or $y' = \dfrac{6x - 3y - 1}{3x + 1}$.

27. $xy^{3/2} = x^2 + y^2$. Differentiating implicitly with respect to x, we obtain

$$y^{3/2} + x\left(\tfrac{3}{2}\right)y^{1/2}\frac{dy}{dx} = 2x + 2y\frac{dy}{dx}$$

$$2y^{3/2} + 3xy^{1/2}\frac{dy}{dx} = 4x + 4y\frac{dy}{dx} \qquad \text{(Multiplying by 2.)}$$

$$(3xy^{1/2} - 4y)\frac{dy}{dx} = 4x - 2y^{3/2}$$

$$\frac{dy}{dx} = \frac{2(2x - y^{3/2})}{3xy^{1/2} - 4y}.$$

29. $(x + y)^3 + x^3 + y^3 = 0$. Differentiating implicitly with respect to x, we obtain

$$3(x + y)^2\left(1 + \frac{dy}{dx}\right) + 3x^2 + 3y^2\frac{dy}{dx} = 0$$

$$(x + y)^2 + (x + y)^2\frac{dy}{dx} + x^2 + y^2\frac{dy}{dx} = 0$$

$$[(x + y)^2 + y^2]\frac{dy}{dx} = -[(x + y)^2 + x^2]$$

$$\frac{dy}{dx} = -\frac{2x^2 + 2xy + y^2}{x^2 + 2xy + 2y^2}.$$

31. $4x^2 + 9y^2 = 36$. Differentiating the equation implicitly, we obtain
$$8x + 18yy' = 0.$$
At the point $(0,2)$, we have $0 + 36y' = 0$ and the slope of the tangent line is 0.
Therefore, an equation of the tangent line is $y = 2$.

33. $x^2y^3 - y^2 + xy - 1 = 0$. Differentiating implicitly with respect to x, we have
$$2xy^3 + 3x^2y^2\frac{dy}{dx} - 2y\frac{dy}{dx} + y + x\frac{dy}{dx} = 0.$$

At $(1,1)$,
$$2 + 3\frac{dy}{dx} - 2\frac{dy}{dx} + 1 + \frac{dy}{dx} = 0$$

$$2\frac{dy}{dx} = -3 \quad \text{and} \quad \frac{dy}{dx} = -\frac{3}{2}.$$

Using the point-slope form of an equation of a line, we have

$$y - 1 = -\tfrac{3}{2}(x - 1)$$

and the equation of the tangent line to the graph of the function f at $(1,1)$ is

$$y = -\tfrac{3}{2}x + \tfrac{5}{2}.$$

35. $xy = 1$. Differentiating implicitly, we have

$$xy' + y = 0, \quad \text{or} \quad y' = -\frac{y}{x}.$$

Differentiating implicitly once again, we have
$$xy'' + y' + y' = 0.$$

Therefore, $\quad y'' = -\dfrac{2y'}{x} = \dfrac{2\left(\dfrac{y}{x}\right)}{x} = \dfrac{2y}{x^2}.$

37. $y^2 - xy = 8$. Differentiating implicitly we have $2yy' - y - xy' = 0$

and $y' = \dfrac{y}{2y - x}$. Differentiating implicitly again, we have

$$2(y')^2 + 2yy'' - y' - y' - xy'' = 0, \quad \text{or} \quad y'' = \frac{2y' - 2(y')^2}{2y - x}.$$

Then $\quad y'' = \dfrac{2\left(\dfrac{y}{2y-x}\right)\left(1 - \dfrac{y}{2y-x}\right)}{2y - x} = \dfrac{2y(2y - x - y)}{(2y - x)^3} = \dfrac{2y(y - x)}{(2y - x)^3}.$

39. a. Differentiating the given equation with respect to t, we obtain

$$\frac{dV}{dt} = \pi r^2 \frac{dh}{dt} + 2\pi r h \frac{dr}{dt} = \pi r \left(r \frac{dh}{dt} + 2h \frac{dr}{dt} \right).$$

b. Substituting $r = 2$, $h = 6$, $\dfrac{dr}{dt} = 0.1$ and $\dfrac{dh}{dt} = 0.3$ into the expression for $\dfrac{dV}{dt}$

we obtain $\dfrac{dV}{dt} = \pi(2)[2(0.3) + 2(6)(0.1)] = 3.6\pi$

and so the volume is increasing at the rate of 3.6π cu in/sec.

41. We are given $\dfrac{dp}{dt} = 2$ and are required to find $\dfrac{dx}{dt}$ when $x = 9$ and $p = 63$.

Differentiating the equation $p + x^2 = 144$ with respect to t, we obtain

$$\frac{dp}{dt} + 2x\frac{dx}{dt} = 0.$$

When $x = 9$, $p = 63$, and $\dfrac{dp}{dt} = 2$,

$$2 + 2(9)\frac{dx}{dt} = 0$$

and

$$\frac{dx}{dt} = -\frac{1}{9} \approx -0.111,$$

or decreasing at the rate of 111 tires per week.

43. $100x^2 + 9p^2 = 3600$. Differentiating the given equation implicitly with respect to t, we have

$$200x\frac{dx}{dt} + 18p\frac{dp}{dt} = 0.$$

Next, when $p = 14$, the given equation yields

$$100x^2 + 9(14)^2 = 3600$$
$$100x^2 = 1836,$$

or $x = 4.2849$. When $p = 14$, $\dfrac{dp}{dt} = -0.15$, and $x = 4.2849$, we have

$$200(4.2849)\frac{dx}{dt} + 18(14)(-0.15) = 0$$

$$\frac{dx}{dt} = 0.0441.$$

So the quantity demanded is increasing at the rate of 44 ten–packs per week.

45. From the results of Problem 44, we have

$$1250p\frac{dp}{dt} - 2x\frac{dx}{dt} = 0.$$

When $p = 1.0770$, $x = 25$, and $\dfrac{dx}{dt} = -1$, we find that

$$1250(1.077)\frac{dp}{dt} - 2(25)(-1) = 0,$$

and

$$\frac{dp}{dt} = -\frac{50}{1250(1.077)} = -0.037.$$

We conclude that the price is decreasing at the rate of 3.7 cents per carton.

47. $p = -0.01x^2 - 0.2x + 8$. Differentiating the given equation implicitly with respect to p, we have

$$1 = -0.02x\frac{dx}{dp} - 0.2\frac{dx}{dp} = [0.02x + 0.2]\frac{dx}{dp}$$

or $\quad \dfrac{dx}{dp} = -\dfrac{1}{0.02x + 0.2}.$

When $x = 15$, $p = -0.01(15)^2 - 0.2(15) + 8 = 2.75$

and $\quad \dfrac{dx}{dp} = -\dfrac{1}{0.02(15) + 0.2} = -2.$

Therefore, $\quad E(p) = -\dfrac{pf'(p)}{f(p)} = -\dfrac{(2.75)(-2)}{15} = 0.37 < 1,$

and the demand is inelastic.

49. $A = \pi r^2$. Differentiating with respect to t, we obtain

$$\frac{dA}{dt} = 2\pi r \frac{dr}{dt}.$$

When the radius of the circle is 40 ft and increasing at the rate of 2 ft/sec,

$$\frac{dA}{dt} = 2\pi(40)(2) = 160\pi \text{ ft}^2 / \text{sec}.$$

51. Let D denote the distance between the two cars, x the distance traveled by the car heading east, and y the distance traveled by the car heading north as shown in the diagram at the right. Then

$$D^2 = x^2 + y^2.$$

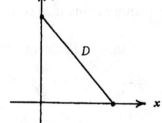

Differentiating with respect to t, we have

$$2D\frac{dD}{dt} = 2x\frac{dx}{dt} + 2y\frac{dy}{dt},$$

or $\quad \dfrac{dD}{dt} = \dfrac{x\dfrac{dx}{dt} + y\dfrac{dy}{dt}}{D}$

When $t = 5$, $x = 30$, $y = 40$, $\dfrac{dx}{dt} = 2(5)+1 = 11$, and $\dfrac{dy}{dt} = 2(5)+3 = 13$.

Therefore, $\dfrac{dD}{dt} = \dfrac{(30)(11)+(40)(13)}{\sqrt{900+1600}} = 17$ ft/sec.

53. Referring to the diagram at the right, we see that
$$D^2 = 120^2 + x^2.$$
Differentiating this last equation with respect to t, we have

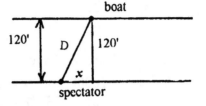

boat

120'   D   120'

x

spectator

$$2D\dfrac{dD}{dt} = 2x\dfrac{dx}{dt} \quad \text{and} \quad \dfrac{dD}{dt} = \dfrac{x\dfrac{dx}{dt}}{D}.$$

When $x = 50$, $D = \sqrt{120^2 + 50^2} = 130$ and $\dfrac{dD}{dt} = \dfrac{(20)(50)}{130} \approx 7.69$, or 7.69 ft/sec.

55. Let V and S denote its volume and surface area. Then we are given that
$\dfrac{dV}{dt} = -kS$, where k is the constant of proportionality. But from $V = \left(\dfrac{4}{3}\right)\pi r^3$,

we find, upon differentiating both sides with respect to t, that
$$\dfrac{dV}{dt} = \left(\dfrac{4}{3}\right)\pi(3\pi r^2)\dfrac{dr}{dt} = 4\pi^2 r^2 \dfrac{dr}{dt}$$
and using the fact stated earlier,
$$\dfrac{dV}{dt} = 4\pi^2 r^2 \dfrac{dr}{dt} = -kS = -k(4\pi r^2).$$

Therefore, $\dfrac{dr}{dt} = -\dfrac{k(4\pi r^2)}{4\pi^2 r^2} = -\dfrac{k}{\pi}$

and this proves that the radius is decreasing at the constant rate of (k/π) units/unit time.

57. Refer to the figure at the right.
We are given that $\dfrac{dx}{dt} = 264$. Using the

Pythagorean Theorem,
$$s^2 = x^2 + 1000^2 = x^2 + 1000000.$$
We want to find $\dfrac{ds}{dt}$ when $s = 1500$.

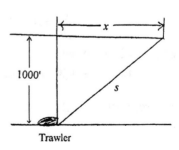

x

1000'

s

Trawler

Differentiating both sides of the equation with respect to t, we have

$$2s\frac{ds}{dt} = 2x\frac{dx}{dt}$$

and so
$$\frac{ds}{dt} = \frac{x\dfrac{dx}{dt}}{s}.$$

Now, when $s = 1500$, we have
$$1500^2 = x^2 + 10000 \quad \text{or} \quad x = \sqrt{1250000}.$$

Therefore,
$$\frac{ds}{dt} = \frac{\sqrt{1250000} \cdot (264)}{1500} \approx 196.8,$$

that is, the aircraft is receding from the trawler at the speed of approximately 196.8 ft/sec.

59. Refer to the diagram at the right.

$$\frac{y}{6} = \frac{y+x}{18}, \quad 18y = 6(y+x)$$

$$3y = y+x, \quad 2y = x, \quad y = \tfrac{1}{2}x.$$

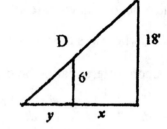

Then $D = y + x = \tfrac{3}{2}x$. Differentiating implicitly, we have

$$\frac{dD}{dt} = \frac{3}{2} \bullet \frac{dx}{dt}$$

and when $\dfrac{dx}{dt} = 6$, $\dfrac{dD}{dt} = \dfrac{3}{2}(6) = 9$, or 9 ft / sec.

61. Differentiating $x^2 + y^2 = 13^2 = 169$ with respect to t gives

$$2x\frac{dx}{dt} + 2y\frac{dy}{dt} = 0.$$

When $x = 12$, we have
$$144 + y^2 = 169 \quad \text{or} \quad y = 5.$$

Therefore, with $x = 12$, $y = 5$, and $\dfrac{dx}{dt} = 8$,

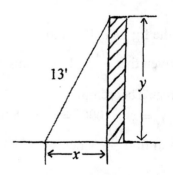

we find $2(12)(8) + 2(5)\dfrac{dy}{dt} = 0$

or $\dfrac{dy}{dt} = -19.2,$ that is, the top of the ladder is sliding down the wall at 19.2 ft/sec.

63. True. Differentiating both sides of the equation with respect to x, we have

$$\frac{d}{dx}[f(x)g(y)] = \frac{d}{dx}(0)$$

$$f(x)g'(y)\frac{dy}{dx} + f'(x)g(y) = 0$$

$$\frac{dy}{dx} = -\frac{f'(x)g(y)}{f(x)g'(y)}$$

provided $f(x) \neq 0$ and $g'(y) \neq 0.$

EXERCISES 3.7, page 268

1. $f(x) = 2x^2$ and $dy = 4x\, dx.$

3. $f(x) = x^3 - x$ and $dy = (3x^2 - 1)\, dx.$

5. $f(x) = \sqrt{x+1} = (x+1)^{1/2}$ and $dy = \dfrac{1}{2}(x+1)^{-1/2}\, dx = \dfrac{dx}{2\sqrt{x+1}}.$

7. $f(x) = 2x^{3/2} + x^{1/2}$ and

$dy = (3x^{1/2} + \frac{1}{2}x^{-1/2})\, dx = \frac{1}{2}x^{-1/2}(6x + 1)dx = \dfrac{6x+1}{2\sqrt{x}}\, dx.$

9. $f(x) = x + \dfrac{2}{x}$ and $dy = \left(1 - \dfrac{2}{x^2}\right)dx = \dfrac{x^2 - 2}{x^2}\, dx.$

11. $f(x) = \dfrac{x-1}{x^2+1}$ and $dy = \dfrac{x^2+1-(x-1)2x}{(x^2+1)^2}\, dx = \dfrac{-x^2+2x+1}{(x^2+1)^2}\, dx.$

13. $f(x) = \sqrt{3x^2 - x} = (3x^2 - x)^{1/2}$ and

$$dy = \frac{1}{2}(3x^2 - x)^{-1/2}(6x - 1)dx = \frac{6x - 1}{2\sqrt{3x^2 - x}}dx.$$

15. $f(x) = x^2 - 1.$
   a. $dy = 2x\, dx.$ b. $dy \approx 2(1)(0.02) = 0.04.$
   c. $\Delta y = [(1.02)^2 - 1] - [1 - 1] = 0.0404.$

17. $f(x) = \frac{1}{x}.$

   a. $dy = -\frac{dx}{x^2}.$ b. $dy \approx -0.05$

   c. $\Delta y = \frac{1}{-0.95} - \frac{1}{-1} = -0.05263.$

19. $y = \sqrt{x}$ and $dy = \frac{dx}{2\sqrt{x}}$. Therefore, $\sqrt{10} = 3 + \frac{1}{2 \cdot \sqrt{9}} = 3.167.$

21. $y = \sqrt{x}$ and $dy = \frac{dx}{2\sqrt{x}}$. Therefore, $\sqrt{49.5} = 7 + \frac{0.5}{2 \cdot 7} = 7.0358.$

23. $y = x^{1/3}$ and $dy = \frac{1}{3}x^{-2/3}\, dx$. Therefore, $\sqrt[3]{7.8} = 2 - \frac{0.2}{3 \cdot 4} = 1.983.$

25. $y = \sqrt{x}$ and $dy = \frac{dx}{2\sqrt{x}}$. Therefore, $\sqrt{0.089} = \frac{1}{10}\sqrt{8.9} = \frac{1}{10}\left[3 - \frac{0.1}{2 \cdot 3}\right] \approx 0.298.$

27. $y = f(x) = \sqrt{x} + \frac{1}{\sqrt{x}} = x^{1/2} + x^{-1/2}$. Therefore,

   $$\frac{dy}{dx} = \frac{1}{2}x^{-1/2} - \frac{1}{2}x^{-3/2}$$

   $$dy = \left(\frac{1}{2x^{1/2}} - \frac{1}{2x^{3/2}}\right)dx.$$

   Letting $x = 4$ and $dx = 0.02$, we find

   $$\sqrt{4.02} + \frac{1}{\sqrt{4.02}} - f(4) = f(4.02) - f(4) = \Delta y \approx dy$$

$$\sqrt{4.02} + \frac{1}{\sqrt{4.02}} = f(4) + dy\Big|_{\substack{x=4 \\ dx=0.02}}$$

$$\approx 2 + \frac{1}{2} + \left(\frac{1}{2 \cdot 2} - \frac{1}{2 \cdot 2\sqrt{2}} \right)(0.02)$$

$$\approx 2.50146.$$

29. The volume of the cube is given by $V = x^3$. Then $dV = 3x^2\, dx$ and
when $x = 12$ and $dx = 0.02$, $dV = 3(144)(\pm 0.02) = \pm 8.64$,
and the possible error that might occur in calculating the volume is ± 8.64 cm^3.

31. The volume of the hemisphere is given by $V = \frac{2}{3}\pi r^3$. The amount of rust-proofer
needed is

$$\Delta V = \frac{2}{3}\pi(r + \Delta r)^3 - \frac{2}{3}\pi r^3$$

$$\approx dV = \left(\frac{2}{3} \right)(3\pi r^2)dr.$$

So, with r $= 60$, and $dr = \frac{1}{12}(0.01)$, we have

$$\Delta V \approx 2\pi(60^2)\left(\frac{1}{12} \right)(0.01) \approx 18.85.$$

So we need approximately 18.85 ft^3 of rust-proofer.

33. $dR = \dfrac{d}{dr}(k\ell r^{-4})dr = -4k\ell r^{-5}\, dr$. With $\dfrac{dr}{r} = 0.1$, we find

$$\frac{dR}{R} = -\frac{4k\ell r^{-5}}{k\ell r^{-4}}dr = -4\frac{dr}{r} = -4(0.1) = -0.4.$$

In other words, the resistance will drop by 40%.

35. $f(n) = 4n\sqrt{n-4} = 4n(n-4)^{1/2}$.
Then $df = 4[(n-4)^{1/2} + \frac{1}{2}n(n-4)^{-1/2}]dn$
When $n = 85$ and $dn = 5$, $df = 4[9 + \frac{85}{2.9}]5 \approx 274$ seconds.

37. $N(r) = \dfrac{7}{1+0.02r^2}$ and $dN = -\dfrac{0.28r}{(1+0.02r^2)^2}dr$. To estimate the decrease in the
number of housing starts when the mortgage rate is increased from 12 to 12.5

percent, we compute

$$dN = -\frac{(0.28)(12)(0.5)}{(3.88)^2} \approx -0.111595 \quad (r = 12, \, dr = 0.5)$$

or 111,595 fewer housing starts.

39. $p = \dfrac{30}{0.02x^2 + 1}$ and $dp = -\dfrac{(1.2x)}{(0.02x^2 + 1)^2}\,dx$. To estimate the change in the price p when the quantity demanded changed from 5000 to 5500 units ($x = 5$ to $x = 5.5$) per week, we compute

$$dp = \frac{(-1.2)(5)(0.5)}{[0.02(25) + 1]^2} \approx -1.33, \text{ or a decrease of } \$1.33.$$

41. $P(x) = -0.000032x^3 + 6x - 100$ and $dP = (-0.000096x^2 + 6)\,dx$. To determine the error in the estimate of Trappee's profits corresponding to a maximum error in the forecast of 15 percent $[dx = \pm 0.15(200)]$, we compute

$$dP = [(-0.000096)(200)^2 + 6]\,(\pm 30) = (2.16)(30) = \pm 64.80$$

or $\pm 64{,}800$.

43. $N(x) = \dfrac{500(400 + 20x)^{1/2}}{(5 + 0.2x)^2}$ and

$$N'(x) = \frac{(5 + 0.2x)^2\,250(400 + 20x)^{-1/2}(20) - 500(400 + 20x)^{1/2}(2)(5 + 0.2x)(0.2)}{(5 + 0.2x)^4}\,dx.$$

To estimate the change in the number of crimes if the level of reinvestment changes from 20 cents per dollars deposited to 22 cents per dollar deposited, we compute

$$dN = \frac{(5 + 4)^2(250)(800)^{-1/2}(20) - 500(400 + 400)^{1/2}(2)(9)(0.2)}{(5 + 4)^4}(2)$$

$$= \frac{(14318.91 - 50911.69)}{9^4}(2) \approx -11$$

or a decrease of approximately 11 crimes per year.

45. True. $dy = f'(x)\,dx = \dfrac{d}{dx}(ax + b)\,dx = a\,dx$. On the other hand,

$$\Delta y = f(x + \Delta x) - f(x) = [a(x + \Delta x) + b] - (ax + b) = a\Delta x = a\,dx.$$

1. $dy = f'(3) dx = 757.87(0.01) \approx 7.5787$.

3. $dy = f'(1) dx = 1.04067285926(0.03) \approx 0.031220185778$.

5. $dy = f'(4)(0.1) = -0.198761598(0.1) = -0.0198761598$.

7. If the interest rate changes from 10% to 10.3% per year, the monthly payment will increase by
   $dP = f'(0.1)(0.003) \approx 26.60279$,
   or approximately $26.60 per month. If the interest rate changes from 10% to 10.4% per year, it will be $35.47 per month. If the interest rate changes from 10% to 10.5% per year, it will be $44.34 per month.

9. $dx = f'(40)(2) \approx -0.625$. That is, the quantity demanded will decrease by 625

   watches per week.

CHAPTER 3 REVIEW, page 274

1. $f'(x) = \dfrac{d}{dx}(3x^5 - 2x^4 + 3x^2 - 2x + 1) = 15x^4 - 8x^3 + 6x - 2$.

3. $g'(x) = \dfrac{d}{dx}(-2x^{-3} + 3x^{-1} + 2) = 6x^{-4} - 3x^{-2}$.

5. $g'(t) = \dfrac{d}{dt}(2t^{-1/2} + 4t^{-3/2} + 2) = -t^{-3/2} - 6t^{-5/2}$.

7. $f'(t) = \dfrac{d}{dt}(t + 2t^{-1} + 3t^{-2}) = 1 - 2t^{-2} - 6t^{-3} = 1 - \dfrac{2}{t^2} - \dfrac{6}{t^3}$.

9. $h'(x) = \dfrac{d}{dx}(x^2 - 2x^{-3/2}) = 2x + 3x^{-5/2} = 2x + \dfrac{3}{x^{5/2}}$.

11. $g(t) = \dfrac{t^2}{2t^2 + 1}$.

$$g'(t) = \frac{(2t^2+1)\frac{d}{dt}(t^2) - t^2\frac{d}{dt}(2t^2+1)}{(2t^2+1)^2}$$

$$= \frac{(2t^2+1)(2t) - t^2(4t)}{(2t^2+1)^2} = \frac{2t}{(2t^2+1)^2}.$$

13. $f(x) = \dfrac{\sqrt{x}-1}{\sqrt{x}+1} = \dfrac{x^{1/2}-1}{x^{1/2}+1}.$

$$f'(x) = \frac{(x^{1/2}+1)(\frac{1}{2}x^{-1/2}) - (x^{1/2}-1)(\frac{1}{2}x^{-1/2})}{(x^{1/2}+1)^2}$$

$$= \frac{\frac{1}{2}+\frac{1}{2}x^{-1/2} - \frac{1}{2}+\frac{1}{2}x^{-1/2}}{(x^{1/2}+1)^2} = \frac{x^{-1/2}}{(x^{1/2}+1)^2} = \frac{1}{\sqrt{x}(\sqrt{x}+1)^2}.$$

15. $f(x) = \dfrac{x^2(x^2+1)}{x^2-1}.$

$$f'(x) = \frac{(x^2-1)\frac{d}{dx}(x^4+x^2) - (x^4+x^2)\frac{d}{dx}(x^2-1)}{(x^2-1)^2}$$

$$= \frac{(x^2-1)(4x^3+2x) - (x^4+x^2)(2x)}{(x^2-1)^2}$$

$$= \frac{4x^5+2x^3-4x^3-2x-2x^5-2x^3}{(x^2-1)^2}$$

$$= \frac{2x^5-4x^3-2x}{(x^2-1)^2} = \frac{2x(x^4-2x^2-1)}{(x^2-1)^2}.$$

17. $f(x) = (3x^3-2)^8;\ f'(x) = 8(3x^3-2)^7(9x^2) = 72x^2(3x^3-2)^7.$

19. $f'(t) = \dfrac{d}{dt}(2t^2+1)^{1/2} = \dfrac{1}{2}(2t^2+1)^{-1/2}\dfrac{d}{dt}(2t^2+1)$

$$= \frac{1}{2}(2t^2+1)^{-1/2}(4t) = \frac{2t}{\sqrt{2t^2+1}}.$$

21. $s(t) = (3t^2 - 2t + 5)^{-2}$

$s'(t) = -2(3t^2 - 2t + 5)^{-3}(6t - 2) = -4(3t^2 - 2t + 5)^{-3}(3t - 1)$

$$= -\frac{4(3t - 1)}{(3t^2 - 2t + 5)^3}.$$

23. $h(x) = \left(x + \dfrac{1}{x}\right)^2 = (x + x^{-1})^2.$

$h'(x) = 2(x + x^{-1})(1 - x^{-2}) = 2\left(x + \dfrac{1}{x}\right)\left(1 - \dfrac{1}{x^2}\right)$

$$= 2\left(\frac{x^2 + 1}{x}\right)\left(\frac{x^2 - 1}{x^2}\right) = \frac{2(x^2 + 1)(x^2 - 1)}{x^3}.$$

25. $h'(t) = (t^2 + t)^4 \dfrac{d}{dt}(2t^2) + 2t^2 \dfrac{d}{dt}(t^2 + t)^4$

$= (t^2 + t)^4 (4t) + 2t^2 \cdot 4(t^2 + t)^3 (2t + 1)$

$= 4t(t^2 + t)^3 [(t^2 + t) + 4t^2 + 2t] = 4t^2(5t + 3)(t^2 + t)^3.$

27. $g(x) = x^{1/2}(x^2 - 1)^3.$

$g'(x) = \dfrac{d}{dx}[x^{1/2}(x^2 - 1)^3] = x^{1/2} \cdot 3(x^2 - 1)^2 (2x) + (x^2 - 1)^3 \cdot \tfrac{1}{2} x^{-1/2}$

$= \tfrac{1}{2} x^{-1/2}(x^2 - 1)^2 [12x^2 + (x^2 - 1)]$

$$= \frac{(13x^2 - 1)(x^2 - 1)^2}{2\sqrt{x}}.$$

29. $h(x) = \dfrac{(3x + 2)^{1/2}}{4x - 3}.$

$h'(x) = \dfrac{(4x - 3)\tfrac{1}{2}(3x + 2)^{-1/2}(3) - (3x + 2)^{1/2}(4)}{(4x - 3)^2}$

$$= \frac{\tfrac{1}{2}(3x + 2)^{-1/2}[3(4x - 3) - 8(3x + 2)]}{(4x - 3)^2} = -\frac{12x + 25}{2\sqrt{3x + 2}(4x - 3)^2}.$$

31. $f(x) = 2x^4 - 3x^3 + 2x^2 + x + 4.$

$$f'(x) = \frac{d}{dx}(2x^4 - 3x^3 + 2x^2 + x + 4) = 8x^3 - 9x^2 + 4x + 1.$$

$$f''(x) = \frac{d}{dx}(8x^3 - 9x^2 + 4x + 1) = 24x^2 - 18x + 4 = 2(12x^2 - 9x + 2).$$

33. $h(t) = \dfrac{t}{t^2 + 4}$. $h'(t) = \dfrac{(t^2 + 4)(1) - t(2t)}{(t^2 + 4)^2} = \dfrac{4 - t^2}{(t^2 + 4)^2}.$

$$h''(t) = \frac{(t^2 + 4)^2(-2t) - (4 - t^2)2(t^2 + 4)(2t)}{(t^2 + 4)^4}$$

$$= \frac{-2t(t^2 + 4)[(t^2 + 4) + 2(4 - t^2)]}{(t^2 + 4)^4} = \frac{2t(t^2 - 12)}{(t^2 + 4)^3}.$$

35. $f'(x) = \dfrac{d}{dx}(2x^2 + 1)^{1/2} = \dfrac{1}{2}(2x^2 + 1)^{-1/2}(4x) = 2x(2x^2 + 1)^{-1/2}.$

$$f''(x) = 2(2x^2 + 1)^{-1/2} + 2x \cdot (-\tfrac{1}{2})(2x^2 + 1)^{-3/2}(4x)$$

$$= 2(2x^2 + 1)^{-3/2}[(2x^2 + 1) - 2x^2] = \frac{2}{(2x^2 + 1)^{3/2}}.$$

37. $6x^2 - 3y^2 = 9$ so $12x - 6y\dfrac{dy}{dx} = 0$ and $-6y\dfrac{dy}{dx} = -12x.$

Therefore, $\dfrac{dy}{dx} = \dfrac{-12x}{-6y} = \dfrac{2x}{y}.$

39. $y^3 + 3x^2 = 3y$, so $3y^2y' + 6x = 3y'$, $3y^2y' - 3y' = -6x$,

and $y'(3y^2 - 3) = -6x$. Therefore, $y' = -\dfrac{6x}{3(y^2 - 1)} = -\dfrac{2x}{y^2 - 1}.$

41. $x^2 - 4xy - y^2 = 12$ so $2x - 4xy' - 4y - 2yy' = 0$ and $y'(-4x - 2y) = -2x + 4y.$

So $y' = \dfrac{-2(x - 2y)}{-2(2x + y)} = \dfrac{x - 2y}{2x + y}.$

43. $f(x) = 2x^3 - 3x^2 - 16x + 3$ and $f'(x) = 6x^2 - 6x - 16.$

a. To find the point(s) on the graph of f where the slope of the tangent line is equal

to –4, we solve
$$6x^2 - 6x - 16 = -4, \; 6x^2 - 6x - 12 = 0, \; 6(x^2 - x - 2) = 0$$
$$6(x-2)(x+1) = 0$$
and $x = 2$ or $x = -1$. Then
$$f(2) = 2(2)^3 - 3(2)^2 - 16(2) + 3 = -25$$
and $f(-1) = 2(-1)^3 - 3(-1)^2 - 16(-1) + 3 = 14$
and the points are $(2,-25)$ and $(-1,14)$.
b. Using the point-slope form of the equation of a line, we find that
$$y - (-25) = -4(x-2), \; y + 25 = -4x + 8, \text{ or } y = -4x - 17$$
and $\qquad y - 14 = -4(x+1), \text{ or } y = -4x + 10$
are the equations of the tangent lines at $(2,-25)$ and $(-1,14)$.

45. $y = (4 - x^2)^{1/2}$. $y' = \frac{1}{2}(4 - x^2)^{-1/2}(-2x) = -\dfrac{x}{\sqrt{4 - x^2}}$.

The slope of the tangent line is obtained by letting $x = 1$, giving
$$m = -\frac{1}{\sqrt{3}} = -\frac{\sqrt{3}}{3}.$$
Therefore, an equation of the tangent line is
$$y - \sqrt{3} = -\frac{\sqrt{3}}{3}(x-1), \text{ or } y = -\frac{\sqrt{3}}{3}x + \frac{4\sqrt{3}}{3}.$$

47. $f(x) = (2x - 1)^{-1}$; $f'(x) = -2(2x - 1)^{-2}$, $f''(x) = 8(2x - 1)^{-3} = \dfrac{8}{(2x-1)^3}$.

$f'''(x) = -48(2x - 1)^4 = -\dfrac{48}{(2x-1)^4}$.

Since $(2x - 1)^4 = 0$ when $x = 1/2$, we see that the domain of f''' is $(-\infty, \frac{1}{2}) \cup (\frac{1}{2}, \infty)$.

49. $N(x) = 1000(1 + 2x)^{1/2}$. $N'(x) = 1000(\frac{1}{2})(1 + 2x)^{-1/2}(2) = \dfrac{1000}{\sqrt{1 + 2x}}$.

The rate of increase at the end of the twelfth week is $N'(12) = \dfrac{1000}{\sqrt{25}} = 200$,

or 200 subscribers/week.

51. a. $R(x) = px = (-0.02x + 600)x = -0.02x^2 + 600x$

b. $R'(x) = -0.04x + 600$

c. $R'(10,000) = -0.04(10,000) + 600 = 200$ and this says that the sale of the 10,001st phone will bring a revenue of $200.

CHAPTER 4

EXERCISES 4.1, page 292

1. f is decreasing on $(-\infty,0)$ and increasing on $(0, \infty)$.

3. f is increasing on $(-\infty,-1) \cup (1,\infty)$, and decreasing on $(-1,1)$.

5. f is increasing on $(0,2)$ and decreasing on $(-\infty,0) \cup (2,\infty)$.

7. f is decreasing on $(-\infty,-1) \cup (1,\infty)$ and increasing on $(-1,1)$.

9. Increasing on $(20.2, 20.6) \cup (21.7, 21.8)$, constant on $(19.6, 20.2) \cup (20.6, 21.1)$, and decreasing on $(21.1, 21.7) \cup (21.8, 22.7)$,

11. $f(x) = 3x + 5$; $f'(x) = 3 > 0$ for all x and so f is increasing on $(-\infty,\infty)$.

13. $f(x) = x^2 - 3x$.
    $f'(x) = 2x - 3$ is continuous everywhere and is equal to zero when $x = 3/2$. From the following sign diagram

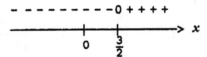

we see that f is decreasing on $(-\infty, \frac{3}{2})$ and increasing on $(\frac{3}{2}, \infty)$.

15. $g(x) = x - x^3$. $g'(x) = 1 - 3x^2$ is continuous everywhere and is equal to zero when $1 - 3x^2 = 0$, or $x = \pm\frac{\sqrt{3}}{3}$. From the following sign diagram

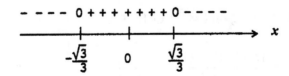

we see that f is decreasing on $(-\infty, -\frac{\sqrt{3}}{3}) \cup (\frac{\sqrt{3}}{3}, \infty)$ and increasing on $(-\frac{\sqrt{3}}{3}, \frac{\sqrt{3}}{3})$.

17. $g(x) = x^3 + 3x^2 + 1;\quad g'(x) = 3x^2 + 6x = 3x(x+2)$.

From the following sign diagram

we see that g is increasing on $(-\infty, -2) \cup (0, \infty)$ and decreasing on $(-2, 0)$.

19. $f(x) = \frac{1}{3}x^3 - 3x^2 + 9x + 20;\quad f'(x) = x^2 - 6x + 9 = (x-3)^2 > 0$ for all x except $x = 3$, at which point $f'(3) = 0$. Therefore, f is increasing on $(-\infty, 3) \cup (3, \infty)$.

21. $h(x) = x^4 - 4x^3 + 10;\ h'(x) = 4x^3 - 12x^2 = 4x^2(x-3)$ if $x = 0$ or 3. From the sign diagram of h',

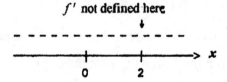

we see that h is increasing on $(3, \infty)$ and decreasing on $(-\infty, 0) \cup (0, 3)$.

23. $f(x) = \dfrac{1}{x-2} = (x-2)^{-1}.\ f'(x) = -1(x-2)^{-2}(1) = -\dfrac{1}{(x-2)^2}$ is discontinuous at $x = 2$ and is continuous everywhere else. From the sign diagram

f' not defined here

we see that f is decreasing on $(-\infty, 2) \cup (2, \infty)$.

25. $h(t) = \dfrac{t}{t-1}.\ h'(t) = \dfrac{(t-1)(1) - t(1)}{(t-1)^2} = -\dfrac{1}{(t-1)^2}$.

From the following sign diagram,

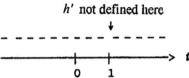

h' not defined here

we see that $h'(t) < 0$ whenever it is defined. We conclude that h is decreasing on $(-\infty,1) \cup (1,\infty)$.

27. $f(x) = x^{3/5}$. $f'(x) = \dfrac{3}{5}x^{-2/5} = \dfrac{3}{5x^{2/5}}$. Observe that $f'(x)$ is not defined at $x = 0$, but is positive everywhere else and therefore increasing on $(-\infty,0) \cup (0,\infty)$.

29. $f(x) = \sqrt{x+1}$. $f'(x) = \dfrac{d}{dx}(x+1)^{1/2} = \dfrac{1}{2}(x+1)^{-1/2} = \dfrac{1}{2\sqrt{x+1}}$ and we see that $f'(x) > 0$ if $x > -1$. Therefore, f is increasing on $(-1,\infty)$.

31. $f(x) = \sqrt{16-x^2} = (16-x^2)^{1/2}$. $f'(x) = \dfrac{1}{2}(16-x^2)^{-1/2}(-2x) = -\dfrac{x}{\sqrt{16-x^2}}$.

Since the domain of f is $[-4,4]$, we consider the sign diagram for f' on this interval. Thus,

f' not defined here

+ + + 0 - - - -

−4 0 4

and we see that f is increasing on $(-4,0)$ and decreasing on $(0,4)$.

33. $f'(x) = \dfrac{d}{dx}(x-x^{-1}) = 1 + \dfrac{1}{x^2} = \dfrac{x^2+1}{x^2}$ and so $f'(x) > 0$ for all $x \neq 0$.

Therefore, f is increasing on $(-\infty,0) \cup (0,\infty)$.

35. $f'(x) = \dfrac{d}{dx}(x-1)^{-2} = -2(x-1)^{-3} = -\dfrac{2}{(x-1)^3}$. From the sign diagram of f'

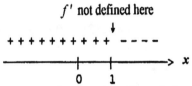

f' not defined here

+ + + + + + + + + + - - - -

0 1

we see that f is increasing on $(-\infty,1)$ and decreasing on $(1,\infty)$.

37. f has a relative maximum of $f(0) = 1$ and relative minima of $f(-1) = 0$ and $f(1) = 0$.

39. f has a relative maximum of $f(-1) = 2$ and a relative minimum of $f(1) = -2$.

41. f has a relative maximum of $f(1) = 3$ and a relative minimum of $f(2) = 2$.

43. f has a relative minimum at $(0,2)$.

45. a   47. d

49. $f(x) = x^2 - 4x$. $f'(x) = 2x - 4 = 2(x - 2)$ has a critical point at $x = 2$. From the following sign diagram

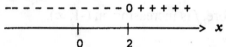

we see that $f(2) = -4$ is a relative minimum by the First Derivative Test.

51. $f(x) = \frac{1}{2}x^2 - 2x + 4$. $f'(x) = x - 2$ giving the critical point $x = 2$. The sign diagram for f' is

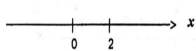

and we see that $f(2) = 2$ is a relative minimum.

53. $f(x) = x^{2/3} + 2$. $f'(x) = \frac{2}{3}x^{-1/3} = \frac{2}{3x^{1/3}}$ and is discontinuous at $x = 0$, a critical point. From the sign diagram

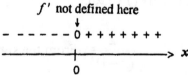

and the First Derivative Test we see that f has a relative minimum at $(0,2)$.

55. $g(x) = x^3 - 3x^2 + 4$. $g'(x) = 3x^2 - 6x = 3x(x - 2) = 0$ if $x = 0$ or 2. From the sign

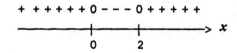

diagram, we see that the critical point $x = 0$ gives a relative maximum, whereas, $x = 2$ gives a relative minimum. The values are $g(0) = 4$ and $g(2) = 8 - 12 + 4 = 0$.

57. $F(x) = \frac{1}{3}x^3 - x^2 - 3x + 4$. Setting $F'(x) = x^2 - 2x - 3 = (x-3)(x+1) = 0$ gives $x = -1$ and $x = 3$ as critical points. From the sign diagram

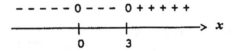

we see that $x = -1$ gives a relative maximum and $x = 3$ gives a relative minimum. The values are
$$F(-1) = -\frac{1}{3} - 1 + 3 + 4 = \frac{17}{3} \quad \text{and} \quad F(3) = 9 - 9 - 9 + 4 = -5,$$
respectively.

59. $g(x) = x^4 - 4x^3 + 8$. Setting $g'(x) = 4x^3 - 12x^2 = 4x^2(x - 3) = 0$ gives $x = 0$ and $x = 3$ as critical points. From the sign diagram

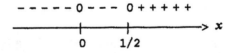

we see that $x = 3$ gives a relative minimum. Its value is $g(3) = 3^4 - 4(3)^3 + 8 = -19$.

61. $f(x) = 3x^4 - 2x^3 + 4$; $f'(x) = 12x^3 - 6x^2 = 6x^2(2x - 1) = 0$ if $x = 0$ or $1/2$. The sign diagram of f' is shown below.

we see that $x=0$ $---$ $0+++++$
$\qquad\qquad 0 \qquad 1/2 \qquad\to x$

and shows that f has a relative minimum at $(\frac{1}{2}, \frac{63}{16})$.

63. $g'(x) = \frac{d}{dx}\left(1 + \frac{1}{x}\right) = -\frac{1}{x^2}$. Observe that g' is never zero for all values of x.

Furthermore, g' is undefined at $x = 0$, but $x = 0$ is not in the domain of g. Therefore g has no critical points and so g has no relative extrema.

65. $f(x) = x + \dfrac{9}{x} + 2.$ Setting $f'(x) = 1 - \dfrac{9}{x^2} = \dfrac{x^2-9}{x^2} = \dfrac{(x+3)(x-3)}{x^2} = 0$

gives $x = -3$ and $x = 3$ as critical points. From the sign diagram

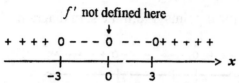

we see that $(-3,-4)$ is a relative maximum and $(3,8)$ is a relative minimum.

67. $f(x) = \dfrac{x}{1+x^2}.$ $f'(x) = \dfrac{(1+x^2)(1)-x(2x)}{(1+x^2)^2} = \dfrac{1-x^2}{(1+x^2)^2} = \dfrac{(1-x)(1+x)}{(1+x^2)^2} = 0$ if $x = \pm 1,$

and these are critical points of f. From the sign diagram of f'

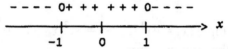

we see that f has a relative minimum at $(-1, -\tfrac{1}{2})$ and a relative maximum at $(1, \tfrac{1}{2})$ ·

69. $f(x) = \dfrac{x^2}{x^2-4}.$ $f'(x) = \dfrac{(x^2-4)(2x)-x^2(2x)}{(x^2-4)^2} = -\dfrac{8x}{(x^2-4)^2}$ is continuous

everywhere except at $x \pm 2$ and has a zero at $x = 0$. Therefore, $x = 0$ is the only critical point of f (the points $x = \pm 2$ do not lie in the domain of f).Using the following sign diagram of f'

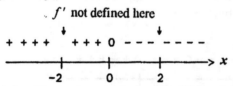

and the First Derivative Test, we conclude that $f(0) = 0$ is a relative maximum of f.

71. $f(x) = (x-1)^{2/3}.$ $f'(x) = \dfrac{2}{3}(x-1)^{-1/3} = \dfrac{2}{3(x-1)^{1/3}}.$

$f'(x)$ is discontinuous at $x = 1$. The sign diagram for f' is

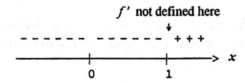

We conclude that $f(1) = 0$ is a relative minimum.

73. $P(x) = -0.001x^2 + 8x - 5000$. $P'(x) = -0.002x + 8 = 0$ if $x = 4000$. Observe that $P'(x) > 0$ if $x < 4000$ and $P'(x) < 0$ if $x > 4000$. So P is increasing on $(0,4000)$ and decreasing on $(4000, \infty)$.

75. $I(t) = \frac{1}{3}t^3 - \frac{5}{2}t^2 + 80$; $I'(t) = t^2 - 5t = t(t - 5) = 0$ if $t = 0$ or 5. From the sign diagram,

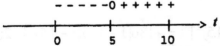

we see that I is decreasing on $(0,5)$ and increasing on $(5,10)$. After declining from 1984 through 1989, the index begins to increase after 1989.

77. $\overline{C}(x) = -0.0001x + 2 + \dfrac{2000}{x}$. $\overline{C}'(x) = -0.0001 - \dfrac{2000}{x^2} < 0$ for all values of x and so $\overline{C}$ is always decreasing.

79. $A(t) = -96.6t^4 + 403.6t^3 + 660.9t^2 + 250$
$A'(t) = -386.4t^3 + 1210.8t^2 + 1321.8t = t(386.4t^2 + 1210.8t + 1321.8)$.
Solving $A'(t) = 0$, we find $t = 0$ and

$$t = \frac{-1210.8 \pm \sqrt{(1210.8)^2 - 4(-386.4)(1321.8)}}{-2(386.4)} = \frac{-1210.8 \pm 1873.2}{-2(386.4)} \approx 4.$$

Since t lies in the interval $[0,5]$, we see that the continuous function A' has zeros at $t = 0$ and $t = 4$. From the sign diagram

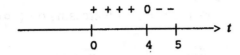

we see that f is increasing on $(0,4)$ and decreasing on $(4,5)$. We conclude that the cash in the Central Provident Trust Funds will be increasing from 1995 to 2035 and decreasing from 2035 to 2045.

81. $C(t) = \dfrac{t^2}{2t^3 + 1}$; $C'(t) = \dfrac{(2t^3 + 1)(2t) - t^2(6t^2)}{(2t^3 + 1)^2} = \dfrac{2t - 2t^4}{(2t^3 + 1)^2} = \dfrac{2t(1 - t^3)}{(2t^3 + 1)^2}$.

From the sign diagram of C' on $(0,\infty)$,

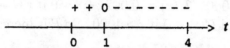

We see that the drug concentration is increasing on $(0,1)$ and decreasing on $(1,4)$.

83. $A(t) = \dfrac{136}{1+0.25(t-4.5)^2} + 28.$

$A'(t) = 136\dfrac{d}{dt}[1+0.25(t-4.5)^2]^{-1} = -136[1+0.25(t-4.5)^2]^{-2}2(0.25)(t-4.5)$

$= -\dfrac{68(t-4.5)}{[1+0.25(t-4.5)^2]^2}.$

Observe that $A'(t) > 0$ if $t < 4.5$ and $A'(t) < 0$ if $t > 4.5$, so the pollution is increasing from 7 A.M. to 11:30 A.M. and decreasing from 11:30 A.M. to 6 P.M.

85. We compute $f'(x) = m$. If $m > 0$, then $f'(x) > 0$ for all x and f is increasing; if $m < 0$, then $f'(x) < 0$ for all x and f is decreasing; if $m = 0$, then $f'(x) = 0$ for all x and f is a constant function.

87. False. The function $f(x) = \begin{cases} -x+1 & x < 0 \\ -\dfrac{1}{2}x+1 & x \geq 0 \end{cases}$

is decreasing on $(-1,1)$, but $f'(0)$ does not exist.

89. False. Let $f(x) = -x$ and $g(x) = -2x$. then both f and g are decreasing on $(-\infty,\infty)$. but $f(x) - g(x) = x - (-2x) = x$ is increasing on $(-\infty,\infty)$.

91. False. Let $f(x) = x^3$. then $f'(0) = 3x^2\big|_{x=0} = 0$. But f does not have a relative extremum at $x = 0$.

93. a. $f'(x) = \begin{cases} -3 & \text{if } x < 0 \\ 2 & \text{if } x > 0 \end{cases}$

$f'(-1) = -3$ and $f'(1) = 2$, so $f'(x)$ changes sign as we move across $x = 0$.

b. No. f does not have a relative minimum at $x = 0$ because $f(0) = 4$ but $f(x) < 4$ if

x is a little less than 4. This does not contradict the First Derivative Test because f is not continuous at $x = 0$.

95. $f(x) = ax^2 + bx + c$. Setting $f'(x) = 2ax + b = 2a\left(x + \frac{b}{2a}\right) = 0$ gives $x = -\frac{b}{2a}$ as the only critical point of f. If $a < 0$, we have the sign diagram

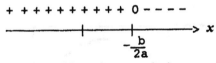

from which we see that $x = -b/2a$ gives a relative maximum. Similarly, you can show that if $a > 0$, then $x = -b/2a$ gives a relative minimum.

97. a. $f'(x) = 3x^2 + 1$ and so $f'(x) > 1$ on the interval $(0,1)$. Therefore, f is increasing on $(0,1)$.
b. $f(0) = -1$ and $f(1) = 1 + 1 - 1 = 1$. So the Intermediate Value Theorem guarantees that there is at least one root of $f(x) = 0$ in $(0,1)$. Since f is increasing on $(0,1)$, the graph of f can cross the x-axis at only one point in $(0,1)$. So $f(x) = 0$ has exactly one root.

USING TECHNOLOGY EXERCISES 4.1, page 298

1. a. f is decreasing on $(-\infty,-0.2934)$ and increasing on $(-0.2934,\infty)$.
   b. Relative minimum: $f(-0.2934) = -2.5435$

3. a. f is increasing on $(-\infty,-1.6144) \cup (0.2390,\infty)$ and decreasing on $(-1.6144, 0.2390)$
   b. Relative maximum: $f(-1.6144) = 26.7991$; relative minimum: $f(0.2390) = 1.6733$

5. a. f is decreasing on $(-\infty,-1) \cup (0.33,\infty)$ and increasing on $(-1,0.33)$
   b. Relative maximum: $f(0.33) = 1.11$; relative minimum: $f(-1) = -0.63$

7. a. f is decreasing on $(-1,-0.71)$ and increasing on $(-0.71,1)$.
   b. f has a relative minimum at $(-0.71,-1.41)$.

9. a.

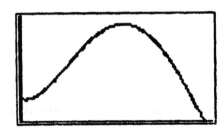

b. f is decreasing on $(0, 0.2398) \cup (6.8758, 12)$ and increasing on $(0.2398, 6.8758)$

c. $(6.8758, 200.14)$; The rate at which the number of banks were failing reached a peak of 200/yr during the latter part of 1988 $(t = 6.8758)$.

11. a.

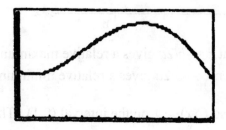

b. f is decreasing on $(0, 0.8343) \cup (7.6726, 12)$ and increasing on $(0.8343, 7.6726)$. The rate at which single-family homes in the greater Boston area were selling was decreasing during most of 1984, but started increasing in late 1984 and continued increasing until mid 1991 when it started decreasing again until 1996.

13. The PSI is increasing on the interval $(0, 4.5)$ and decreasing on $(4.5, 11)$. It is highest when $t = 4.5$ (11:30 A.M.) and has value 164.

EXERCISES 4.2, page 313

1. f is concave downward on $(-\infty, 0)$ and concave upward on $(0, \infty)$. f has an inflection point at $(0, 0)$.

3. f is concave downward on $(-\infty, 0) \cup (0, \infty)$.

5. f is concave upward on $(-\infty, 0) \cup (1, \infty)$ and concave downward on $(0, 1)$. $(0, 0)$ and $(1, -1)$ are inflection points of f.

7. f is concave downward on $(-\infty, -2) \cup (-2, 2) \cup (2, \infty)$.

9. a

11. b

13. a. $D_1'(t) > 0$, $D_2'(t) > 0$, $D_1''(t) > 0$, and $D_2''(t) < 0$ on $(0, 12)$.

b. With or without the proposed promotional campaign, the deposits will increase, but with the promotion, the deposits will increase at an increasing rate whereas without the promotion, the deposits will increase at a decreasing rate.

15. The significance of the inflection point Q is that at the time t_0, corresponding to its t-coordinate, the restoration process is working at its peak.

17. $g(x) = x^4 + \frac{1}{2}x^2 + 6x + 10$; $g'(x) = 4x^3 + x + 6$ and $g''(x) = 12x^2 + 1$. We see that $g''(x) \geq 1$ for all values of x and so g is concave upward everywhere.

19. $f(x) = \dfrac{1}{x^4} = x^{-4}$; $f'(x) = -\dfrac{4}{x^5}$ and $f''(x) = \dfrac{20}{x^6} > 0$ for all values of x in
$(-\infty, 0) \cup (0, \infty)$ and so f is concave upward everywhere.

21. $g'(x) = \dfrac{d}{dx}\left[-(4-x^2)^{1/2}\right] = -\frac{1}{2}(4-x^2)^{-1/2}(-2x) = x(4-x^2)^{-1/2}$.

$g''(x) = (4-x^2)^{-1/2} + x(-\frac{1}{2})(4-x^2)^{-3/2}(-2x)$

$= (4-x^2)^{-3/2}[(4-x^2)+x^2] = \dfrac{4}{(4-x^2)^{3/2}} > 0,$

whenever it is defined and so g is concave upward wherever it is defined.

23. $f(x) = 2x^2 - 3x + 4$; $f'(x) = 4x - 3$ and $f''(x) = 4 > 0$ for all values of x. So f is concave upward on $(-\infty, \infty)$.

25. $f(x) = x^3 - 1$. $f'(x) = 3x^2$ and $f''(x) = 6x$. The sign diagram of f'' follows.

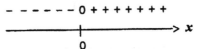

We see that f is concave downward on $(-\infty, 0)$ and concave upward on $(0, \infty)$.

27. $f(x) = x^4 - 6x^3 + 2x + 8$; $f'(x) = 4x^3 - 18x^2 + 2$ and $f''(x) = 12x^2 - 36x = 12x(x - 3)$. The sign diagram of f''

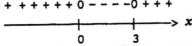

shows that f is concave upward on $(-\infty, 0) \cup (3, \infty)$ and is concave downward on $(0, 3)$.

29. $f(x) = x^{4/7}$. $f'(x) = \dfrac{4}{7}x^{-3/7}$ and $f''(x) = -\dfrac{12}{49}x^{-10/7} = -\dfrac{12}{49x^{10/7}}$.

Observe that $f''(x) < 0$ for all x different from zero. So f is concave downward on $(-\infty,0) \cup (0,\infty)$.

31. $f(x) = (4-x)^{1/2}$. $f'(x) = \dfrac{1}{2}(4-x)^{-1/2}(-1) = -\dfrac{1}{2}(4-x)^{-1/2}$;

$f''(x) = \dfrac{1}{4}(4-x)^{-3/2}(-1) = -\dfrac{1}{4(4-x)^{3/2}} < 0$.

whenever it is defined. So f is concave downward on $(-\infty,4)$.

33. $f'(x) = \dfrac{d}{dx}(x-2)^{-1} = -(x-2)^{-2}$ and $f''(x) = 2(x-2)^{-3} = \dfrac{2}{(x-2)^3}$.

The sign diagram of f'' shows that f is concave downward on $(-\infty,2)$ and concave upward on $(2,\infty)$.

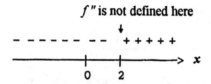

35. $f'(x) = \dfrac{d}{dx}(2+x^2)^{-1} = -(2+x^2)^{-2}(2x) = -2x(2+x^2)^{-2}$ and

$f''(x) = -2(2+x^2)^{-2} - 2x(-2)(2+x^2)^{-3}(2x)$

$= 2(2+x^2)^{-3}[-(2+x^2)+4x^2] = \dfrac{2(3x^2-2)}{(2+x^2)^3} = 0$ if $x = \pm\sqrt{2/3}$.

From the sign diagram of f''

```
  + + + + 0  - - - - - -  0+ + + +
 ─────────┼──────────┼──────────────> x
       -√2/3        0        √2/3
```

we see that f is concave upward on $(-\infty,-\sqrt{2/3}) \cup (\sqrt{2/3}, \infty)$ and concave downward on $(-\sqrt{2/3}, \sqrt{2/3})$.

37. $h(t) = \dfrac{t^2}{t-1}$; $h'(t) = \dfrac{(t-1)(2t)-t^2(1)}{(t-1)^2} = \dfrac{t^2-2t}{(t-1)^2}$;

$$h''(t) = \frac{(t-1)^2(2t-2) - (t^2-2t)2(t-1)}{(t-1)^4}$$

$$= \frac{(t-1)(2t^2 - 4t + 2 - 2t^2 + 4t)}{(t-1)^4} = \frac{2}{(t-1)^3}.$$

The sign diagram of h'' is

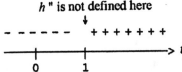

h'' is not defined here

− − − − − − + + + + + + +

0   1   → t

and tells us that h is concave downward on $(-\infty,1)$ and concave upward on $(1,\infty)$.

39. $g'(x) = 1 - 2x^{-3}$ and $g''(x) = 6x^{-4} = \dfrac{6}{x^4} > 0$ whenever $x \neq 0$. Therefore, g is

concave upward on $(-\infty,0) \cup (0,\infty)$.

41. $g(t) = (2t - 4)^{1/3}$. $g'(t) = = \dfrac{1}{3}(2t-4)^{-2/3}(2) = \dfrac{2}{3}(2t-4)^{-2/3}$.

$$g''(t) = -\frac{4}{9}(2t-4)^{-5/3} = -\frac{4}{9(2t-4)^{5/3}}.$$

The sign diagram of g''

g'' is not defined here

+ + + + + + + + + + + − − − −

0   2   → x

tells us that g is concave upward on $(-\infty,2)$ and concave downward on $(2,\infty)$.

43. $f(x) = \dfrac{x^2}{x^2-1}$. $f'(x) = \dfrac{(x^2-1)(2x) - x^2(2x)}{(x^2-1)^2} = -\dfrac{2x}{(x^2-1)^2}$.

$$f''(x) = -\frac{(x^2-1)^2(2) - (-2x)(2)(x^2-1)(2x)}{(x^2-1)^4} = -\frac{2(x^2-1)[x^2-1)+4x^2]}{(x^2-1)^4}$$

$$= \frac{2(3x^2+1)}{(x^2-1)^3}.$$

The sign diagram of f'' is

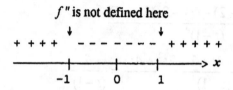

f" is not defined here

$$+ + + + \quad - - - - - - - \quad + + + + +$$

$$\xrightarrow{\qquad} x$$

-1   0   1

and we see that *f* is concave upward on $(-\infty,-1) \cup (1,\infty)$ and concave downward on $(-1,1)$.

45. $f(x) = x^3 - 2. f'(x) = 3x^2$ and $f''(x) = 6x. f''(x)$ is continuous everywhere and has a zero at $x = 0$. From the sign diagram of f''

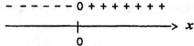

$$- - - - - - 0 + + + + + + +$$

$$\xrightarrow{\qquad} x$$

0

we conclude that $(0,-2)$ is an inflection point of *f*.

47. $f(x) = 6x^3 - 18x^2 + 12x - 15; f'(x) = 18x^2 - 36x + 12$ and $f''(x) = 36x - 36 = 36(x - 1) = 0$ if $x = 1$. The sign diagram of f''

$$- - - - - - - - 0 + + + + + +$$

$$\xrightarrow{\qquad} x$$

0   1

tells us that *f* has an inflection point at $(1,-15)$.

49. $f(x) = 3x^4 - 4x^3 + 1. f'(x) = 12x^3 - 12x^2$ and $f''(x) = 36x^2 - 24x = 12x(3x - 2) = 0$ if $x = 0$ or $2/3$. These are candidates for inflection points. The sign diagram of f''

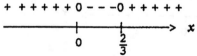

$$+ + + + + + 0 - - - 0 + + + + +$$

$$\xrightarrow{\qquad} x$$

0     $\frac{2}{3}$

shows that $(0,1)$ and $(\frac{2}{3}, \frac{11}{27})$ are inflection points of *f*.

51. $g(t) = t^{1/3}, g'(t) = \frac{1}{3}t^{-2/3}$ and $g''(t) = -\frac{2}{9}t^{-5/3} = -\dfrac{2}{9t^{5/3}}$. Observe that $t = 0$ is in the domain of *g*. Next, since $g''(t) > 0$ if $t < 0$ and $g''(t) < 0$, if $t > 0$, we see that $(0,0)$ is an inflection point of *g*.

53. $f(x) = (x - 1)^3 + 2. f'(x) = 3(x - 1)^2$ and $f''(x) = 6(x - 1)$. Observe that $f''(x) < 0$ if $x < 1$ and $f''(x) > 0$ if $x > 1$ and so $(1,2)$ is an inflection point of *f*.

55. $f(x) = \dfrac{2}{1+x^2} = 2(1+x^2)^{-1}$. $f'(x) = -2(1+x^2)^{-2}(2x) = -4x(1+x^2)^{-2}$.

$f''(x) = -4(1+x^2)^{-2} - 4x(-2)(1+x^2)^{-3}(2x)$

$\qquad = 4(1+x^2)^{-3}[-(1+x^2)+4x^2] = \dfrac{4(3x^2 - 1)}{(1+x^2)^3}$,

is continuous everywhere and has zeros at $x = \pm\frac{\sqrt{3}}{3}$. From the sign diagram of f''

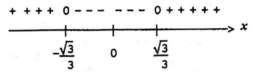

we conclude that $(-\frac{\sqrt{3}}{3}, \frac{3}{2})$ and $(\frac{\sqrt{3}}{3}, \frac{3}{2})$ are inflection points of f.

57. $f(x) = -x^2 + 2x + 4$ and $f'(x) = -2x + 2$. The critical point of f is $x = 1$. Since $f''(x) = -2$ and $f''(1) = -2 < 0$, we conclude that $f(1) = 5$ is a relative maximum of f.

59. $f(x) = 2x^3 + 1$; $f'(x) = 6x^2 = 0$ if $x = 0$ and this is a critical point of f. Next, $f''(x) = 12x$ and so $f''(0) = 0$. Thus, the Second Derivative Test fails. But the First Derivative Test shows that $(0,0)$ is not a relative extremum.

61. $f(x) = \frac{1}{3}x^3 - 2x^2 - 5x - 10$. $f'(x) = x^2 - 4x - 5 = (x-5)(x+1)$ and this gives $x = -1$ and $x = 5$ as critical points of f. Next, $f''(x) = 2x - 4$. Since $f''(-1) = -6 < 0$, we see that $(-1, -\frac{22}{3})$ is a relative maximum. Next, $f''(5) = 6 > 0$ and this shows that $(5, -\frac{130}{3})$ is a relative minimum.

63. $g(t) = t + \dfrac{9}{t}$. $g'(t) = 1 - \dfrac{9}{t^2} = \dfrac{t^2 - 9}{t^2} = \dfrac{(t+3)(t-3)}{t^2}$ and this shows that $t = \pm 3$ are

critical points of g. Now, $g''(t) = 18t^{-3} = \dfrac{18}{t^3}$. Since $g''(-3) = -\dfrac{18}{27} < 0$ the Second

Derivative Test implies that g has a relative maximum at $(-3, -6)$. Also,

$g''(3) = \dfrac{18}{27} > 0$ and so g has a relative minimum at $(3, 6)$.

65. $f(x) = \dfrac{x}{1-x}$. $f'(x) = \dfrac{(1-x)(1) - x(-1)}{(1-x)^2} = \dfrac{1}{(1-x)^2}$ is never zero.

So there are no critical points and f has no relative extrema.

67. $f(t) = t^2 - \dfrac{16}{t}$. $f'(t) = 2t + \dfrac{16}{t^2} = \dfrac{2t^3 + 16}{t^2} = \dfrac{2(t^3 + 8)}{t^2}$. Setting
$f'(t) = 0$ gives $t = -2$ as a critical point. Next, we compute
$f''(t) = \dfrac{d}{dt}(2t + 16t^{-2}) = 2 - 32t^{-3} = 2 - \dfrac{32}{t^3}$. Since $f''(-2) = 2 - \dfrac{32}{(-8)} = 6 > 0$, we
see that $(-2, 12)$ is a relative minimum.

69. $g(s) = \dfrac{s}{1+s^2}$; $g'(s) = \dfrac{(1+s^2)(1) - s(2s)}{(1+s^2)^2} = \dfrac{1-s^2}{(1+s^2)^2} = 0$ gives $s = -1$ and $s = 1$

as critical points of g. Next, we compute
$$g''(s) = \frac{(1+s^2)^2(-2s) - (1-s^2)2(1+s^2)(2s)}{(1+s^2)^4}$$
$$= \frac{2s(1+s^2)(-1-s^2 - 2 + 2s^2)}{(1+s^2)^4} = \frac{2s(s^2 - 3)}{(1+s^2)^3}.$$
Now, $g''(-1) = \frac{1}{2} > 0$ and so $g(-1) = -\frac{1}{2}$ is a relative minimum of g. Next,
$g''(1) = -\frac{1}{2} < 0$ and so $g(1) = \frac{1}{2}$ is a relative maximum of g.

71. $f(x) = \dfrac{x^4}{x-1}$.
$$f'(x) = \frac{(x-1)(4x^3) - x^4(1)}{(x-1)^2} = \frac{4x^4 - 4x^3 - x^4}{(x-1)^2}$$
$$= \frac{3x^4 - 4x^3}{(x-1)^2} = \frac{x^3(3x-4)}{(x-1)^2}$$
and so $x = 0$ and $x = 4/3$ are critical points of f. Next,
$$f''(x) = \frac{(x-1)^2(12x^3 - 12x^2) - (3x^4 - 4x^3)(2)(x-1)}{(x-1)^4}$$
$$= \frac{(x-1)(12x^4 - 12x^3 - 12x^3 + 12x^2 - 6x^4 + 8x^3)}{(x-1)^4}$$
$$= \frac{6x^4 - 16x^3 + 12x^2}{(x-1)^3} = \frac{2x^2(3x^2 - 8x + 6)}{(x-1)^3}.$$

Since $f''(\tfrac{4}{3}) > 0$, we see that $f(\tfrac{4}{3}) = \tfrac{256}{27}$ is a relative minimum. Since $f''(0) = 0$, the Second Derivative Test fails. Using the sign diagram for f',

f' is not defined here

$$+ + + +\ 0 - - - \downarrow - 0 + + + + +$$

$$\xrightarrow{\hspace{1cm}}\ x$$

$$0 \qquad 1 \quad \tfrac{4}{3}$$

and the First Derivative Test, we see that $f(0) = 0$ is a relative maximum.

73. $g(x) = \dfrac{2-x}{(x+2)^3}$.

$$g'(x) = \dfrac{(x+2)^3(-1) - (2-x)(3)(x+2)^2}{(x+2)^6}$$

$$= \dfrac{(x+2)^2[-(x+2) - 3(2-x)]}{(x+2)^6} = \dfrac{2(x-4)}{(x+2)^4}.$$

We see that $x = 4$ is a critical point of g.

$$g''(x) = \dfrac{(x+2)^4(2) - 2(x-4)(4)(x+2)^3}{(x+2)^8}$$

$$= \dfrac{2(x+2)^3[(x+2) - 4(x-4)]}{(x+2)^8} = -\dfrac{2(3x-18)}{(x+2)^5}.$$

Since $g''(4) = \dfrac{12}{(6)^5} > 0$, we see that g has a relative minimum at $(4, -\tfrac{1}{108})$.

75. a. $N'(t)$ is positive because N is increasing on $(0,12)$.

b. $N''(t) < 0$ on $(0,6)$ and $N''(t) > 0$ on $(6,12)$.

c. The rate of growth of the number of help-wanted advertisements was decreasing over the first six months of the year and increasing over the last six months.

77. The behavior of $f(t)$ is just the opposite of that given in the solution to exercise 76. $f(t)$ increases at a decreasing rate until the water level reaches the middle of the vase (and this corresponds to the inflection point of f). After that $f(t)$ increases until the vase is filled and does so at an increasing rate (see the following figure).

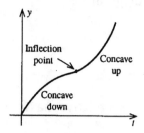

79. $S(x) = -0.002x^3 + 0.6x^2 + x + 500$; $S'(x) = -0.006x^2 + 1.2x + 1$;
$S''(x) = -0.012x + 1.2$.
$x = 100$ is a candidate for an inflection point of S. The sign diagram for S'' is

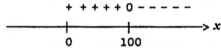

We see that $(100, 4600)$ is an inflection point of S.

81. We wish to find the inflection point of the function $N(t) = -t^3 + 6t^2 + 15t$. Now, $N'(t) = -3t^2 + 12t + 15$ and $N''(t) = -6t + 12 = -6(t - 2)$ giving $t = 2$ as the only candidate for an inflection point of N. From the sign diagram

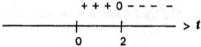

for N'', we conclude that $t = 2$ gives an inflection point of N. Therefore, the average worker is performing at peak efficiency at 10 A.M.

83. $s = f(t) = -t^3 + 54t^2 + 480t + 6$. The velocity of the rocket is
$$v = f'(t) = -3t^2 + 108t + 480 \text{ and}$$
its acceleration is
$$a = f''(t) = -6t + 108 = -6(t - 18).$$
From the sign diagram

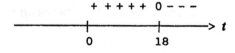

we see that $(18, 20{,}310)$ is an inflection point of f. Our computations reveal that the maximum velocity of the rocket is attained when $t = 18$. The maximum velocity is

$$f'(18) = -3(18)^2 + 108(18) + 480$$
$$= 1452, \text{ or } 1452 \text{ ft/sec.}$$

85. True. If f' is increasing on (a,b), then $-f'$ is decreasing on (a,b), and so if the graph of f is concave upward on (a,b), the graph of f must be downward on (a,b).

87. True. The given conditions imply that $f''(0) < 0$ and the Second Derivative Test gives the desired conclusion.

89. Yes. First of all, since $f''(a) = 0$ or fails to exist, we see that a is a critical point of f. Next, $f''(x)$ changes sign as we move across $x = a$. Therefore, f' must have a relative extremum at $x = a$.

91. a. $f'(x) = 3x^2$, $g'(x) = 4x^3$, and $h'(x) = -4x^3$. Setting $f'(x) = 0$, $g'(x) = 0$, and $h'(x) = 0$, respectively, gives $x = 0$ as a critical point of each function.
    b. $f''(x) = 6x$, $g''(x) = 12x^2$, and $h''(x) = -12x^2$, so that $f''(0) = 0$, $g''(0) = 0$, and $h''(0) = 0$. Thus, the second derivative test yields no conclusion in these cases.
    c. Since $f'(x) > 0$ for both $x > 0$ and $x < 0$, $f'(x)$ does not change sign as we move across the critical point $x = 0$ by the First Derivative Test. Next, $g'(x)$ changes sign from negative to positive as we move. Finally, we see that $h'(x) < 0$ for $x > 0$, so h has a relative maximum at $x = 0$.

USING TECHNOLOGY EXERCISES 4.2, page 319

1. a. f is concave upward on $(-\infty,0) \cup (1.1667,\infty)$ and concave downward on $(0, 1.1667)$.
   b. $(1.1667, 1.1153); (0,2)$

3. a. f is concave downward on $(-\infty,0)$ and concave upward on $(0, \infty)$.
   b. $(0,2)$

5. a. f is concave downward on $(-\infty,0)$ and concave upward on $(0, \infty)$.
    b. $(0,0)$

7. a. f is concave downward on $(-\infty,-2.4495) \cup (0, 2.4495)$; f is concave upward on $(-2.4495, 0) \cup (2.4495, \infty)$.
    b. $(-2.4495, -0.3402)$; $(2.4495, 0.3402)$

9. a.

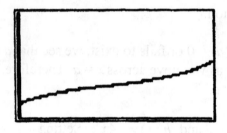

    b. $(5.5318, 35.9483)$ c. $t = 5.5318$

11. a.

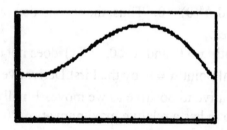

    b. $(3.9024, 77.0919)$; sales of houses were increasing at the fastest rate in late 1988.

EXERCISES 4.3, page 331

1. $y = 0$ is a horizontal asymptote.

3. $y = 0$ is a horizontal asymptote and $x = 0$ is a vertical asymptote.

5. $y = 0$ is a horizontal asymptote and $x = -1$ and $x = 1$ are vertical asymptotes.

7. $y = 3$ is a horizontal asymptote and $x = 0$ is a vertical asymptote.

9. $y = 1$ and $y = -1$ are horizontal asymptotes.

11. $\lim\limits_{x \to \infty} \dfrac{1}{x} = 0$ and so $y = 0$ is a horizontal asymptote. Next, since the numerator of the rational expression is not equal to zero and the denominator is zero at $x = 0$, we see that $x = 0$ is a vertical asymptote.

13. $f(x) = -\dfrac{2}{x^2}$. $\lim\limits_{x \to \infty} -\dfrac{2}{x^2} = 0$, so $y = 0$ is a horizontal asymptote. Next, the denominator of $f(x)$ is equal to zero at $x = 0$. Since the numerator of $f(x)$ is not equal to zero at $x = 0$, we see that $x = 0$ is a vertical asymptote.

15. $\lim\limits_{x \to \infty} \dfrac{x-1}{x+1} = \lim\limits_{x \to \infty} \dfrac{1-\frac{1}{x}}{1+\frac{1}{x}} = 1$, and so $y = 1$ is a horizontal asymptote. Next, the denominator is equal to zero at $x = -1$ and the numerator is not equal to zero at this point, so $x = -1$ is a vertical asymptote.

17. $h(x) = x^3 - 3x^2 + x + 1$. $h(x)$ is a polynomial function and, therefore, it does not have any horizontal or vertical asymptotes.

19. $\lim\limits_{t \to \infty} \dfrac{t^2}{t^2 - 9} = \lim\limits_{t \to \infty} \dfrac{1}{1 - \frac{9}{t^2}} = 1$, and so $y = 1$ is a horizontal asymptote. Next, observe that the denominator of the rational expression $t^2 - 9 = (t+3)(t-3) = 0$ if $t = -3$ and $t = 3$. But the numerator is not equal to zero at these points. Therefore, $t = -3$ and $t = 3$ are vertical asymptotes.

21. $\lim\limits_{x \to \infty} \dfrac{3x}{x^2 - x - 6} = \lim\limits_{x \to \infty} \dfrac{\frac{3}{x}}{1 - \frac{1}{x} - \frac{6}{x^2}} = 0$ and so $y = 0$ is a horizontal asymptote. Next, observe that the denominator $x^2 - x - 6 = (x-3)(x+2) = 0$ if $x = -2$ or $x = 3$. But the numerator $3x$ is not equal to zero at these points. Therefore, $x = -2$ and $x = 3$ are vertical asymptotes.

23. $\lim_{t \to \infty} \left[2 + \dfrac{5}{(t-2)^2} \right] = 2$, and so $y = 2$ is a horizontal asymptote. Next observe that

$$\lim_{t \to 2^+} g(t) = \lim_{t \to 2^-} \left[2 + \dfrac{5}{(t-2)^2} \right] = \infty$$

and so $t = 2$ is a vertical asymptote.

25. $\lim_{x \to \infty} \dfrac{x^2 - 2}{x^2 - 4} = \lim_{x \to \infty} \dfrac{1 - \frac{2}{x^2}}{1 - \frac{4}{x^2}} = 1$ and so $y = 1$ is a horizontal asymptote. Next, observe

that the denominator $x^2 - 4 = (x+2)(x-2) = 0$ if $x = -2$ or 2. Since the numerator $x^2 - 2$ is not equal to zero at these points, the lines $x = -2$ and $x = 2$ are vertical asymptotes.

27. $g(x) = \dfrac{x^3 - x}{x(x+1)}$; Rewrite $g(x)$ as $g(x) = \dfrac{x^2 - 1}{x+1}$ $(x \neq 0)$ and note that

$\lim_{x \to -\infty} g(x) = \lim_{x \to -\infty} \dfrac{x - \frac{1}{x}}{1 + \frac{1}{x}} = -\infty$ and $\lim_{x \to \infty} g(x) = \infty$. Therefore, there are no horizontal

asymptotes. Next, note that the denominator of $g(x)$ is equal to zero at $x = 0$ and $x = -1$. However, since the numerator of $g(x)$ is also equal to zero when $x = 0$, we see that $x = 0$ is not a vertical asymptote. Also, the numerator of $g(x)$ is equal to zero when $x = -1$, so $x = -1$ is not a vertical asymptote.

29. f is the derivative function of the function g. Observe that at a relative maximum (relative minimum) of g, $f(x) = 0$.

31.   33.

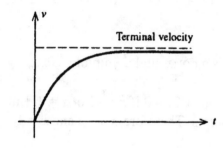

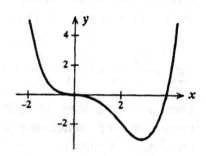

35.

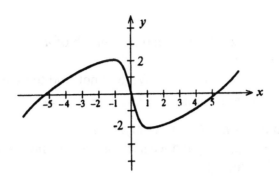

37. $g(x) = 4 - 3x - 2x^3$.

   We first gather the following information on the graph of f.

   1. The domain of f is $(-\infty, \infty)$.

   2. Setting $x = 0$ gives $y = 4$ as the y-intercept. Setting $y = g(x) = 0$ gives a cubic equation which is not easily solved and we will not attempt to find the x-intercepts.

   3. $\lim\limits_{x \to -\infty} g(x) = \infty$ and $\lim\limits_{x \to \infty} g(x) = -\infty$.

   4. There are no asymptotes of g.

   5. $g'(x) = -3 - 6x^2 = -3(2x^2 + 1) < 0$ for all values of x and so g is decreasing on $(-\infty, \infty)$.

   6. The results of 5 show that g has no critical points and hence has no relative extrema.

   7. $g''(x) = -12x$. Since $g''(x) > 0$ for $x < 0$ and $g''(x) < 0$ for $x > 0$, we see that g is concave upward on $(-\infty, 0)$ and concave downward on $(0, \infty)$.

   8. From the results of (7), we see that $(0,4)$ is an inflection point of g.
   The graph of g follows.

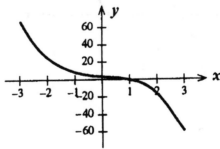

39. $h(x) = x^3 - 3x + 1$

We first gather the following information on the graph of h.
1. The domain of h is $(-\infty, \infty)$.
2. Setting $x = 0$ gives 1 as the y-intercept. We will not find the x-intercept.
3. $\lim\limits_{x \to -\infty} (x^3 - 3x + 1) = -\infty$ and $\lim\limits_{x \to \infty} (x^3 - 3x + 1) = \infty$
4. There are no asymptotes since $h(x)$ is a polynomial.
5. $h'(x) = 3x^2 - 3 = 3(x + 1)(x - 1)$, and we see that $x = -1$ and $x = 1$ are critical points. From the sign diagram

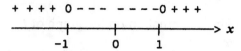

we see that h is increasing on $(-\infty, -1) \cup (1, \infty)$ and decreasing on $(-1, 1)$.
6. The results of (5) shows that $(-1, 3)$ is a relative maximum and $(1, -1)$ is a relative minimum.
7. $h''(x) = 6x$ and $h''(x) < 0$ if $x < 0$ and $h''(x) > 0$ if $x > 0$. So the graph of h is concave downward on $(-\infty, 0)$ and concave upward on $(0, \infty)$.
8. The results of (7) show that $(0, 1)$ is an inflection point of h.

The graph of h follows.

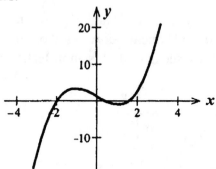

41. $f(x) = -2x^3 + 3x^2 + 12x + 2$
We first gather the following information on the graph of f.
1. The domain of f is $(-\infty, \infty)$.
2. Setting $x = 0$ gives 2 as the y-intercept.

3. $\lim_{x \to -\infty}(-2x^3 + 3x^2 + 12x + 2) = \infty$ and $\lim_{x \to \infty}(-2x^3 + 3x^2 + 12x + 2) = -\infty$

4. There are no asymptotes because $f(x)$ is a polynomial function.

5. $f'(x) = -6x^2 + 6x + 12 = -6(x^2 - x - 2) = -6(x - 2)(x + 1) = 0$ if $x = -1$ or $x = 2$, the critical points of f. From the sign diagram

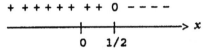

we see that f is decreasing on $(-\infty, -1) \cup (2, \infty)$ and increasing on $(-1, 2)$.

6. The results of (5) show that $(-1, -5)$ is a relative minimum and $(2, 22)$ is a relative maximum.

7. $f''(x) = -12x + 6 = 0$ if $x = 1/2$. The sign diagram of f''

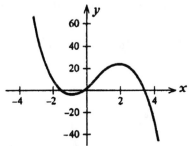

shows that the graph of f is concave upward on $(-\infty, 1/2)$ and concave downward on $(1/2, \infty)$.

8. The results of (7) show that $(\frac{1}{2}, \frac{17}{2})$ is an inflection point.

The graph of f follows.

43. $h(x) = \frac{3}{2}x^4 - 2x^3 - 6x^2 + 8$

We first gather the following information on the graph of h.

1. The domain of h is $(-\infty, \infty)$.

2. Setting $x = 0$ gives 8 as the y-intercept.

3. $\lim_{x \to -\infty} h(x) = \lim_{x \to \infty} h(x) = \infty$

4. There are no asymptotes.

5. $h'(x) = 6x^3 - 6x^2 - 12x = 6x(x^2 - x - 2) = 6x(x-2)(x+1) = 0$ if $x = -1, 0$, or 2, and these are the critical points of h. The sign diagram of h' is

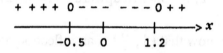

and this tells us that h is increasing on $(-1, 0) \cup (2, \infty)$ and decreasing on $(-\infty, -1) \cup (0, 2)$.

6. The results of (5) show that $(-1, \frac{11}{2})$ and $(2, -8)$ are relative minima of h and $(0, 8)$ is a relative maximum of h.

7. $h''(x) = 18x^2 - 12x - 12 = 6(3x^2 - 2x - 2)$. The zeros of h'' are

$$x = \frac{2 \pm \sqrt{4 + 24}}{6} \approx -0.5 \text{ or } 1.2.$$

The sign diagram of h'' is

and tells us that the graph of h is concave upward on $(-\infty, -0.5) \cup (1.2, \infty)$ and is concave downward on $(0.5, 1.2)$.

8. The results of (7) also show that $(-0.5, 6.8)$ and $(1.2, -1)$ are inflection points. The graph of h follows.

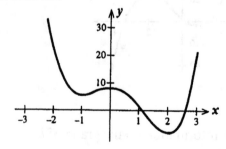

45. $f(t) = \sqrt{t^2 - 4}$.

We first gather the following information on f.

1. The domain of f is found by solving $t^2 - 4 \geq 0$ giving it as $(-\infty, -2] \cup [2, \infty)$.

2. Since $t \neq 0$, there is no y-intercept. Next, setting $y = f(t) = 0$ gives the t-intercepts as -2 and 2.

3. $\lim_{t \to -\infty} f(t) = \lim_{t \to \infty} f(t) = \infty$

4. There are no asymptotes.

5. $f'(t) = \frac{1}{2}(t^2 - 4)^{-1/2}(2t) = t(t^2 - 4)^{-1/2} = \frac{t}{\sqrt{t^2 - 4}}$.

Setting $f'(t) = 0$ gives $t = 0$. But $t = 0$ is not in the domain of f and so there are no critical points. The sign diagram for f' is

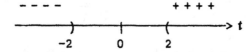

We see that f is increasing on $(2, \infty)$ and decreasing on $(-\infty, -2)$.

6. From the results of (5) we see that there are no relative extrema.

7. $f''(t) = (t^2 - 4)^{-1/2} + t(-\frac{1}{2})(t^2 - 4)^{-3/2}(2t) = (t^2 - 4)^{-3/2}(t^2 - 4 - t^2)$

$$= -\frac{4}{(t^2 - 4)^{3/2}}.$$

8. Since $f''(t) < 0$ for all t in the domain of f, we see that f is concave downward everywhere. From the results of (7), we see that there are no inflection points.

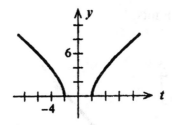

The graph of f follows.

47. $g(x) = \frac{1}{2}x - \sqrt{x}$.

We first gather the following information on g.

1. The domain of g is $[0, \infty)$.

2. The y-intercept is 0. To find the x-intercept, set $y = 0$, giving

$$\tfrac{1}{2}x - \sqrt{x} = 0$$

$$x = 2\sqrt{x}$$

$$x^2 = 4x$$

$$x(x-4) = 0, \text{ and } x = 0 \text{ or } x = 4$$

3. $\lim\limits_{x\to\infty} (\tfrac{1}{2}x - \sqrt{x}) = \lim\limits_{x\to\infty} \tfrac{1}{2}x(1 - \tfrac{2}{\sqrt{x}}) = \infty.$

4. There are no asymptotes.

5. $g'(x) = \tfrac{1}{2} - \tfrac{1}{2}x^{-1/2} = \tfrac{1}{2}x^{-1/2}(x^{1/2} - 1) = \dfrac{\sqrt{x}-1}{2\sqrt{x}}$

which is zero when $x = 1$. From the sign diagram for g'

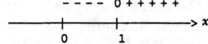

we see that g is decreasing on $(0,1)$ and increasing on $(1,\infty)$.

6. From the sign diagram of g', we see that $g(1) = -1/2$ is a relative minimum.

7. $g''(x) = (-\tfrac{1}{2})(-\tfrac{1}{2})x^{-3/2} = \dfrac{1}{4x^{3/2}} > 0$ for $x > 0$, and so g is concave upward on $(0,\infty)$.

8. There are no inflection points.
   The graph of g follows.

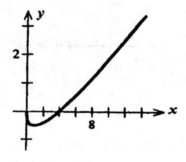

49. $g(x) = \dfrac{2}{x-1}.$

We first gather the following information on g.

1. The domain of g is $(-\infty,1) \cup (1,\infty)$.
2. Setting $x = 0$ gives -2 as the y-intercept. There are no x-intercepts since $\dfrac{2}{x-1} \neq 0$ for all values of x.
3. $\displaystyle\lim_{x\to-\infty} \dfrac{2}{x-1} = 0$ and $\displaystyle\lim_{x\to\infty} \dfrac{2}{x-1} = 0$.

4. The results of (3) show that $y = 0$ is a horizontal asymptote. Furthermore, the denominator of $g(x)$ is equal to zero at $x = 1$ but the numerator is not equal to zero there. Therefore, $x = 1$ is a vertical asymptote.
5. $g'(x) = -2(x-1)^{-2} = -\dfrac{2}{(x-1)^2} < 0$ for all $x \neq 1$ and so g is decreasing on $(-\infty,1)$ and $(1,\infty)$.
6. Since g has no critical points, there are no relative extrema.
7. $g''(x) = \dfrac{4}{(x-1)^3}$ and so $g''(x) < 0$ if $x < 1$ and $g''(x) > 0$ if $x > 1$.

Therefore, the graph of g is concave downward on $(-\infty,1)$ and concave upward on $(1,\infty)$.
8. Since $g''(x) \neq 0$, there are no inflection points.

The graph of g follows.

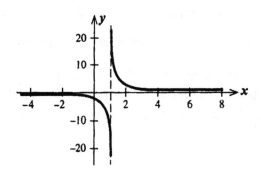

51. $h(x) = \dfrac{x+2}{x-2}$.

We first gather the following information on the graph of h.

1. The domain of h is $(-\infty,2) \cup (2,\infty)$.
2. Setting $x = 0$ gives $y = -1$ as the y-intercept. Next, setting $y = 0$ gives $x = -2$ as the x-intercept.

3. $\displaystyle\lim_{x\to\infty} h(x) = \lim_{x\to-\infty} \frac{1+\dfrac{2}{x}}{1-\dfrac{2}{x}} = \lim_{x\to-\infty} h(x) = 1.$

4. Setting $x - 2 = 0$ gives $x = 2$. Furthermore,

$$\lim_{x\to2^+} \frac{x+2}{x-2} = \infty \quad \text{and} \quad \lim_{x\to2^+} \frac{x+2}{x-2} = -\infty$$

So $x = 2$ is a vertical asymptote of h. Also, from the resultsof (3), we see that $y = 1$ is a horizontal asymptote of h.

5. $h'(x) = \dfrac{(x-2)(1)-(x+2)(1)}{(x-2)^2} = -\dfrac{4}{(x-2)^2}.$

We see that there are no critical points of h. (Note $x = 2$ does not belong to the domain of h.) The sign diagram of h' follows.

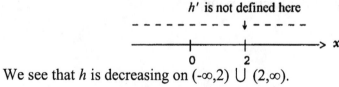

We see that h is decreasing on $(-\infty,2) \cup (2,\infty)$.

6. From the results of (5), we see that there is no relative extremum.

7. $h''(x) = \dfrac{8}{(x-2)^3}.$ Note that $x = 2$ is not a candidate for an inflection point because $h(2)$ is not defined. Since $h''(x) < 0$ for $x < 2$ and $h''(x) > 0$ for $x > 2$, we see that h is concave downward on $(-\infty,2)$ and concave upward on $(2,\infty)$.

8. From the results of (7), we see that there are no inflection points. The graph of h follows.

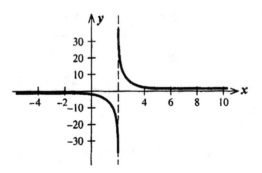

53. $f(t) = \dfrac{t^2}{1+t^2}$.

We first gather the following information on the graph of f.
1. The domain of f is $(-\infty, \infty)$.
2. Setting $t = 0$ gives the y-intercept as 0. Similarly, setting $y = 0$ gives the t-intercept as 0.

3. $\displaystyle\lim_{t\to-\infty}\dfrac{t^2}{1+t^2} = \lim_{t\to\infty}\dfrac{t^2}{1+t^2} = 1$.

4. The results of (3) show that $y = 1$ is a horizontal asymptote. There are no vertical asymptotes since the denominator is not equal to zero.

5. $f'(t) = \dfrac{(1+t^2)(2t) - t^2(2t)}{(1+t^2)^2} = \dfrac{2t}{(1+t^2)^2} = 0$, if $t = 0$, the only critical point of f.

Since $f'(t) < 0$ if $t < 0$ and $f'(t) > 0$ if $t > 0$, we see that f is decreasing on $(-\infty, 0)$ and increasing on $(0, \infty)$.
6. The results of (5) show that $(0,0)$ is a relative minimum.

7. $f''(t) = \dfrac{(1+t^2)^2(2) - 2t(2)(1+t^2)(2t)}{(1+t^2)^4} = \dfrac{2(1+t^2)[(1+t^2) - 4t^2]}{(1+t^2)^4}$

$= \dfrac{2(1-3t^2)}{(1+t^2)^3} = 0$ if $t = \pm\dfrac{\sqrt{3}}{3}$.

The sign diagram of f'' is

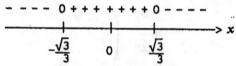

and shows that f is concave downward on $(-\infty, -\frac{\sqrt{3}}{3}) \cup (\frac{\sqrt{3}}{3}, \infty)$ and concave

upward on $(-\frac{\sqrt{3}}{3}, \frac{\sqrt{3}}{3})$.

8. The results of (7) show that $(-\frac{\sqrt{3}}{3}, \frac{1}{4})$ and $(\frac{\sqrt{3}}{3}, \frac{1}{4})$ are inflection points.
The graph of f follows.

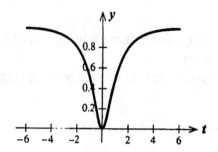

55. $g(t) = -\dfrac{t^2 - 2}{t - 1}$.

First we obtain the following information on g.
1. The domain of g is $(-\infty, 1) \cup (1, \infty)$.
2. Setting $t = 0$ gives -2 as the y-intercept.
3. $\displaystyle\lim_{t \to -\infty} -\frac{t^2 - 2}{t - 1} = \infty$ and $\displaystyle\lim_{t \to \infty} -\frac{t^2 - 2}{t - 1} = -\infty$.
4. There are no horizontal asymptotes. The denominator is equal to zero at $t = 1$ at which point the numerator is not equal to zero. Therefore $t = 1$ is a vertical asymptote.
5. $g'(t) = -\dfrac{(t-1)(2t) - (t^2 - 2)(1)}{(t-1)^2} = -\dfrac{t^2 - 2t + 2}{(t-1)^2} \neq 0$

for all values of t. The sign diagram of g'

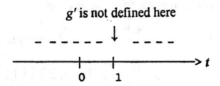

g' is not defined here

shows that g is decreasing on $(-\infty,1) \cup (1, \infty)$.

6. Since there are no critical points, g has no relative extrema.

7. $g''(t) = -\dfrac{(t-1)^2(2t-2)-(t^2-2t+2)(2)(t-1)}{(t-1)^4}$

$= \dfrac{-2(t-1)(t^2-2t+1-t^2+2t-2)}{(t-1)^4} = \dfrac{2}{(t-1)^3}.$

The sign diagram of g''

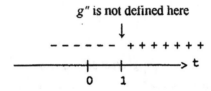

g" is not defined here

shows that the graph of g is concave upward on $(1,\infty)$ and concave downward on $(-\infty,1)$.

8. There are no inflection points since $g''(x) \neq 0$ for all x.

The graph of g follows.

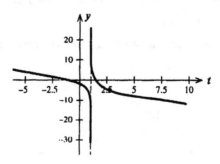

57. $g(t) = \dfrac{t+1}{t^2-2t-1}.$

We first gather some information on the graph of g.

1. Since $t^2 - 2t - 1 = 0$ if $t = \dfrac{2 \pm \sqrt{4+4}}{2} = 1 \pm \sqrt{2}$, we see that

the domain of g is $(-\infty, 1 - \sqrt{2}) \cup (1 - \sqrt{2}, 1 + \sqrt{2}) \cup (1 + \sqrt{2}, \infty)$.

2. Setting $t = 0$ gives -1 as the y-intercept. Setting $y = 0$ gives -1 as the t-intercept.

3. $\displaystyle\lim_{t \to -\infty} g(t) = \lim_{t \to \infty} g(t) = 0.$

4. The results of (3) show that $y = 0$ is a horizontal asymptote. Since the denominator (but not the numerator) is zero at $t = 1 \pm \sqrt{2}$, we see that $t = 1 - \sqrt{2}$ and $t = 1 + \sqrt{2}$ are vertical asymptotes.

5. $g'(t) = \dfrac{(t^2 - 2t - 1)(1) - (t+1)(2t-2)}{(t^2 - 2t - 1)^2} = -\dfrac{(t^2 + 2t - 1)}{(t^2 - 2t - 1)^2} = 0$

if $t = \dfrac{-2 \pm \sqrt{4+4}}{2} = -1 \pm \sqrt{2}.$

The sign diagram of g' is

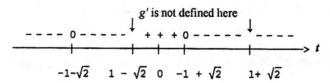

We see that g is decreasing on

$(-\infty, -1 - \sqrt{2}) \cup (-1 - \sqrt{2}, 1 - \sqrt{2}) \cup (-1 + \sqrt{2}, 1 + \sqrt{2}) \cup (1 + \sqrt{2}, \infty)$

and increasing on $(1 - \sqrt{2}, -1 + \sqrt{2})$.

6. From the results of (5), we see that g has a relative maximum at $t = -1 + \sqrt{2}$.

The graph of g follows.

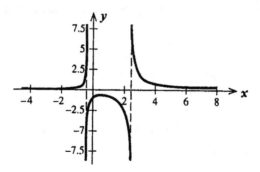

59. $h(x) = (x-1)^{2/3} + 1$.

We begin by obtaining the following information on h.

1. The domain of h is $(-\infty, \infty)$.

2. Setting $x = 0$ gives 2 as the y-intercept; setting $h(x) = 0$ gives 2 as the x-intercept.

3. $\lim_{x\to\infty} [(x-1)^{2/3} + 1] = \infty$. Similarly, $\lim_{x\to-\infty} [(x-1)^{2/3} + 1] = \infty$.

4. There are no asymptotes.

5. $h'(x) = \frac{2}{3}(x-1)^{-1/3}$ and is positive if $x > 1$ and negative if $x < 1$. So h is increasing on $(1,\infty)$, and decreasing on $(-\infty,1)$.

6. From (5), we see that h has a relative minimum at $(1,1)$.

7. $h''(x) = \frac{2}{3}(-\frac{1}{3})(x-1)^{-4/3} = -\frac{2}{9}(x-1)^{-4/3} = -\frac{2}{(x-1)^{4/3}}$. Since $h''(x) < 0$ on

$(-\infty,1) \cup (1,\infty)$, we see that h is concave downward on $(-\infty,1) \cup (1,\infty)$. Note that $h''(x)$ is not defined at $x = 1$.

8. From the results of (7), we see h has no inflection points.

The graph of h follows.

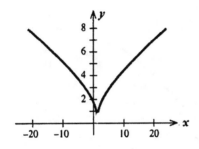

61. a. $\lim\limits_{x\to\infty} \overline{C}(x) = \lim\limits_{x\to\infty} (2.2 + \dfrac{2500}{x}) = 2.2$, and so $y = 2.2$ is a horizontal asymptote.

b. The limiting value is 2.2, or \$2.20 per disc.

63. a. $\lim\limits_{x\to\infty} \dfrac{ax}{x+b} = \lim\limits_{x\to\infty} \dfrac{a}{1+\dfrac{b}{x}} = a.$

b. The initial speed of the reaction approaches a moles/liter/sec as the amount of substrate becomes arbitrarily large.

65. $G(t) = -0.2t^3 + 2.4t^2 + 60.$
We first gather the following information on the graph of G.
1. The domain of G is $(0,\infty)$.
2. Setting $t = 0$ gives 60 as the y-intercept.
Note that Step 3 is not necessary in this case because of the restricted domain.
4. There are no asymptotes since G is a polynomial function.
5. $G'(t) = -0.6t^2 + 4.8t = -0.6t(t - 8) = 0$, if $t = 0$ or $t = 8$. But these points do not lie in the interval $(0,8)$, so they are not critical points. The sign diagram of G' shows that G is increasing on $(0,8)$.

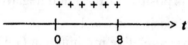

6. The results of (5) tell us that there are no relative extrema.
7. $G''(t) = -1.2t + 4.8 = -1.2(t - 4)$. The sign diagram of G'' is

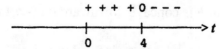

and shows that G is concave upward on $(0,4)$ and concave downward on $(4,8)$.
8. The results of (7) shows that $(4,85.6)$ is an inflection point.
The graph of G follows.

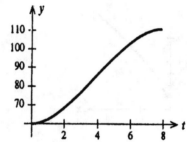

67. $f(t) = 100\left(\dfrac{t^2 - 4t + 4}{t^2 + 4}\right)$. We first gather the following information..

1. The domain of f is restricted to $[0,\infty)$.
2. Setting $t = 0$ gives $y = 100$. Next, setting $y = 0$ gives $t = 2$.
3. $\displaystyle\lim_{t\to\infty} 100\left[\dfrac{t^2 - 4t + 4}{t^2 + 4}\right] = 100\lim_{t\to\infty}\left[\dfrac{1 - \frac{4}{t} + \frac{4}{t^2}}{1 + \frac{4}{t^2}}\right] = 100$

4. From the results of (3), we see that $y = 100$ is a horizontal asymptote. There are no vertical asymptotes.

5. $f'(t) = 100\left[\dfrac{(t^2 + 4)(2t - 4) - (t^2 - 4t + 4)(2t)}{(t^2 + 4)^2}\right]$

$= \dfrac{400(t^2 - 4)}{(t^2 + 4)^2} = \dfrac{400(t - 2)(t + 2)}{(t^2 + 4)^2}.$

Setting $f'(t) = 0$ gives $t = 2$ as a critical point of f. ($t = -2$ is not in the domain of f.) Since $f'(t) < 0$ when $t < 2$ and $f'(t) > 0$ when $t > 2$, we see that f is decreasing on $(0,2)$ and increasing on $(2,\infty)$.

6. The results of (5) imply that $f(2) = 0$ is a relative minimum.

7. $f''(t) = 400\left[\dfrac{(t^2 + 4)^2(2t) - (t^2 - 4)2(t^2 + 4)(2t)}{(t^2 + 4)^4}\right]$

$= 400\left[\dfrac{(2t)(t^2 + 4)(t^2 + 4 - 2t^2 + 8)}{(t^2 + 4)^4}\right] = -\dfrac{800t(t^2 - 12)}{(t^2 + 4)^3}.$

Setting $f''(t) = 0$ gives $t = 2\sqrt{3}$ as a candidate for a point of inflection. The sign diagram for f'' is

$$\begin{array}{c} +\ +\ +\ \ +\ 0\ -\ -\ - \\ \xrightarrow{\hspace{1cm}|\hspace{1cm}|\hspace{1cm}} t \\ \ \ \ 0 \quad\quad 2\sqrt{3} \end{array}$$

We see that f is concave upward on $(0, 2\sqrt{3})$ and concave downward on $(2\sqrt{3}, \infty)$.

8. From the results of (7), we see that $(2\sqrt{3}, 50(2 - \sqrt{3}))$ is an inflection point of f. The graph of f follows.

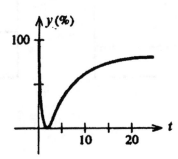

69. $C(x) = \dfrac{0.5x}{100 - x}$.

We first gather the following information on the graph of C.
1. The domain of C is $[0,100)$.
2. Setting $x = 0$ gives the y-intercept as 0.

Because of the restricted domain, we omit steps 3 and 4.

5. $C'(x) = 0.5 \left[\dfrac{(100-x)(1) - x(-1)}{(100-x)^2} \right] = \dfrac{50}{(100-x)^2} > 0$ for all $x \neq 100$. Therefore C

is increasing on $(0,100)$.

6. There are no relative extrema.
The graph of C follows.

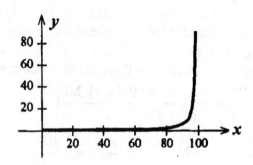

USING TECHNOLOGY EXERCISES 4.3, page 337

1.

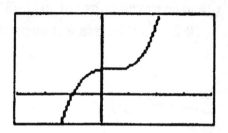

3.

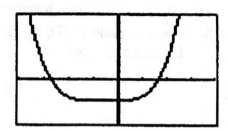

5.   -0.9733; 2.3165, 4.6569   7.   -1.1310; 2.9267   9.   1.5142

EXERCISES 4.4, page 350

1. f has no absolute extrema.   3. f has an absolute minimum at $(0,0)$.

5. f has an absolute minimum at $(0,-2)$ and an absolute maximum at $(1,3)$.

7. f has an absolute minimum at $(\frac{3}{2}, -\frac{27}{16})$ and an absolute maximum at $(-1,3)$.

9.   The graph of $f(x) = 2x^2 + 3x - 4$ is a parabola that opens upward. Therefore, the
     vertex of the parabola is the absolute minimum of f. To find the vertex, we solve
     the equation
     $$f'(x) = 4x + 3 = 0$$
     giving $x = -3/4$. We conclude that the absolute minimum value is $f(-\frac{3}{4}) = -\frac{41}{8}$.

11.   Since $\lim\limits_{x\to-\infty} x^{1/3} = -\infty$ and $\lim\limits_{x\to\infty} x^{1/3} = \infty$, we see that h is unbounded. Therefore it has
      no absolute extrema.

13. $f(x) = \dfrac{1}{1+x^2}$.
     Using the techniques of graphing, we sketch the graph of f (see Fig. 4.41, page
     297, in the text). The absolute maximum of f is $f(0) = 1$. Alternatively, observe
     that $1 + x^2 \geq 1$ for all real values of x. Therefore, $f(x) \leq 1$ for all x, and we see that
     the absolute maximum is attained when $x = 0$.

15. $f(x) = x^2 - 2x - 3$ and $f'(x) = 2x - 2 = 0$, so $x = 1$ is a critical point. From the table,

x	-2	1	3
$f(x)$	5	-4	0

     we conclude that the absolute maximum value is $f(-2) = 5$ and the absolute
     minimum value is $f(1) = -4$.

17. $f(x) = -x^2 + 4x + 6$; The function f is continuous and defined on the closed interval

[0,5]. $f'(x) = -2x + 4$ and $x = 2$ is a critical point. From the table

x	0	2	5
$f(x)$	6	10	1

we conclude that $f(2) = 10$ is the absolute maximum value and $f(5) = 1$ is the absolute minimum value.

19. The function $f(x) = x^3 + 3x^2 - 1$ is continuous and defined on the closed interval [-3,2] and differentiable in (-3,2). The critical points of f are found by solving
$$f'(x) = 3x^2 + 6x = 3x(x + 2)$$
giving $x = -2$ and $x = 0$. Next, we compute the values of f given in the following table.

x	-3	-2	0	2
$f(x)$	-1	3	-1	19

From the table, we see that the absolute maximum value of f is $f(2) = 19$ and the absolute minimum value is $f(-3) = -1$ and $f(0) = -1$.

21. The function $g(x) = 3x^4 + 4x^3$ is continuous and differentiable on the closed interval [-2,1] and differentiable in (-2,1). The critical points of g are found by solving

$$g'(x) = 12x^3 + 12x^2 = 12x^2(x + 1)$$
giving $x = 0$ and $x = -1$. We next compute the values of g shown in the following table.

x	-2	-1	0	1
$g(x)$	16	-1	0	7

From the table we see that $g(-2) = 16$ is the absolute maximum value of g and $g(-1) = -1$ is the absolute minimum value of g.

23. $f(x) = \dfrac{x+1}{x-1}$ on [2,4]. Next, we compute,

$$f'(x) = \frac{(x-1)(1)-(x+1)(1)}{(x-1)^2} = -\frac{2}{(x-1)^2}.$$

Since there are no critical points, ($x = 1$ is not in the domain of f), we need only test the endpoints. From the table

x	2	4
$g(x)$	3	5/3

we conclude that $f(4) = 5/3$ is the absolute minimum value and $f(2) = 3$ is the absolute maximum value.

25. $f(x) = 4x + \dfrac{1}{x}$ is continuous on $[1,3]$ and differentiable in $(1,3)$. To find the

critical points of f, we solve $f'(x) = 4 - \dfrac{1}{x^2} = 0$, obtaining $x = \pm\frac{1}{2}$. Since these

critical points lie outside the interval $[1,3]$, they are not candidates for the absolute extrema of f. Evaluating f at the endpoints of the interval $[1,3]$, we find that the absolute maximum value of f is $f(3) = \frac{37}{3}$, and the absolute minimum of f is $f(1) = 5$.

27. $f(x) = \frac{1}{2}x^2 - 2\sqrt{x} = \frac{1}{2}x^2 - 2x^{1/2}$. To find the critical points of f, we solve
$$f'(x) = x - x^{-1/2} = 0, \quad \text{or} \quad x^{3/2} - 1 = 0,$$
obtaining $x = 1$. From the table

x	0	1	3
$f(x)$	0	$-\frac{3}{2}$	$\frac{9}{2} - 2\sqrt{3} \approx 1.04$

we conclude that $f(3) \approx 1.04$ is the absolute maximum value and $f(1) = -3/2$ is the absolute minimum value.

29. The graph of $f(x) = 1/x$ over the interval $(0,\infty)$ follows.

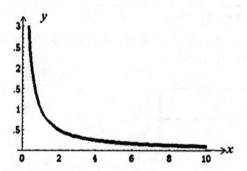

From the graph of f, we conclude that f has no absolute extrema.

31. $f(x) = 3x^{2/3} - 2x$. The function f is continuous on [0,3] and differentiable on (0,3). To find the critical points of f, we solve
$$f'(x) = 2x^{-1/3} - 2 = 0$$
obtaining $x = 1$ as the critical point. From the table,

x	0	1	3
$f(x)$	0	1	$3^{5/3} - 6 \approx 0.24$

we conclude that the absolute maximum value is $f(1) = 1$ and the absolute minimum value is $f(0) = 0$.

33. $f(x) = x^{2/3}(x^2 - 4)$. $f'(x) = x^{2/3}(2x) + \frac{2}{3}x^{-1/3}(x^2 - 4) = \frac{2}{3}x^{-1/3}[3x^2 + (x^2 - 4)]$
$$= \frac{8(x^2 - 1)}{3x^{1/3}} = 0.$$
Observe that f' is not defined at $x = 0$. Furthermore, $f'(x) = 0$ at $x \pm 1$. So the critical points of f are -1, 0, 1. From the following table,

x	-1	0	1	2
$f(x)$	-3	0	-3	0

we see that f has an absolute minimum at (-1,-3) and (1,-3) and absolute maxima at (0,0) and (2,0).

35. $f(x) = \dfrac{x}{x^2+2}$. To find the critical points of f, we solve

$$f'(x) = \frac{(x^2+2) - x(2x)}{(x^2+2)^2} = \frac{2-x^2}{(x^2+2)^2} = 0$$

obtaining $x = \pm\sqrt{2}$. Since $x = -\sqrt{2}$ lies outside $[-1,2]$, $x = \sqrt{2}$ is the only critical point in the given interval. From the table

x	-1	$\sqrt{2}$	2
$f(x)$	$-\frac{1}{3}$	$\sqrt{2}/4 \approx 0.35$	$\frac{1}{3}$

we conclude that $f(\sqrt{2})) = \sqrt{2}/4 \approx 0.35$ is the absolute maximum value and $f(-1) = -1/3$ is the absolute minimum value.

37. The function $f(x) = \dfrac{x}{\sqrt{x^2+1}} = \dfrac{x}{(x^2+1)^{1/2}}$ is continuous and defined on the closed interval $[-1,1]$ and differentiable on $(-1,1)$. To find the critical points of f, we first compute

$$f'(x) = \frac{(x^2+1)^{1/2}(1) - x(\frac{1}{2})(x^2+1)^{-1/2}(2x)}{[(x^2+1)^{1/2}]^2}$$

$$= \frac{(x^2+1)^{-1/2}[x^2+1-x^2]}{x^2+1} = \frac{1}{(x^2+1)^{3/2}}$$

which is never equal to zero. Next, we compute the values of f shown in the following table.

x	-1	1
$f(x)$	$-\sqrt{2}/2$	$\sqrt{2}/2$

We conclude that $f(-1) = -\sqrt{2}/2$ is the absolute minimum value and $f(1) = \sqrt{2}/2$ is the absolute maximum value.

39. $h(t) = -16t^2 + 64t + 80$. To find the maximum value of h, we solve
$h'(t) = -32t + 64 = -32(t - 2) = 0$

giving $t = 2$ as the critical point of h. Furthermore, this value of t gives rise to the absolute maximum value of h since the graph of h is parabola that opens downward. The maximum height is given by

$$h(2) = -16(4) + 64(2) + 80 = 144, \text{ or } 144 \text{ feet.}$$

41.   We compute $P'(x) = -0.08x + 240$. Setting $P'(x) = 0$ gives $x = 3000$. The graph of P is a parabola that opens downward and so $x = 3000$ gives rise to the absolute maximum of P. Thus, to maximize profits, the company should produce 3000 cameras per month.

43. $N'(t) = 0.81 - 1.14(\frac{1}{2}t^{-1/2}) = 0.81 - \dfrac{0.57}{t^{1/2}}$. Setting $N'(t) = 0$ gives $t^{1/2} = \dfrac{0.57}{0.81}$, or

$t = 0.4952$ as a critical point of N. Evaluating $N(t)$ at the endpoints $t = 0$ and $t = 6$ as well as at the critical point, we have

t	0	0.4952	6
$N(t)$	1.53	1.13	3.60

From the table, we see that the absolute maximum of N occurs at $t = 6$ and the absolute minimum occurs at $t \approx 0.5$. Our results tell us that the number of nonfarm full-time self-employed women over the time interval from 1963 to 1993 was the highest in 1993 and stood at approximately 3.6 million.

45.   The revenue is $R(x) = px = -0.00042x^2 + 6x$. Therefore, the profit is
$$P(x) = R(x) - C(x) = -0.00042x^2 + 6x - (600 + 2x - 0.00002x^2)$$
$$= -0.0004x^2 + 4x - 600.$$
$$P'(x) = -0.0008x + 4 = 0$$
if $x = 5000$, a critical point of P. From the following table

x	0	5000	12000
$P(x)$	-600	9400	-10200

we see that Phonola should produce 5000 discs/month.

47. $R(x) = px = -0.05x^2 + 600x$
$P(x) = R(x) - C(x) = -0.05x^2 + 600x - (0.000002x^3 - 0.03x^2 + 400x + 80000)$

$$= -0.000002x^3 - 0.02x^2 + 200x - 80000.$$

We want to maximize P on $[0, 12{,}000]$. $P'(x) = -0.000006x^2 - 0.04x + 200$.

Setting $P'(x) = 0$ gives $3x^2 + 20{,}000x - 100{,}000{,}000 = 0$

or $\quad x = \dfrac{-20{,}000 \pm \sqrt{20{,}000^2 + 1{,}200{,}000{,}000}}{6} = -10{,}000$, or $3{,}333.3$

So $x = 3{,}333.3$ is a critical point in the interval $[0, 12{,}000]$.

x	0	3,333	12,000
$f(x)$	-80,000	290,370	-4,016,000

From the table, we see that a level of production of 3,333 units will yield a maximum profit.

49. a. $\overline{C}(x) = \dfrac{C(x)}{x} = 0.0025x + 80 + \dfrac{10{,}000}{x}$.

b. $\overline{C}'(x) = 0.0025 - \dfrac{10{,}000}{x^2} = 0$ if $0.0025x^2 = 10{,}000$, or $x = 2000$.

Since $\overline{C}''(x) = \dfrac{20{,}000}{x^3}$, we see that $\overline{C}''(x) > 0$ for $x > 0$ and so $\overline{C}$ is concave upward on $(0,\infty)$. Therefore, $x = 2000$ yields a minimum.

c. We solve $\overline{C}(x) = C'(x)$. $0.0025x + 80 + \dfrac{10{,}000}{x} = 0.005x + 80$,

$0.0025x^2 = 10{,}000$, or $x = 2000$.

d. It appears that we can solve the problem in two ways.
REMARK This can be proved.

51. The demand equation is $p = \sqrt{800 - x} = (800 - x)^{1/2}$.

The revenue function is $R(x) = xp = x(800 - x)^{1/2}$.

To find the maximum of R, we compute

$$R'(x) = \tfrac{1}{2}(800 - x)^{-1/2}(-1)(x) + (800 - x)^{1/2}$$
$$= \tfrac{1}{2}(800 - x)^{-1/2}[-x + 2(800 - x)]$$
$$= \tfrac{1}{2}(800 - x)^{-1/2}(1600 - 3x).$$

Next, $R'(x) = 0$ implies $x = 800$ or $x = 1600/3$ are critical points of R. Next, we compute the values of R given in the following table.

x	0	800	1600/3
$R(x)$	0	0	8709

We conclude that $R(\frac{1600}{3}) = 8709$ is the absolute maximum value. Therefore, the revenue is maximized by producing $1600/3 \approx 533$ dresses.

53. $f(t) = 100\left[\dfrac{t^2 - 4t + 4}{t^2 + 4}\right]$.

    a. $f'(t) = 100\left[\dfrac{(t^2 + 4)(2t - 4) - (t^2 - 4t + 4)(2t)}{(t^2 + 4)^2}\right] = \dfrac{400(t^2 - 4)}{(t^2 + 4)^2}$

$$= \dfrac{400(t - 2)(t + 2)}{(t^2 + 4)^2}.$$

From the sign diagram for f'

we see that $t = 2$ gives a relative minimum, and we conclude that the oxygen content is the lowest 2 days after the organic waste has been dumped into the pond.

    b.

$$f''(t) = 400\left[\dfrac{(t^2 + 4)^2(2t) - (t^2 - 4)2(t^2 + 4)(2t)}{(t + 4)^4}\right] = 400\left[\dfrac{(2t)(t^2 + 4)(t^2 + 4 - 2t^2 + 8)}{(t^2 + 4)^4}\right]$$

$$= -\dfrac{800t(t^2 - 12)}{(t^2 + 4)^3}$$

and $f''(t) = 0$ when $t = 0$ and $t = \pm 2\sqrt{3}$. We reject $t = 0$ and $t = -2\sqrt{3}$. From the sign diagram for f'',

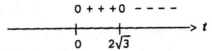

$$0 + + + 0 \ - - - -$$

we see that $f'(2\sqrt{3})$ gives an inflection point of f and we conclude that this is an absolute maximum. Therefore, the rate of oxygen regeneration is greatest 3.5 days after the organic waste has been dumped into the pond.

55. We compute $\overline{R}'(x) = \dfrac{xR'(x) - R(x)}{x^2}$. Setting $\overline{R}'(x) = 0$ gives $xR'(x) - R(x) = 0$

or $R'(x) = \dfrac{R(x)}{x} = \overline{R}(x)$, so a critical point of $\overline{R}$ occurs when $\overline{R}(x) = R'(x)$.

Next, we compute $\overline{R}''(x) = \dfrac{x^2[R'(x) + xR''(x) - R'(x)] - [xR'(x) - R(x)](2x)}{x^4}$

$= \dfrac{R''(x)}{x} < 0.$

So, by the Second Derivative Test, the critical point does give a maximum revenue.

57. The growth rate is $G'(t) = -0.6t^2 + 4.8t$. To find the maximum growth rate, we compute
$$G''(t) = -1.2t + 4.8.$$
Setting $G''(t) = 0$ gives $t = 4$ as a critical point.

t	0	4	8
$G'(t)$	0	9.6	0

From the table, we see that G is maximal at $t = 4$; that is, the growth rate is greatest in 1992.

59. $f(t) = -0.0129t^4 + 0.3087t^3 + 2.1760t^2 + 62.8466t + 506.2955$
   To find the maximum of $f(t)$, we first compute
   $$f'(t) = -0.0516t^3 + 0.9261t^2 + 4.352t + 62.8466.$$

   Then
   $$f'(23.6811) = -0.0516(23.6811)^3 + 0.9261(23.6811)^2 + 4.352(23.6811) + 62.8466$$
   $$\approx 0$$

Next, we compute $f'(23) \approx 25.03$ and $f'(24) = -15.18$. Since f is a polynomial function it is continuous. We conclude that $f(t)$ is maximized when $t = 23.6811$ since f' changes sign from positive to negative as we move across the critical point $t = 23.6811$. These results may be confirmed by graphing the derivitive function f' on your graphing calculator.

61. $R = D^2\left(\dfrac{k}{2} - \dfrac{D}{3}\right) = \dfrac{kD^2}{2} - \dfrac{D^3}{3}$. $\dfrac{dR}{dD} = \dfrac{2kD}{2} - \dfrac{3D^2}{3} = kD - D^2 = D(k - D)$

Setting $\dfrac{dR}{dD} = 0$, we have $D = 0$ or $k = D$. We only consider $k = D$

(since $D > 0$). If $k > 0$, $\dfrac{dR}{dD} > 0$ and if $k < 0$, $\dfrac{dR}{dD} < 0$. Therefore $k = D$ provides a relative maximum. The nature of the problem suggests that $k = D$ gives the absolute maximum of R. We can also verify this by graphing R.

63. False. Let $f(x) = \begin{cases} |x| & \text{if } x \neq 0 \\ 1 & \text{if } x = 0 \end{cases}$ on [-1, 1].

65. False. Let $f(x) = \begin{cases} -x & \text{if } -1 \leq x < 0 \\ \dfrac{1}{2} & \text{if } 0 \leq x < 1 \end{cases}$. Then f is discontinuous at $x = 0$. But f has an absolute maximum value of 1 attained at $x = -1$.

67. Since $f(x) = c$ for all x, the function f satisfies $f(x) \leq c$ for all x and so f has an absolute maximum at all points of x. Similarly, f has an absolute minimum at all points of x.

69. a. g is not continuous at $x = 0$ because $\lim\limits_{x \to 0} g(x)$ does not exist.

b. $\lim\limits_{x \to 0} g(x) = \lim\limits_{x \to 0^-} \dfrac{1}{x} = -\infty$ and $\lim\limits_{x \to 0^+} g(x) = \lim\limits_{x \to 0^+} \dfrac{1}{x} = \infty$

c.

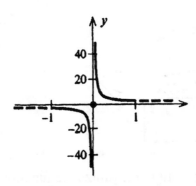

USING TECHNOLOGY EXERCISES 4.4, page 355

1. Absolute maximum value: 145.8985; absolute minimum value: -4.3834

3. Absolute maximum value: 16; absolute minimum value: -0.1257

5. Absolute maximum value: 2.8889; absolute minimum value: 0

7. a.

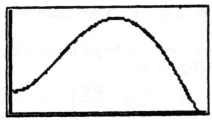

   b. 200.1410 banks/yr

9. a.

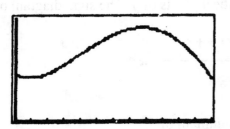

   b. Absolute maximum value: 108.8756;

   absolute minimum value: 49.7773

EXERCISES 4.5, page 365

1. Refer to the following figure.

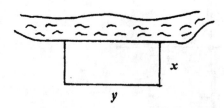

We have $2x + y = 3000$ and we want to maximize the function
$$A = f(x) = xy = x(3000 - 2x) = 3000x - 2x^2$$
on the interval $[0,1500]$. The critical point of A is obtained by solving
$f'(x) = 3000 - 4x = 0$, giving $x = 750$. From the table of values

x	0	750	1500
$f(x)$	0	1,125,000	0

we conclude that $x = 750$ yields the absolute maximum value of A. Thus, the required dimensions are 750×1500 yards. The maximum area is 1,125,000 square yards.

3. Let x denote the length of the side made of wood and y the length of the side made of steel. The cost of construction will be $C = 6(2x) + 3y$.

But $xy = 800$. So $y = 800/x$ and therefore $C = f(x) = 12x + 3\left(\dfrac{800}{x}\right) = 12x + \dfrac{2400}{x}$.

To minimize C, we compute $f'(x) = 12 - \dfrac{2400}{x^2} = \dfrac{12x^2 - 2400}{x^2} = \dfrac{12(x^2 - 200)}{x^2}$.

Setting $f'(x) = 0$ gives $x = \pm\sqrt{200}$ as critical points of f. The sign diagram of f'

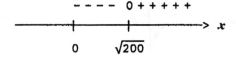

shows that $x = \pm\sqrt{200}$ gives a relative minimum of f.

$$f''(x) = \frac{4800}{x^3} > 0$$

if $x > 0$ and so f is concave upward for $x > 0$. Therefore $x = \sqrt{200} = 10\sqrt{2}$ actually yields the absolute minimum. So the dimensions of the enclosure should be

$$10\sqrt{2} \text{ ft} \times \frac{800}{10\sqrt{2}} \text{ ft, or } 14.1 \text{ ft} \times 56.6 \text{ ft.}$$

5. Let the dimensions of each square that is cut out be $x'' \times x''$. Refer to the following diagram.

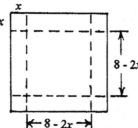 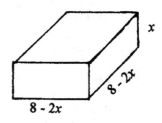

Then the dimensions of the box will be $(8 - 2x)''$ by $(8 - 2x)''$ by x''. Its volume will be $V = f(x) = x(8 - 2x)^2$. We want to maximize f on $[0,4]$.

$$f'(x) = (8 - 2x)^2 + x(2)(8 - 2x)(-2) \qquad \text{[Using the Product Rule.]}$$
$$= (8 - 2x)[(8 - 2x) - 4x] = (8 - 2x)(8 - 6x) = 0$$

if $x = 4$ or $4/3$. The latter is a critical point in $(0,4)$.

x	0	4/3	4
$f(x)$	0	1024/27	0

We see that $x = 4/3$ yields an absolute maximum for f. So the dimensions of the box should be $\frac{16}{3}'' \times \frac{16}{3}'' \times \frac{4}{3}''$.

7. Let x denote the length of the sides of the box and y denote its height. Referring to the following figure, we see that the volume of the box is given by $x^2 y = 128$. The

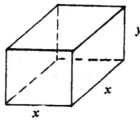

amount of material used is given by

$$S = f(x) = 2x^2 + 4xy$$
$$= 2x^2 + 4x\left(\frac{128}{x^2}\right)$$
$$= 2x^2 + \frac{512}{x} \text{ square inches.}$$

We want to minimize f subject to the condition that $x > 0$. Now
$$f'(x) = 4x - \frac{512}{x^2} = \frac{4x^3 - 512}{x^2} = \frac{4(x^3 - 128)}{x^2}.$$
Setting $f'(x) = 0$ yields $x = 5.04$, a critical point of f. Next,
$$f''(x) = 4 + \frac{1024}{x^3} > 0$$
for all $x > 0$. Thus, the graph of f is concave upward and so $x = 5.04$ yields an absolute minimum of f. Thus, the required dimensions are 5.04" × 5.04" × 5.04".

9. The length plus the girth of the box is $4x + h = 108$ and $h = 108 - 4x$. Then
$$V = x^2h = x^2(108 - 4x) = 108x^2 - 4x^3$$
and $V' = 216x - 12x^2$. We want to maximize V on the interval $[0,27]$. Setting $V'(x) = 0$ and solving for x, we obtain $x = 18$ and $x = 0$. Evaluating $V(x)$ at $x = 0$, $x = 18$, and $x = 27$, we obtain
$$V(0) = 0, \ V(18) = 11{,}664, \text{ and } V(27) = 0$$
Thus, the dimensions of the box are 18" × 18" × 36" and its maximum volume is approximately 11664 cu in.

11. We take $2\pi r + \ell = 108$. We want to maximize
$$V = \pi r^2 \ell = \pi r^2(-2\pi r + 108) = -2\pi^2 r^3 + 108\pi r^2$$
subject to the condition that $0 \le r \le \frac{54}{\pi}$. Now
$$V'(r) = -6\pi^2 r^2 + 216\pi r = -6\pi r(\pi r - 36).$$
Since $V' = 0$, we find $r = 0$ or $r = 36/\pi$, the critical points of V. From the table

r	0	$36/\pi$	$54/\pi$
V	0	$46{,}656/\pi$	0

we conclude that the maximum volume occurs when $r = 36/\pi \approx 11.5$ inches and $\ell = 108 - 2\pi\left(\frac{36}{\pi}\right) = 36$ inches and its volume is $46{,}656/\pi$ cu in .

13. Let y denote the height and x the width of the cabinet. Then $y = (3/2)x$. Since the volume is to be 2.4 cu ft, we have $xyd = 2.4$, where d is the depth of the cabinet.

We have $\quad x\left(\dfrac{3}{2}x\right)d = 2.4 \quad$ or $\quad d = \dfrac{2.4(2)}{3x^2} = \dfrac{1.6}{x^2}$.

The cost for constructing the cabinet is

$$C = 40(2xd + 2yd) + 20(2xy) = 80\left[\dfrac{1.6}{x} + \left(\dfrac{3}{2}x\right)\left(\dfrac{1.6}{x^2}\right)\right] + 40x\left(\dfrac{3}{2}x\right)$$

$$= \dfrac{320}{x} + 60x^2.$$

$$C'(x) = -\dfrac{320}{x^2} + 120x = \dfrac{120x^3 - 320}{x^2} = 0 \ \text{ if } x = \sqrt[3]{\dfrac{8}{3}} = \dfrac{2}{\sqrt[3]{3}} = \dfrac{2}{3}\sqrt[3]{9}$$

and $x = \frac{2}{3}\sqrt[3]{9}$ is a critical point of C. The sign diagram

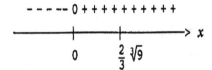

shows that $x = \frac{2}{3}\sqrt[3]{9}$ gives a relative minimum. Next,

$$C''(x) = \dfrac{640}{x^3} + 120 > 0$$

for all $x > 0$ tells us that the graph of C is concave upward. So $x = \dfrac{2}{3}\sqrt[3]{9}$ yields an

absolute minimum. The required dimensions are $\frac{2}{3}\sqrt[3]{9}' \times \sqrt[3]{9}' \times \frac{2}{5}\sqrt[3]{9}'$.

15. We want to maximize the function
$$R(x) = (200 + x)(300 - x) = -x^2 + 100x + 60000.$$
$$R'(x) = -2x + 100 = 0$$
gives $x = 50$ and this is a critical point of R. Since $R''(x) = -2 < 0$, we see that $x = 50$ gives an absolute maximum of R. Therefore, the number of passengers should be 250. The fare will then be \$250/passenger and the revenue will be \$62,500.

17. We want to maximize $S = kh^2w$. But $h^2 + w^2 = 24^2$ or $h^2 = 576 - w^2$. So
$$S = f(w) = kw(576 - w^2) = k(576w - w^3). \text{ Now, setting}$$
$$f'(w) = k(576 - 3w^2) = 0$$
gives $w = \pm\sqrt{192} \approx \pm 13.86$. Only the positive root is a critical point of interest.

Next, we find $f''(w) = -6kw$, and in particular,

$$f''(\sqrt{192}) = -6\sqrt{192}\, k < 0,$$

so that $w = \pm\sqrt{192} \approx \pm 13.86$ gives a relative maximum of f. Since $f''(w) < 0$ for $w > 0$, we see that the graph of f is concave downward on $(0,\infty)$ and so, $w = \sqrt{192}$ gives an absolute maximum of f. We find $h^2 = 576 - 192 = 384$ or $h \approx 19.60$. So the width and height of the log should be approximately 13.86 inches and 19.60 inches, respectively.

19. We want to minimize $C(x) = (10{,}000 - x) + 3\sqrt{3000^2 + x^2}$ subject to $0 \le x \le 10{,}000$. Now

$$C'(x) = -1 + 3(\tfrac{1}{2})(9{,}000{,}000 + x^2)^{-1/2}(2x)$$

$$= -1 + \frac{3x}{\sqrt{9{,}000{,}000 + x^2}} = 0.$$

$$C'(x) = 0 \Rightarrow 3x = \sqrt{9{,}000{,}000 + x^2}$$

$$9x^2 = 9{,}000{,}000 + x^2$$

or
$$x = \frac{3000}{\sqrt{8}} \approx 750\sqrt{2}.$$

x	0	$750\sqrt{2}$	10000
$f(x)$	19000	18485	31321

From the table, we see that $x = 750\sqrt{2}$ gives the absolute minimum.

21. The fuel cost is $x/400$ dollars per mile, and the labor cost is $8/x$ dollars per mile. Therefore, the total cost is $C(x) = \dfrac{8}{x} + \dfrac{x}{400}$; $C'(x) = -\dfrac{8}{x^2} + \dfrac{1}{400}$.

Setting $C'(x) = 0$ gives $-\dfrac{8}{x^2} = -\dfrac{1}{400}$; $x^2 = 3200$, and $x = 56.57$.

Next, $C''(x) = \dfrac{16}{x^3} > 0$ for all $x > 0$ so C is concave upward. Therefore, $x = 56.57$ gives the absolute minimum. So the most economical speed is 56.57 mph.

23. Let x denote the number of bottles in each order. We want to minimize

$$C(x) = 200\left(\frac{2,000,000}{x}\right) + \frac{x}{2}(0.40) = \frac{400,000,000}{x} + 0.2x.$$

We compute $C'(x) = -\dfrac{400,000,000}{x^2} + 0.2.$ Setting $C'(x) = 0$ gives

$$x^2 = \frac{400,000,000}{0.2} = 2,000,000,000$$

or $x = 44,721$, a critical point of C.

$$C'(x) = \frac{800,000,000}{x^3} > 0 \text{ for all } x > 0,$$

and we see that the graph of C is concave upward and so $x = 44,721$ gives an absolute minimum of C. Therefore, there should be $2,000,000/x \approx 45$ orders per year (since we can not have fractions of an order.) Then each order should be for $2,000,000/4.5 \approx 44,445$ bottles.

CHAPTER 4 REVIEW EXERCISES, page 369

1. a. $f(x) = \frac{1}{3}x^3 - x^2 + x - 6.$ $f'(x) = x^2 - 2x + 1 = (x - 1)^2.$ $f'(x) = 0$ gives $x = 1$, the critical point of f. Now, $f'(x) > 0$ for all $x \neq 1$. Thus, f is increasing on $(-\infty, 1) \cup (1, \infty)$.
    b. Since $f'(x)$ does not change sign as we move across the critical point $x = 1$, the First Derivative Test implies that $x = 1$ does not give rise to a relative extremum of f.
    c. $f''(x) = 2(x - 1)$. Setting $f''(x) = 0$ gives $x = 1$ as a candidate for an inflection point of f. Since $f''(x) < 0$ for $x < 1$, and $f''(x) > 0$ for $x > 1$, we see that f is concave downward on $(-\infty, 1)$ and concave upward on $(1, \infty)$.
    d. The results of (c) imply that $(1, -\frac{17}{3})$ is an inflection point.

3. a. $f(x) = x^4 - 2x^2.$ $f'(x) = 4x^3 - 4x = 4x(x^2 - 1) = 4x(x + 1)(x - 1)$. The sign diagram of f'

```
      - -   -0 + + 0 - -  0 + + +
    ─────────┼────┼────┼─────────→ x
            -1    0    1
```

shows that f is decreasing on $(-\infty, -1) \cup (0, 1)$ and increasing on $(-1, 0) \cup (1, \infty)$.
    b. The results of (a) and the First Derivative Test show that $(-1, -1)$ and $(1, -1)$ are relative minima and $(0, 0)$ is a relative maximum.

4 Applications of the Derivative

c. $f''(x) = 12x^2 - 4 = 4(3x^2 - 1) = 0$ if $x = \pm\sqrt{3}/3$. The sign diagram

shows that f is concave upward on $(-\infty, -\sqrt{3}/3) \cup (\sqrt{3}/3, \infty)$ and concave downward on $(-\sqrt{3}/3, \sqrt{3}/3)$.

d. The results of (c) show that $(-\sqrt{3}/3, -5/9)$ and $(\sqrt{3}/3, -5/9)$ are inflection points.

5. a. $f(x) = \dfrac{x^2}{x-1}$. $f'(x) = \dfrac{(x-1)(2x) - x^2(1)}{(x-1)^2} = \dfrac{x^2 - 2x}{(x-1)^2} = \dfrac{x(x-2)}{(x-1)^2}$.

The sign diagram of f'

shows that f is increasing on $(-\infty, 0) \cup (2, \infty)$ and decreasing on $(0,1) \cup (1,2)$.

b. The results of (a) show that $(0,0)$ is a relative maximum and $(2,4)$ is a relative minimum.

c. $f''(x) = \dfrac{(x-1)^2(2x-2) - x(x-2)2(x-1)}{(x-1)^4} = \dfrac{2(x-1)[(x-1)^2 - x(x-2)]}{(x-1)^4}$

$= \dfrac{2}{(x-1)^3}$.

Since $f''(x) < 0$ if $x < 1$ and $f''(x) > 0$ if $x > 1$, we see that f is concave downward on $(-\infty, 1)$ and concave upward on $(1, \infty)$.

d. Since $x = 1$ is not in the domain of f, there are no inflection points.

7. $f(x) = (1-x)^{1/3}$. $f'(x) = -\dfrac{1}{3}(1-x)^{-2/3} = -\dfrac{1}{3(1-x)^{2/3}}$.

The sign diagram for f' is

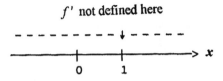

f' not defined here

a. f is decreasing on $(-\infty,1) \cup (1,\infty)$.

b. There are no relative extrema.

c. Next, we compute $f''(x) = -\dfrac{2}{9}(1-x)^{-5/3} = -\dfrac{2}{9(1-x)^{5/3}}$.

The sign diagram for f'' is

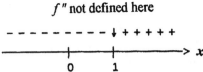

f'' not defined here

We find f is concave downward on $(-\infty,1)$ and concave upward on $(1,\infty)$.

d. $x = 1$ is a candidate for an inflection point of f. Referring to the sign diagram for f'', we see that $(1,0)$ is an inflection point.

9. a. $f(x) = \dfrac{2x}{x+1}$. $f'(x) = \dfrac{(x+1)(2) - 2x(1)}{(x+1)^2} = \dfrac{2}{(x+1)^2} > 0$ if $x \neq -1$.

Therefore f is increasing on $(-\infty,-1) \cup (-1,\infty)$.

b. Since there are no critical points, f has no relative extrema.

c. $f''(x) = -4(x+1)^{-3} = -\dfrac{4}{(x+1)^3}$.

Since $f''(x) > 0$ if $x < -1$ and $f''(x) < 0$ if $x > -1$, we see that f is concave upward on $(-\infty,-1)$ and concave downward on $(-1,\infty)$.

d. There are no inflection points since $f''(x) \neq 0$ for all x in the domain of f.

11. $f(x) = x^2 - 5x + 5$
    1. The domain of f is $(-\infty, \infty)$.
    2. Setting $x = 0$ gives 5 as the y-intercept.
    3. $\lim\limits_{x \to -\infty} (x^2 - 5x + 5) = \lim\limits_{x \to \infty} (x^2 - 5x + 5) = \infty$.
    4. There are no asymptotes because f is a quadratic function.

5. $f'(x) = 2x - 5 = 0$ if $x = 5/2$. The sign diagram

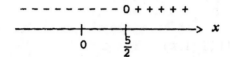

shows that f is increasing on $(\frac{5}{2}, \infty)$ and decreasing on $(-\infty, \frac{5}{2})$.

6. The First Derivative Test implies that $(\frac{5}{2}, -\frac{5}{4})$ is a relative minimum.

7. $f''(x) = 2 > 0$ and so f is concave upward on $(-\infty, \infty)$.

8. There are no inflection points.

The graph of f follows.

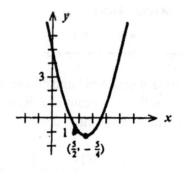

13. $g(x) = 2x^3 - 6x^2 + 6x + 1$.

1. The domain of g is $(-\infty, \infty)$.

2. Setting $x = 0$ gives 1 as the y-intercept.

3. $\lim\limits_{x \to -\infty} g(x) = -\infty$, $\lim\limits_{x \to \infty} g(x) = \infty$.

4. There are no vertical or horizontal asymptotes.

5. $g'(x) = 6x^2 - 12x + 6 = 6(x^2 - 2x + 1) = 6(x - 1)^2$. Since $g'(x) > 0$ for all $x \neq 1$, we see that g is increasing on $(-\infty, 1) \cup (1, \infty)$.

6. $g'(x)$ does not change sign as we move across the critical point $x = 1$, so there is no extremum.

7. $g''(x) = 12x - 12 = 12(x - 1)$. Since $g''(x) < 0$ if $x < 1$ and $g''(x) > 0$ if $x > 1$, we see that g is concave upward on $(1, \infty)$ and concave downward on $(-\infty, 1)$.

8. The point $x = 1$ gives rise to the inflection point $(1, 3)$.

9. The graph of g follows.

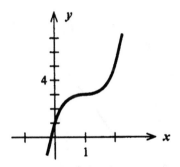

15. $h(x) = x\sqrt{x-2}$.

1. The domain of h is $[2,\infty)$.
2. There are no y-intercepts. Next, setting $y = 0$ gives 2 as the x-intercept.
3. $\lim\limits_{x\to\infty} x\sqrt{x-2} = \infty$.
4. There are no asymptotes.
5. $h'(x) = (x-2)^{1/2} + x(\tfrac{1}{2})(x-2)^{-1/2} = \tfrac{1}{2}(x-2)^{-1/2}[2(x-2)+x]$

$$= \frac{3x-4}{2\sqrt{x-2}} > 0 \quad \text{on } [2,\infty)$$

and so h is increasing on $[2,\infty)$.
6. Since h has no critical points in $(2,\infty)$, there are no relative extrema.

7. $h''(x) = \dfrac{1}{2}\left[\dfrac{(x-2)^{1/2}(3) - (3x-4)\tfrac{1}{2}(x-2)^{-1/2}}{x-2} \right]$

$$= \frac{(x-2)^{-1/2}[6(x-2)-(3x-4)]}{4(x-2)} = \frac{3x-8}{4(x-2)^{3/2}}.$$

The sign diagram for h''

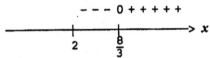

shows that h is concave downward on $(2,\tfrac{8}{3})$ and concave upward on $(\tfrac{8}{3},\infty)$.

8. The results of (7) tell us that $(\tfrac{8}{3}, \tfrac{8\sqrt{6}}{9})$ is an inflection point.

4 Applications of the Derivative

The graph of h follows.

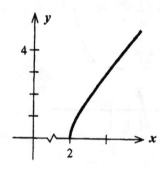

17. $f(x) = \dfrac{x-2}{x+2}$:

1. The domain of f is $(-\infty,-2) \cup (-2,\infty)$.
2. Setting $x = 0$ gives -1 as the y-intercept. Setting $y = 0$ gives 2 as the x-intercept.
3. $\displaystyle\lim_{x\to-\infty} \dfrac{x-2}{x+2} = \lim_{x\to\infty} \dfrac{x-2}{x+2} = 1.$
4. The results of (3) tell us that $y = 1$ is a horizontal asymptote. Next, observe that the denominator of $f(x)$ is equal to zero at $x = -2$, but its numerator is not equal to zero there. Therefore, $x = -2$ is a vertical asymptote.
5. $\qquad f'(x) = \dfrac{(x+2)(1) - (x-2)(1)}{(x+2)^2} = \dfrac{4}{(x+2)^2}.$

The sign diagram of f'

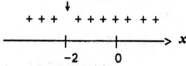

tells us that f is increasing on $(-\infty,-2) \cup (-2,\infty)$.
6. The results of (5) tells us that there are no relative extrema.
7. $f''(x) = -\dfrac{8}{(x+2)^3}.$ The sign diagram of f'' follows

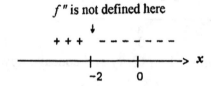

and it shows that f is concave upward on $(-\infty,-2)$ and concave downward on $(-2,\infty)$.

8. There are no inflection points.

The graph of f follows.

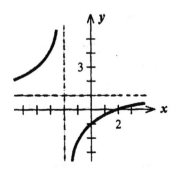

19. $\lim\limits_{x\to-\infty}\dfrac{1}{2x+3} = \lim\limits_{x\to\infty}\dfrac{1}{2x+3} = 0$ and so $y = 0$ is a horizontal asymptote. Since the denominator is equal to zero at $x = -3/2$, but the numerator is not equal to zero there, we see that $x = -3/2$ is a vertical asymptote.

21. $\lim\limits_{x\to-\infty}\dfrac{5x}{x^2-2x-8} = \lim\limits_{x\to\infty}\dfrac{5x}{x^2-2x-8} = 0$ and so $y = 0$ is a horizontal asymptote. Next, note that the denominator is zero if $x^2 - 2x - 8 = (x - 4)(x + 2) = 0$, or $x = -2$ or $x = 4$. Since the numerator is not equal to zero at these points, we see that $x = -2$ and $x = 4$ are vertical asymptotes.

23. $f(x) = 2x^2 + 3x - 2$; $f'(x) = 4x + 3$. Setting $f'(x) = 0$ gives $x = -3/4$ as a critical point of f. Next, $f''(x) = 4 > 0$ for all x, so f is concave upward on $(-\infty, \infty)$. Therefore, $f(-\frac{3}{4}) = -\frac{25}{8}$ is an absolute minimum of f. There is no absolute maximum.

25. $g(t) = \sqrt{25-t^2} = (25-t^2)^{1/2}$. Differentiating $g(t)$, we have
$$g'(t) = \tfrac{1}{2}(25-t^2)^{-1/2}(-2t) = -\frac{t}{\sqrt{25-t^2}}.$$
Setting $g'(t) = 0$ gives $t = 0$ as a critical point of g. The domain of g is given by solving the inequality $25 - t^2 \geq 0$ or $(5 - t)(5 + t) \geq 0$ which implies that $t \in [-5,5]$.

From the table

t	-5	0	5
$g(t)$	0	5	0

we conclude that $g(0) = 5$ is the absolute maximum of g and $g(-5) = 0$ and $g(5) = 0$ is the absolute minimum value of g.

27. $h(t) = t^3 - 6t^2$. $h'(t) = 3t^2 - 12t = 3t(t - 4) = 0$ if $t = 0$ or $t = 4$, critical points of h. But only $t = 4$ lies in $(2,5)$.

t	2	4	5
$h(t)$	-16	-32	-25

From the table, we see that there is an absolute minimum at $(4,-32)$ and an absolute maximum at $(2,-16)$.

29. $f(x) = x - \dfrac{1}{x}$ on $[1,3]$. $f'(x) = 1 + \dfrac{1}{x^2}$. Since $f'(x)$ is never zero, f has no critical point.

x	1	3
$f(x)$	0	$\frac{8}{3}$

We see that $f(1) = 0$ is the absolute minimum value and $f(3) = 8/3$ is the absolute maximum value.

31. $f(s) = s\sqrt{1-s^2}$ on $[-1,1]$. The function f is continuous on $[-1,1]$ and differentiable on $(-1,1)$. Next,

$$f'(s) = (1-s^2)^{1/2} + s(\tfrac{1}{2})(1-s^2)^{-1/2}(-2s) = \frac{1-2s^2}{\sqrt{1-s^2}}.$$

Setting $f'(s) = 0$, we have $s = \pm\sqrt{2}/2$, giving the critical points of f. From the table

x	-1	$-\sqrt{2}/2$	$\sqrt{2}/2$	1
$f(x)$	0	-1/2	1/2	0

we see that $f(-\sqrt{2}/2) = -1/2$ is the absolute minimum value and $f(\sqrt{2}/2) = 1/2$ is the absolute maximum value of f.

33. We want to maximize $P(x) = -x^2 + 8x + 20$. Now, $P'(x) = -2x + 8 = 0$ if $x = 4$, a critical point of P. Since $P''(x) = -2 < 0$, the graph of P is concave downward. Therefore, the critical point $x = 4$ yields an absolute maximum. So, to maximize profit, the company should spend $4000 on advertising per month.

35. The revenue is $R(x) = px = x(-0.0005x^2 + 60) = -0.0005x^3 + 60x$. Therefore, the total profit is $P(x) = R(x) - C(x) = -0.0005x^3 + 0.001x^2 + 42x - 4000$.
$$P'(x) = -0.0015x^2 + 0.002x + 42.$$
Setting $P'(x) = 0$, we have $3x^2 - 4x - 84,000 = 0$. Solving for x, we find
$$x = \frac{4 \pm \sqrt{16 - 4(3)(84,000)}}{2(3)} = \frac{4 \pm 1004}{6} = 168, \text{ or } -167.$$
We reject the negative root. Next,
$$P''(x) = -0.003x + 0.002 \text{ and } P''(168) = -0.003(168) + 0.002 = -0.502 < 0.$$
By the Second Derivative Test, $x = 168$ gives a relative maximum. Therefore, the required level of production is 168 video discs.

37. $N(t) = -2t^3 + 12t^2 + 2t$. We wish to find the inflection point of the function N. Now, $N'(t) = -6t^2 + 24t + 2$ and $N''(t) = -12t + 24 = -12(t - 2)$.
Setting $N''(t) = 0$ gives $t = 2$. Furthermore, $N''(t) > 0$ when $t < 2$ and $N''(t) < 0$ when $t > 2$. Therefore, $t = 2$ is an inflection point of N. Thus, the average worker is performing at peak efficiency at 10 A.M.

39. Let x denote the number of cases in each order. Then the average number of cases of beer in storage during the year is $x/2$. The storage cost is $2(x/2)$, or x dollars. Next, we see that the number of orders required is $800,000/x$, and so the ordering cost is
$$\frac{500(800,000)}{x} = \frac{400,000,000}{x}$$

dollars. Thus, the total cost incurred by the company per year is given by

$$C(x) = x + \frac{400,000,000}{x}.$$

We want to minimize C in the interval $(0, \infty)$. Now

$$C'(x) = 1 - \frac{400,000,000}{x^2}.$$

Setting $C'(x) = 0$ gives $x^2 = 400,000,000$, or $x = 20,000$ (we reject $x = -20,000$).

Next, $\quad C''(x) = \dfrac{800,000,000}{x^3} > 0 \quad$ for all x, so C is concave upward. Thus,

$x = 20,000$ gives rise to the absolute minimum of C. Thus, the company should order 20,000 cases of beer per order.

CHAPTER 5

EXERCISES 5.1, page 379

1. a. $4^{-3} \times 4^5 = 4^{-3+5} = 4^2 = 16$ b. $3^{-3} \times 3^6 = 3^{6-3} = 3^3 = 27.$

3. a. $9(9)^{-1/2} = \dfrac{9}{9^{1/2}} = \dfrac{9}{3} = 3.$ b. $5(5)^{-1/2} = 5^{1/2} = \sqrt{5}.$

5. a. $\dfrac{(-3)^4(-3)^5}{(-3)^8} = (-3)^{4+5-8} = (-3)^1 = -3.$

   b. $\dfrac{(2^{-4})(2^6)}{2^{-1}} = 2^{-4+6+1} = 2^3 = 8.$

7. a. $\dfrac{5^{3.3} \cdot 5^{-1.6}}{5^{-0.3}} = \dfrac{5^{3.3-1.6}}{5^{-0.3}} = 5^{1.7+(0.3)} = 5^2 = 25.$

   b. $\dfrac{4^{2.7} \cdot 4^{-1.3}}{4^{-0.4}} = 4^{2.7-1.3+0.4} = 4^{1.8} \approx 12.1257.$

9. a. $(64x^9)^{1/3} = 64^{1/3}(x^{9/3}) = 4x^3.$
   b. $(25x^3y^4)^{1/2} = 25^{1/2}(x^{3/2})(y^{4/2}) = 5x^{3/2}y^2 = 5xy^2\sqrt{x}.$

11. a. $\dfrac{6a^{-5}}{3a^{-3}} = 2a^{-5+3} = 2a^{-2} = \dfrac{2}{a^2}.$ b. $\dfrac{4b^{-4}}{12b^{-6}} = \dfrac{1}{3}b^{-4+6} = \dfrac{1}{3}b^2.$

13. a. $(2x^3y^2)^3 = 2^3 \times x^{3(3)} \times y^{2(3)} = 8x^9y^6.$
    b. $(4x^2y^2z^3)^2 = 4^2 \times x^{2(2)} \times y^{2(2)} \times z^{3(2)} = 16x^4y^4z^6.$

15. a. $\dfrac{5^0}{(2^{-3}x^{-3}y^2)^2} = \dfrac{1}{2^{-3(2)}x^{-3(2)}y^{2(2)}} = \dfrac{2^6x^6}{y^4} = \dfrac{64x^6}{y^4}.$

    b. $\dfrac{(x+y)(x-y)}{(x-y)^0} = (x+y)(x-y).$

17. $6^{2x} = 6^4$ if and only if $2x = 4$ or $x = 2$.

19. $3^{3x-4} = 3^5$ if and only if $3x - 4 = 5$, $3x = 9$, or $x = 3$.

21. $(2.1)^{x+2} = (2.1)^5$ if and only if $x + 2 = 5$, or $x = 3$.

23. $8^x = (\frac{1}{32})^{x-2}$, $(2^3)^x = (32)^{2-x} = (2^5)^{2-x}$, so $2^{3x} = 2^{5(2-x)}$, $3x = 10 - 5x$, $8x = 10$, or $x = 5/4$.

25. Let $y = 3^x$, then the given equation is equivalent to
$$y^2 - 12y + 2y = 0$$
$$(y-9)(y-3) = 0$$
giving $y = 3$ or 9. So $3^x = 3$ or $3^x = 9$, and therefore, $x = 1$ or $x = 2$.

27. $y = 2^x$, $y = 3^x$, and $y = 4^x$

29. $y = 2^{-x}$, $y = 3^{-x}$, and $y = 4^{-x}$

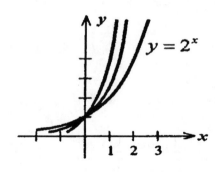

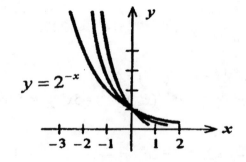

31. $y = 4^{0.5x}$, $y = 4x$, and $y = 4^{2x}$

33. $y = e^{0.5x}$, $y = e^x$, $y = e^{1.5x}$

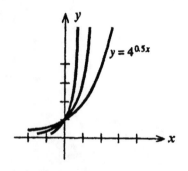

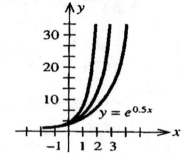

35. $y = 0.5e^{-x}$, $y = e^{-x}$, and $y = 2e^{-x}$

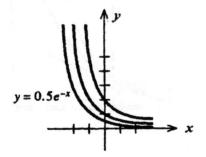

$y = 0.5e^{-x}$

37. False. $e^x e^y = e^{x+y}$

39. True. If $a < b < 1$, then $f(x) = b^x$ is a decreasing function of x and so if $x < y$, then $f(x) > f(y)$ or $b^x > b^y$.

USING TECHNOLOGY EXERCISES 5.1, page 381

1.

3.

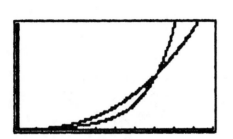

5.

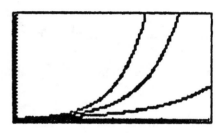

7.   9.

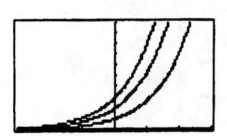   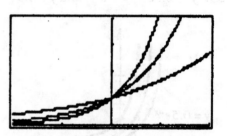

EXERCISES 5.2 , page 389

1. $\log_2 64 = 6$ 3. $\log_3 \dfrac{1}{9} = -2$ 5. $\log_{1/3} \dfrac{1}{3} = 1$

7. $\log_{32} 8 = \dfrac{3}{5}$ 9. $\log_{10} 0.001 = -3$

11. $\log 12 = \log 4 \times 3 = \log 4 + \log 3 = 0.6021 + 0.4771 = 1.0792.$

13. $\log 16 = \log 4^2 = 2 \log 4 = 2(0.6021) = 1.2042.$

15. $\log 48 = \log 3 \times 4^2 = \log 3 + 2 \log 4 = 0.4771 + 2(0.6021) = 1.6813.$

17. $\log x(x+1)^4 = \log x + \log (x+1)^4 = \log x + 4 \log (x+1).$

19. $\log \dfrac{\sqrt{x+1}}{x^2+1} = \log (x+1)^{1/2} - \log(x^2+1) = \frac{1}{2} \log (x+1) - \log (x^2+1)$

21. $\ln xe^{-x^2} = \ln x - x^2.$

23. $\ln \left(\dfrac{x^{1/2}}{x^2\sqrt{1+x^2}} \right) = \ln x^{1/2} - \ln x^2 - \ln (1+x^2)^{1/2}$

           $= \frac{1}{2} \ln x - 2 \ln x - \frac{1}{2} \ln (1+x^2) = -\frac{3}{2} \ln x - \frac{1}{2} \ln (1+x^2).$

25. $\ln x^x = x \ln x.$

27. $y = \log_3 x$

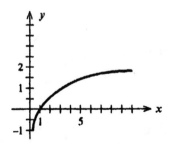

29. $y = \ln 2x$

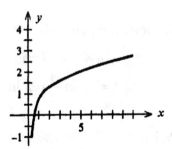

31. $y = 2^x$ and $y = \log_2 x$

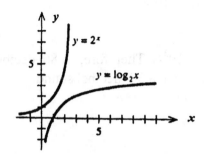

33. $e^{0.4t} = 8$, $0.4t \ln e = \ln 8$, and $0.4t = \ln 8$ ($\ln e = 1$.) So, $t = \dfrac{\ln 8}{0.4} = 5.1986$.

35. $5e^{-2t} = 6$, $e^{-2t} = \frac{6}{5} = 1.2$. Taking the logarithm, we have
$$-2t \ln e = \ln 1.2, \text{ or } t = -\frac{\ln 1.2}{2} \approx -0.0912.$$

37. $2e^{-0.2t} - 4 = 6$, $2e^{-0.2t} = 10$. Taking the logarithm, we have
$$\ln e^{-0.2t} = \ln 5, \ -0.2t \ln e = \ln 5, \ -0.2t = \ln 5,$$
and $\quad t = -\dfrac{\ln 5}{0.2} \approx -8.0472.$

39. $\dfrac{50}{1 + 4e^{0.2t}} = 20$, $\ 1 + 4e^{0.2t} = \dfrac{50}{20} = 2.5$, $\ 4e^{0.2t} = 1.5$,

$$e^{0.2t} = \frac{1.5}{4} = 0.375, \ln e^{0.2t} = \ln 0.375, 0.2t = \ln 0.375. \text{ So } t = \frac{\ln 0.375}{0.2} \approx -4.9041.$$

41. Taking the logarithm on both sides, we obtain

$$\ln A = \ln Be^{-t/2}, \ \ln A = \ln B + \ln e^{-t/2}, \ \ln A - \ln B = -t/2 \ln e,$$

$$\ln \frac{A}{B} = -\frac{t}{2} \text{ or } t = -2 \ln \frac{A}{B} = 2 \ln \frac{B}{A}$$

43. $p(x) = 19.4 \ln x + 18$. For a child weighing 92 lb, we find
$p(92) = 19.4 \ln 92 + 18 = 105.72$ millimeters of mercury.

45. a. $30 = 10 \log \frac{I}{I_0}$; $3 = \log \frac{I}{I_0}$; $\frac{I}{I_0} = 10^3 = 1000$.

So $I = 1000 I_0$.
b. When $D = 80$, $I = 10^8 I_0$ and when $D = 30$, $I = 10^3 I_0$. Therefore, an 80–decibel
sound is $10^8/10^3$ or $10^5 = 100,000$ times louder than a 30–decibel sound.
c. It is $10^{15}/10^8 = 10^7$, or 10,000,000, times louder.

47. With $T_0 = 70$, $T_1 = 98.6$, and $T = 80$, we have
$$80 = 70 + (98.6 - 70)(0.97)^t$$
$$28.6(0.97)^t = 10,$$
$$(0.97)^t = 0.34965.$$
Taking logarithms, we have
$$\ln (0.97)^t = \ln 0.34965, \text{ or } t = \frac{\ln 0.34965}{\ln 0.97} \approx 34.50.$$
So he was killed 34½ hours earlier at 1:30 P.M.

49. False. Take $a = 2e$ and $b = e$. Then
$$\ln a - \ln b = \ln 2e - \ln e = \ln 2 + \ln e - \ln e = \ln 2.$$
But $\ln(a - b) = \ln(2e - e) = \ln e = 1$.

51. True. If $x > 0$, then $|x| = x$ and $f(x) = \ln|x| = \ln x$ is continuous on $(0, \infty)$. If $x < 0$,
then $|x| = -x$ and so $f(x) = \ln|x| = \ln(-x)$ is continuous on $(-\infty, 0)$.

53. a. Let $p = \log_b m$ and $q = \log_b n$ so that $m = b^p$ and $n = b^q$. Then $mn = b^p b^q = b^{p+q}$
and by definition, $p + q = \log_b mn$; that is, $\log_b mn = \log_b m + \log_b n$.

b. $\dfrac{m}{n} = \dfrac{b^p}{b^q} = b^{p-q}$. So, by definition, $p - q = \log_b \dfrac{m}{n}$; that is

$\log_b \dfrac{m}{n} = \log_b m - \log_b n$.

55. a. By definition $\log_b 1 = 0$ means $1 = b^0 = 1$.
   b. By definition $\log_b b = 1$ means $b = b^1 = b$.

EXERCISES 5.3, page 401

1. $A = 2500\left(1 + \dfrac{0.07}{2}\right)^{20} = 4974.47$, or $4974.47.

3. $A = 150{,}000\left(1 + \dfrac{0.1}{12}\right)^{48} = 223{,}403.11$, or $223,403.11

5. a. Using the formula
$$r_{eff} = \left(1 + \dfrac{r}{m}\right)^m - 1$$
with $r = 0.10$ and $m = 2$, we have
$$r_{eff} = \left(1 + \dfrac{0.10}{2}\right)^2 - 1 = 0.1025, \quad \text{or } 10.25 \text{ percent}..$$

   b. Using the formula
$$r_{eff} = \left(1 + \dfrac{r}{m}\right)^m - 1$$
with $r = 0.09$ and $m = 4$, we have
$$r_{eff} = \left(1 + \dfrac{0.09}{4}\right)^4 - 1 = 0.09308, \text{ or } 9.308 \text{ percent per year.}$$

7. a. The present value is given by
$$P = 40{,}000\left(1 + \dfrac{0.08}{2}\right)^{-8} = 29{,}227.61, \quad \text{or } $29{,}227.61.$$
   b. The present value is given by

$$P = 40,000\left(1+\frac{0.08}{4}\right)^{-16} = 29,137.83 \qquad \text{or } \$29,137.83.$$

9. $A = 5000e^{0.08(4)} \approx 6885.64$, or \$6,885.64.

11. The Estradas can expect to pay $80,000(1+0.09)^4$, or approximately \$112,926.52.

13. The investment will be worth

$$A = 1.5\left(1+\frac{0.095}{2}\right)^{20} = 3.794651 \text{ , or approximately \$3.8 million}$$

dollars.

15. The present value of the \$8000 loan due in 3 years is given by

$$P = 8000\left(1+\frac{0.10}{2}\right)^{-6} = 5969.72, \text{ or } \$5969.72.$$

The present value of the \$15,000 loan due in 6 years is given by

$$P = 15,000\left(1+\frac{0.10}{2}\right)^{-12} = 8352.56, \text{ or } \$8352.56.$$

Therefore, the amount the proprietors of the inn will be required to pay at the end of 5 years is given by

$$A = 14,322.28\left(1+\frac{0.10}{2}\right)^{10} = 23,329.48, \text{ or } \$23,329.48.$$

17. We solve the equation
$$2 = 1(1+0.075)^t$$
for t.   Taking the logarithm on both sides, we have
$$\ln 2 = \ln(1.075)^t \approx t \ln 1.075.$$
So $t = \dfrac{\ln 2}{\ln 1.075} \approx 9.58$, or approximately 9.6 years.

19. The effective annual rate of return on his investment is found by solving the equation
$$(1+r)^2 = \frac{32100}{25250}$$

$$1+r = \left(\frac{32100}{25250}\right)^{1/2}$$

$$1+r \approx 1.1275$$

and $r = 0.1275$, or 12.75 percent.

21. $P = Ae^{-rt} = 59673e^{-(0.08)5} \approx 40,000.008$, or approximately \$40,000.

23. a. If they invest the money at 10.5 percent compounded quarterly, they should set aside

$$P = 70,000\left(1+\frac{0.105}{4}\right)^{-28} \approx 33,885.14, \text{ or } \$33,885.14.$$

b. If they invest the money at 10.5 percent compounded continuously, they should set aside

$$P = 70,000e^{-0.735} = 33,565.38, \text{ or } \$33,565.38.$$

25. a. If inflation over the next 15 years is 6 percent, then Eleni's first year's pension will be worth

$$P = 40,000e^{-0.9} = 16,262.79, \text{ or } \$16,262.79.$$

b. If inflation over the next 15 years is 8 percent, then Eleni's first year's pension will be worth

$$P = 40,000e^{-1.2} = 12,047.77, \text{ or } \$12,047.77.$$

c. If inflation over the next 15 years is 12 percent, then Eleni's first year's pension will be worth

$$P = 40,000e^{-1.8} = 6611.96, \text{ or } \$6,611.96.$$

27. $r_{eff} = \lim\limits_{m\to\infty}\left(1+\dfrac{r}{m}\right)^{m} - 1 = e^{r} - 1.$

29. The effective rate of interest at Bank A is given by

$$R = \left(1+\frac{0.07}{4}\right)^{4} - 1 = 0.07186,$$

or 7.186 percent. The effective rate at Bank B is given by

$$R = e^{r} - 1 = e^{0.07125} - 1 = 0.07385$$

or 7.385 percent. We conclude that Bank B has the higher effective rate of interest.

31. The nominal rate of interest that, when compounded continuously, yields an effective rate of interest of 10 percent per year is found by solving the equation

$R = e^r - 1,\ 0.10 = e^r - 1,\ 1.10 = e^r.$

$\ln 1.10 = r \ln e,\ r = \ln 1.10 \approx 0.09531,$

or 9.531 percent.

EXERCISES 5.4 , page 411

1. $f(x) = e^{3x};\ f'(x) = 3e^{3x}$

3. $g(t) = e^{-t};\ g'(t) = -e^{-t}$

5. $f(x) = e^x + x;\ f'(x) = e^x + 1$

7. $f(x) = x^3 e^x,\ f'(x) = x^3 e^x + e^x(3x^2) = x^2 e^x(x + 3).$

9. $f(x) = \dfrac{2e^x}{x},\ f'(x) = \dfrac{x(2e^x) - 2e^x(1)}{x^2} = \dfrac{2e^x(x-1)}{x^2}.$

11. $f(x) = 3(e^x + e^{-x});\ f'(x) = 3(e^x - e^{-x}).$

13. $f(w) = \dfrac{e^w + 1}{e^w} = 1 + \dfrac{1}{e^w} = 1 + e^{-w}.\ f'(w) = -e^{-w} = -\dfrac{1}{e^w}.$

15. $f(x) = 2e^{3x-1},\ f'(x) = 2e^{3x-1}(3) = 6e^{3x-1}.$

17. $h(x) = e^{-x^2};\ h'(x) = e^{-x^2}(-2x) = -2xe^{-x^2}.$

19. $f(x) = 3e^{-1/x};\ f'(x) = 3e^{-1/x} \cdot \dfrac{d}{dx}\left(-\dfrac{1}{x}\right) = 3e^{-1/x}\left(\dfrac{1}{x^2}\right) = \dfrac{3e^{-1/x}}{x^2}.$

21. $f(x) = (e^x + 1)^{25},\ f'(x) = 25(e^x + 1)^{24}e^x = 25e^x(e^x + 1)^{24}.$

23. $f(x) = e^{\sqrt{x}}; \ f'(x) = e^{\sqrt{x}} \dfrac{d}{dx} x^{1/2} = e^{\sqrt{x}} \dfrac{1}{2} x^{-1/2} = \dfrac{e^{\sqrt{x}}}{2\sqrt{x}}.$

25. $f(x) = (x-1)e^{3x+2}; f'(x) = (x-1)(3)e^{3x+2} + e^{3x+2} = e^{3x+2}(3x - 3 + 1) = e^{3x+2}(3x - 2).$

27. $f(x) = \dfrac{e^x - 1}{e^x + 1}; \ f'(x) = \dfrac{(e^x + 1)(e^x) - (e^x - 1)(e^x)}{(e^x + 1)^2} = \dfrac{e^x(e^x + 1 - e^x + 1)}{(e^x + 1)^2} = \dfrac{2e^x}{(e^x + 1)^2}.$

29. $f(x) = e^{-4x} + 2e^{3x}; f'(x) = -4e^{-4x} + 6e^{3x}$ and
$f''(x) = 16e^{-4x} + 18e^{3x} = 2(8e^{-4x} + 9e^{3x}).$

31. $f(x) = 2xe^{3x}; f'(x) = 2e^{3x} + 2xe^{3x}(3) = 2(3x + 1)e^{3x}.$
$f''(x) = 6e^{3x} + 2(3x + 1)e^{3x}(3) = 6(3x + 2)e^{3x}.$

33. $y = f(x) = e^{2x-3}$. $f'(x) = 2e^{2x-3}$. To find the slope of the tangent line to the graph of f at $x = 3/2$, we compute $f'(\tfrac{3}{2}) = 2e^{3-3} = 2$.

Next, using the point–slope form of the equation of a line, we find that
$$y - 1 = 2(x - \tfrac{3}{2})$$
$$= 2x - 3, \quad \text{or} \quad y = 2x - 2.$$

35. $f(x) = e^{-x^2/2}$, $f'(x) = e^{-x^2/2}(-x) = -xe^{-x^2/2}$. Setting $f'(x) = 0$, gives $x = 0$ as the only critical point of f. From the sign diagram,

```
+ + + + + + + 0 - - - - -
─────────────┼─────────────> x
             0
```

we conclude that f is increasing on $(-\infty, 0)$ and decreasing on $(0, \infty)$.

37. $f(x) = \tfrac{1}{2}e^x - \tfrac{1}{2}e^{-x}$, $f'(x) = \tfrac{1}{2}(e^x + e^{-x}), f''(x) = \tfrac{1}{2}(e^x - e^{-x})$. Setting $f''(x) = 0$, gives $e^x = e^{-x}$ or $e^{2x} = 1$, and $x = 0$. From the sign diagram for f'',

```
- - - - - - - -0 + + + ++ + + +
─────────────┼─────────────> x
             0
```

we conclude that f is concave upward on $(0, \infty)$ and concave downward on $(-\infty, 0)$.

39. $f(x) = xe^{-2x}$. $f'(x) = e^{-2x} + xe^{-2x}(-2) = (1 - 2x)e^{-2x}$.

$f''(x) = -2e^{-2x} + (1 - 2x)e^{-2x}(-2) = 4(x - 1)e^{-2x}$.

Observe that $f''(x) = 0$ if $x = 1$. The sign diagram of f''

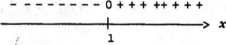

shows that $(1, e^{-2})$ is an inflection point.

41. $f(x) = e^{-x^2}$. $f'(x) = -2xe^{-x^2} = 0$ if $x = 0$, the only critical point of f.

x	-1	0	1
$f(x)$	e^{-1}	1	e^{-1}

From the table, we see that f has an absolute minimum value of e^{-1} attained at $x = -1$ and $x = 1$. It has an absolute maximum at $(0,1)$.

43. $g(x) = (2x - 1)e^{-x}$; $g'(x) = 2e^{-x} + (2x - 1)e^{-x}(-1) = (3 - 2x)e^{-x} = 0$, if $x = 3/2$. The graph of g shows that $(\frac{3}{2}, 2e^{-3/2})$ is an absolute maximum, and $(0,-1)$ is an absolute minimum.

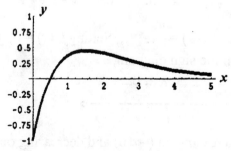

45. $f(t) = e^t - t$;

We first gather the following information on f.

1. The domain of f is $(-\infty,\infty)$.
2. Setting $t = 0$ gives 1 as the y-intercept.
3. $\lim_{t \to -\infty} (e^t - t) = \infty$ and $\lim_{t \to \infty} (e^t - t) = \infty$.
4. There are no asymptotes.

5. $f'(t) = e^t - 1$ if $t = 0$, a critical point of f. From the sign diagram for f'

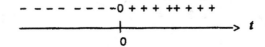

we see that f is decreasing on $(-\infty, 0)$ and increasing on $(0, \infty)$.
6. From the results of (5), we see that $(0,1)$ is a relative minimum of f.
7. $f''(t) = e^t > 0$ for all t in $(-\infty, \infty)$. So the graph of f is concave upward on $(-\infty, \infty)$.
8. There are no inflection points.
The graph of f follows.

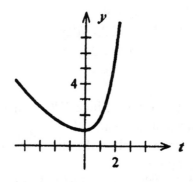

47. $f(x) = 2 - e^{-x}$.
We first gather the following information on f.
1. The domain of f is $(-\infty, \infty)$.
2. Setting $x = 0$ gives 1 as the y-intercept.
3. $\lim\limits_{x \to -\infty} (2 - e^{-x}) = -\infty$ and $\lim\limits_{x \to \infty} (2 - e^{-x}) = 2$,
4. From the results of (3), we see that $y = 2$ is a horizontal asymptote of f.
5. $f'(x) = e^{-x}$. Observe that $f'(x) > 0$ for all x in $(-\infty, \infty)$ and so f is increasing on $(-\infty, \infty)$.
6. Since there are no critical points, f has no relative extrema.
7. $f''(x) = -e^{-x} < 0$ for all x in $(-\infty, \infty)$ and so the graph of f is concave downward on $(-\infty, \infty)$.
8. There are no inflection points
The graph of f follows.

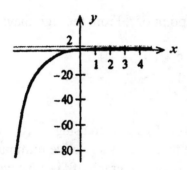

49. $S(t) = 20,000(1 + e^{-0.5t});$ $S'(t) = 20,000(-0.5e^{-0.5t}) = -10,000e^{-0.5t};$
    $S'(1) = -10,000e^{-0.5} = -6065,$ or $-\$6065/day.$
    $S'(2) = -10,000e^{-1} = -3679,$ or $-\$3679/day.$
    $S'(3) = -10,000(e^{-1.5}) = -2231,$ or $-\$2231/day.$
    $S'(4) = -10,000e^{-2} = -1353,$ or $-\$1353/day.$

51. $N(t) = 5.3e^{0.095t^2-0.85t}$
    a. $N'(t) = 5.3e^{0.095t^2-0.85t}(0.19t - 0.85).$ Since $N'(t)$ is negative for $(0 \le t \le 4)$, we see that $N(t)$ is decreasing over that interval.
    b. To find the rate at which the number of polio cases was decreasing at the beginning of 1959, we compute
    $$N'(0) = 5.3e^{0.095(0^2)-0.85(0)}(0.85) = 5.3(-0.85) = -4.505$$
    (t is measured in thousands), or 4,505 cases per year. To find the rate at which the number of polio cases was decreasing at the beginning of 1962, we compute
    $$N'(3) = 5.3e^{0.095(9)-0.85(3)}(0.57 - 0.85)$$
    $$= (-0.28)(0.9731) \approx -0.273, \text{ or } 273 \text{ cases per year.}$$

53. The demand equation is
    $$p(x) = 100e^{-0.0002x} + 150.$$
    Next, $p'(x) = 100(-0.0002)e^{-0.0002x} = -0.02e^{-0.0002x}.$
    a. To find the rate of change of the price per bottle when $x = 1000$, we compute
    $$p'(1000) = -0.02e^{-0.0002(1000)} = -0.02e^{-0.2} \approx -0.0163, \text{ or } -1.63 \text{ cents per bottle.}$$
    To find the rate of change of the price per bottle when $x = 2000$, we compute
    $$p'(2000) = -0.02e^{-0.0002(2000)} = -0.02e^{-0.4} \approx -0.0134,$$
    or -1.34 cents per bottle.
    b. The price per bottle when $x = 1000$ is given by
    $$p(1000) = 100e^{-0.0002(1000)} + 150 \approx 231.87,$$

or \$231.87/bottle. The price per bottle when $x = 2000$ is given by
$$p(2000) = 100e^{-0.0002(2000)} + 150 \approx 217.03,$$
or \$217.03/bottle.

55. a. $N(0) = \dfrac{3000}{1+99} = 30.$

b. $N'(x) = 3000 \dfrac{d}{dx}(1 + 99e^{-x})^{-1} = -3000(1 + 99e^{-x})^{-2}(-99e^{-x}) = \dfrac{297{,}000e^{-x}}{(1+99e^{-x})^{2}}.$

Since $N'(x) > 0$ for all x in $(0,\infty)$, we see that N is increasing on $(0,\infty)$.

c. The graph of N follows.

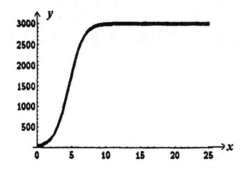

The total number of students who contracted influenza during that particular epidemic is approximately
$$\lim_{x \to \infty} \dfrac{3000}{1+99e^{-x}} = 3000.$$

57. $P(t) = 80{,}000\, e^{\sqrt{t}/2 - 0.09t} = 80{,}000\, e^{\frac{1}{2}t^{1/2} - 0.09t}.$

$P'(t) = 80{,}000(\tfrac{1}{4}t^{-1/2} - 0.09)e^{\frac{1}{2}t^{1/2} - 0.09t}.$

Setting $P'(t) = 0$, we have

$$\tfrac{1}{4}t^{-1/2} = 0.09, \quad t^{-1/2} = 0.36, \quad \dfrac{1}{\sqrt{t}} = 0.36, \quad t = \left(\dfrac{1}{0.36}\right)^{2} \approx 7.72.$$

Evaluating $P(t)$ at each of its endpoints and at the point $t = 7.72$, we find

t	$P(t)$
0	80,000
7.72	160,207.69
8	160,170.71

We conclude that P is optimized at $t = 7.72$. The optimal price is $160,207.69.

59. $P'(t) = 20.6(-0.009)e^{-0.009t} = -0.1854e^{-0.009t}$
$P'(10) = -0.1694$, $P'(20) = -0.1549$, and $P'(30) = -0.1415$,
and this tells us that the percentage of the total population relocating was
decreasing at the rate of 0.17% in 1970, 0.15% in 1980, and 0.14% in 1990.

61. a. The price at $t = 0$ is $8 + 4$, or 12, dollars per unit.

b. $\dfrac{dp}{dt} = -8e^{-2t} + e^{-2t} - 2te^{-2t}$.

$\left. \dfrac{dp}{dt} \right|_{t=0} = -8e^{-2t} + e^{-2t} - 2te^{-2t} \Big|_{t=0} = -8 + 1 = -7.$

That is, the price is decreasing at the rate of $7/week.

c. The equilibrium price is $\lim_{t \to \infty}(8 + 4e^{-2t} + te^{-2t}) = 8 + 0 + 0$

or $8 per unit.

63. We are given that

$$c(1 - e^{-at/V}) < m$$

$$1 - e^{-at/V} < \frac{m}{c}$$

$$-e^{-at/V} < \frac{m}{c} - 1$$

$$e^{-at/V} > 1 - \frac{m}{c}.$$

Taking the log of both sides of the inequality, we have

$$-\frac{at}{V}\ln e > \ln\frac{c-m}{c}$$

$$-\frac{at}{V} > \ln\frac{c-m}{c}$$

$$-t > \frac{V}{a}\ln\frac{c-m}{c}$$

or
$$t < \frac{V}{a}\left(-\ln\frac{c-m}{c}\right) = \frac{V}{a}\ln\left(\frac{c}{c-m}\right).$$

Therefore the liquid must not be allowed to enter the organ for a time longer than

$$t = \frac{V}{a}\ln\left(\frac{c}{c-m}\right) \text{ minutes.}$$

65. False. $f(x) = e^{\pi}$ is a constant function and so $f'(x) = 0$.

67. True. Differentiating both sides of the equation with respect to x, we have

$$\frac{d}{dx}(x^2 + e^y) = \frac{d}{dx}(10)$$

$$2x + e^y\frac{dy}{dx} = 0$$

and so
$$\frac{dy}{dx} = -\frac{2x}{e^y}.$$

USING TECHNOLOGY EXERCISES 5.4, page 414

1. 5.4366   3. 12.3929   5. 0.1861

7. a. The initial population of crocodiles is $P(0) = \frac{300}{6} = 50$.

   b. $\displaystyle\lim_{t\to 0} P(t) = \lim_{t\to 0}\frac{300e^{-0.024t}}{5e^{-0.024t}+1} = \frac{0}{0+1} = 0.$

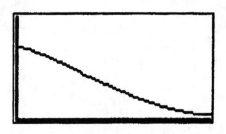

c.

9. a.   b. 4.2720 billion/half century

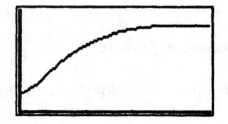

11. a. Using the function evaluation capabilities of a graphing utility, we find
          $f(11) = 153.024$ and $g(11) = 235.180977624$
    and this tells us that the number of violent-crime arrests will be 153,024 at the
    beginning of the year 2000, but if trends like inner-city drug use and wider
    availability of guns continue, then the number of arrests will be 235,181.
    b. Using the differentiation capability of a graphing utility, we find
          $f'(11) = -0.634$ and $g'(11) = 18.4005596893$
    and this tells us that the number of violent-crime arrests will be decreasing at the
    rate of 634 per year at the beginning of the year 2000. But if the trends like inner-
    city drug use and wider availability of guns continues, then the number of arrests
    will be increasing at the rate of 18,400 per year at the beginning of the year 2000.

13. a. $P(10) = \dfrac{74}{1+2.6e^{-0.166(10)+0.04536(10)^2-0.0066(10)^3}} \approx 69.63$ percent.

    b. $P'(10) = 5.09361$, or 5.09361%/decade

EXERCISES 5.5, page 423

1. $f(x) = 5 \ln x; f'(x) = 5\left(\dfrac{1}{x}\right) = \dfrac{5}{x}.$

3. $f(x) = \ln (x + 1); f'(x) = \dfrac{1}{x+1}.$

5. $f(x) = \ln x^8; f'(x) = \dfrac{8x^7}{x^8} = \dfrac{8}{x}.$

7. $f(x) = \ln x^{1/2} ; \quad f'(x) = \dfrac{\frac{1}{2}x^{-1/2}}{x^{1/2}} = \dfrac{1}{2x}.$

9. $f(x) = \ln\left(\dfrac{1}{x^2}\right) = \ln x^{-2} = -2 \ln x; \quad f'(x) = -\dfrac{2}{x}.$

11. $f(x) = \ln (4x^2 - 6x + 3); \quad f'(x) = \dfrac{8x-6}{4x^2 - 6x + 3} = \dfrac{2(4x-3)}{4x^2 - 6x + 3}.$

13. $f(x) = \ln\left(\dfrac{2x}{x+1}\right) = \ln 2x - \ln (x + 1).$

$$f'(x) = \dfrac{2}{2x} - \dfrac{1}{x+1} = \dfrac{2(x+1)-2x}{2x(x+1)} = \dfrac{2x+2-2x}{2x(x+1)}$$

$$= \dfrac{2}{2x(x+1)} = \dfrac{1}{x(x+1)}.$$

15. $f(x) = x^2 \ln x ; \; f'(x) = x^2\left(\dfrac{1}{x}\right) + (\ln x)(2x) \; = x + 2x \ln x = x(1 + 2 \ln x)$

17. $f(x) = \dfrac{2 \ln x}{x}. \quad f'(x) = \dfrac{x\left(\frac{2}{x}\right) - 2 \ln x}{x^2} = \dfrac{2(1 - \ln x)}{x^2}.$

19. $f(u) = \ln (u - 2)^3; f'(u) = \dfrac{3(u-2)^2}{(u-2)^3} = \dfrac{3}{u-2}.$

21. $f(x) = (\ln x)^{1/2}$ and $f'(x) = \frac{1}{2}(\ln x)^{-1/2}\left(\frac{1}{x}\right) = \frac{1}{2x\sqrt{\ln x}}$.

23. $f(x) = (\ln x)^3$; $f'(x) = 3(\ln x)^2\left(\frac{1}{x}\right) = \frac{3(\ln x)^2}{x}$.

25. $f(x) = \ln(x^3 + 1)$; $f'(x) = \frac{3x^2}{x^3 + 1}$.

27. $f(x) = e^x \ln x$. $f'(x) = e^x \ln x + e^x\left(\frac{1}{x}\right) = \frac{e^x(x \ln x + 1)}{x}$.

29. $f(t) = e^{2t} \ln(t + 1)$

$f'(t) = e^{2t}\left(\frac{1}{t+1}\right) + \ln(t+1)\cdot(2e^{2t})$

$= \frac{[2(t+1)\ln(t+1) + 1]e^{2t}}{t+1}$.

31. $f(x) \quad \frac{\ln x}{x}$. $f'(x) = \frac{x(\frac{1}{x}) - \ln x}{x^2} = \frac{1 - \ln x}{x^2}$.

33. $f(x) = \ln 2 + \ln x$; So $f'(x) = \frac{1}{x}$ and $f''(x) = -\frac{1}{x^2}$.

35. $f(x) = \ln(x^2 + 2)$; $f'(x) = \frac{2x}{(x^2 + 2)}$ and

$f''(x) = \frac{(x^2 + 2)(2) - 2x(2x)}{(x^2 + 2)^2} = \frac{2(2 - x^2)}{(x^2 + 2)^2}$.

37. $y = (x + 1)^2(x + 2)^3$

$\ln y = \ln(x + 1)^2(x + 2)^3 = \ln(x + 1)^2 + \ln(x + 2)^3$

$= 2\ln(x + 1) + 3\ln(x + 2)$.

$\frac{y'}{y} = \frac{2}{x+1} + \frac{3}{x+2} = \frac{2(x+2) + 3(x+1)}{(x+1)(x+2)} = \frac{5x+7}{(x+1)(x+2)}$.

$$y' = \frac{(5x+7)(x+1)^2(x+2)^3}{(x+1)(x+2)} = (5x+7)(x+1)(x+2)^2.$$

39. $y = (x-1)^2(x+1)^3(x+3)^4$

$\ln y = 2 \ln (x-1) + 3 \ln (x+1) + 4 \ln (x+3)$

$$\frac{y'}{y} = \frac{2}{x-1} + \frac{3}{x+1} + \frac{4}{x+3}$$

$$= \frac{2(x+1)(x+3) + 3(x-1)(x+3) + 4(x-1)(x+1)}{(x-1)(x+1)(x+3)}$$

$$= \frac{2x^2 + 8x + 6 + 3x^2 + 6x - 9 + 4x^2 - 4}{(x-1)(x+1)(x+3)} = \frac{9x^2 + 14x - 7}{(x-1)(x+1)(x+3)}.$$

Therefore,

$$y' = \frac{9x^2 + 14x - 7}{(x-1)(x+1)(x+3)} \cdot y$$

$$= \frac{(9x^2 + 14x - 7)(x-1)^2(x+1)^3(x+3)^4}{(x-1)(x+1)(x+3)}$$

$$= (9x^2 + 14x - 7)(x-1)(x+1)^2(x+3)^3.$$

41. $y = \dfrac{(2x^2-1)^5}{\sqrt{x+1}}.$

$$\ln y = \ln \frac{(2x^2-1)^5}{(x+1)^{1/2}} = 5 \ln(2x^2-1) - \frac{1}{2}\ln(x+1)$$

So $\dfrac{y'}{y} = \dfrac{20x}{2x^2-1} - \dfrac{1}{2(x+1)} = \dfrac{40x(x+1) - (2x^2-1)}{2(2x^2-1)(x+1)}$

$$= \frac{38x^2 + 40x + 1}{2(2x^2-1)(x+1)}.$$

$$y' = \frac{38x^2 + 40x + 1}{2(2x^2-1)(x+1)} \cdot \frac{(2x^2-1)^5}{\sqrt{x+1}} = \frac{(38x^2 + 40x + 1)(2x^2-1)^4}{2(x+1)^{3/2}}.$$

43. $y = 3^x$ $\qquad \ln y = x \ln 3$

$$\frac{1}{y} \cdot \frac{dy}{dx} = \ln 3$$

$$\frac{dy}{dx} = y \ln 3 = 3^x \ln 3.$$

45. $y = (x^2 + 1)^x$; $\ln y = \ln (x^2 + 1)^x = x \ln (x^2 + 1)$. So

$$\frac{y'}{y} = \ln(x^2 + 1) + x\left(\frac{2x}{x^2 + 1}\right) = \frac{(x^2 + 1)\ln(x^2 + 1) + 2x^2}{x^2 + 1}.$$

$$y' = \frac{[(x^2 + 1)\ln(x^2 + 1) + 2x^2](x^2 + 1)^x}{x^2 + 1}$$

47. $y = x \ln x$. The slope of the tangent line at any point is

$$y' = \ln x + x\left(\frac{1}{x}\right) = \ln x + 1.$$

In particular, the slope of the tangent line at $(1,0)$ where $x = 1$ is $m = \ln 1 + 1 = 1$. So, an equation of the tangent line is

$$y - 0 = 1(x - 1) \quad \text{or} \quad y = x - 1.$$

49. $f(x) = \ln x^2 = 2 \ln x$ and so $f'(x) = 2/x$. Since $f'(x) < 0$ if $x < 0$, and $f'(x) > 0$ if $x > 0$, we see that f is decreasing on $(-\infty,0)$ and increasing on $(0,\infty)$.

51. $f(x) = x^2 + \ln x^2$; $f'(x) = 2x + \frac{2x}{x^2} = 2x + \frac{2}{x}$.

$$f''(x) = 2 - \frac{2}{x^2}.$$

To find the intervals of concavity for f, we first set $f''(x) = 0$ giving

$$2 - \frac{2}{x^2} = 0, \quad 2 = \frac{2}{x^2}, \quad 2x^2 = 2$$

or $\qquad\qquad\qquad x^2 = 1 \text{ and } x = \pm 1.$

Next, we construct the sign diagram for f''

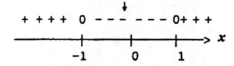

and conclude that f is concave upward on $(-\infty,-1) \cup (1,\infty)$ and concave downward on $(-1,0) \cup (0,1)$.

53. $f(x) = \ln(x^2 + 1)$.

$$f'(x) = \frac{2x}{x^2 + 1}; \; f''(x) = \frac{(x^2 + 1)(2) - (2x)(2x)}{(x^2 + 1)^2} = -\frac{2(x^2 - 1)}{(x^2 + 1)^2}.$$

Setting $f''(x) = 0$ gives $x = \pm 1$ as candidates for inflection points of f. From the sign diagram for f''

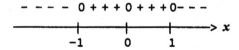

we see that $(-1, \ln 2)$ and $(1, \ln 2)$ are inflection points of f.

55. $f(x) = x - \ln x; \quad f'(x) = 1 - \dfrac{1}{x} = \dfrac{x - 1}{x} = 0$ if $x = 1$, a critical point of f.

x	1/2	1	3
$f(x)$	$1/2 + \ln 2$	1	$3 - \ln 3$

From the table, we see that f has an absolute minimum at $(1,1)$ and an absolute maximum at $(3, 3 - \ln 3)$.

57. $f(x) = \ln(x - 1)$.
   1. The domain of f is obtained by requiring that $x - 1 > 0$. We find the domain to be $(1, \infty)$.
   2. Since $x \neq 0$, there are no y-intercepts. Next, setting $y = 0$ gives $x - 1 = 1$ or $x = 2$ as the x-intercept.
   3. $\lim\limits_{x \to 1^+} \ln(x - 1) = -\infty$.
   4. There are no horizontal asymptotes. Observe that $\lim\limits_{x \to 1^+} \ln(x - 1) = -\infty$ so $x = 1$
   is a vertical asymptote.
   5. $f'(x) = \dfrac{1}{x - 1}.$
   The sign diagram for f' is

f' is not defined here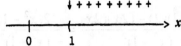

We conclude that *f* is increasing on $(1,\infty)$.

6. The results of (5) show that *f* is increasing on $(1,\infty)$.

7. $f''(x) = -\dfrac{1}{(x-1)^2}$. Since $f''(x) < 0$ for $x > 1$, we see that *f* is concave downward on $(1,\infty)$.

8. From the results of (7), we see that *f* has no inflection points. The graph of *f* follows.

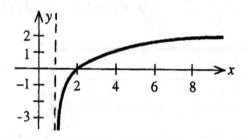

59. False. ln 5 is a constant function and $f'(x) = 0..$

61. If $x \le 0$, then $|x| = -x$. Therefore, $\ln |x| = \ln(-x)$. Writing $f(x) = \ln |x|$ we have
$$|x| = -x = e^{f(x)}.$$
Differentiating both sides with respect to *x* and using the Chain Rule, we have
$$-1 = e^{f(x)} \cdot f'(x) \quad \text{or} \quad f'(x) = -\frac{1}{e^{f(x)}} = -\frac{1}{-x} = \frac{1}{x}.$$

EXERCISES 5.6 , page 433

1. a. The growth constant is $k = 0.05$. b. Initially, the quantity present is 400 units.
    c.

t	0	10	20	100	1000
Q	400	659	1087	59365	2.07×10^{24}

3. a. $Q(t) = Q_0 e^{kt}$. Here $Q_0 = 100$ and so $Q(t) = 100 e^{kt}$. Since the number of cells doubles in 20 minutes, we have
$$Q(20) = 100 e^{20k} = 200, \quad e^{20k} = 2, \quad 20k = \ln 2, \text{ or } k = \tfrac{1}{20}\ln 2 \approx 0.03466.$$
$$Q(t) = 100 e^{0.03466t}$$
    b. We solve the equation $100 e^{0.03466t} = 1{,}000{,}000$. We obtain
$$e^{0.03466t} = 10000 \text{ or } 0.03466t = \ln 10000,$$
$$t = \frac{\ln 10{,}000}{0.03466} \approx 266, \quad \text{or } 266 \text{ minutes.}$$
    c. $Q(t) = 1000 e^{0.03466t}$.

5. a. We solve the equation
$$5.3 e^{0.0198t} = 3(5.3) \text{ or } e^{0.0198t} = 3,$$
or $\qquad 0.0198t = \ln 3 \text{ and } t = \dfrac{\ln 3}{0.0198} \approx 55.5.$
    So the world population will triple in approximately 55.5 years.
    b. If the growth rate is 1.8 percent, then proceeding as before, we find
$$k = \ln 1.018 \approx 0.0178.$$
    So $N(t) = 5.3 e^{0.0178t}$. If $t = 55.5$, the population would be
$$N(55.5) = 5.3 e^{0.0178(55.5)} \approx 14.25,$$
    or approximately 14.25 billion.

7. $P(h) = p_0 e^{-kh}$, $P(0) = 15$, therefore, $p_0 = 15$.
$$P(4000) = 15 e^{-4000k} = 12.5$$
$$e^{-4000k} = \frac{12.5}{15},$$
$$-4000k = \ln\left(\frac{12.5}{15}\right) \quad \text{and } k = 0.00004558.$$
    Therefore, $P(12{,}000) = 15 e^{-0.00004558(12{,}000)} = 8.68$, or 8.7 lb/sq in.
    The rate of change of the atmospheric pressure with respect to altitude is given by
$$P'(h) = \frac{d}{dh}(15 e^{-0.00004558h}) = -0.0006837 e^{-0.00004558h}.$$
    So, the rate of change of the atmospheric pressure with respect to altitude when the altitude is 12,000 feet is
$$P'(12{,}000) = -0.0006837 e^{-0.00004558(12{,}000)} \approx -0.00039566.$$

That is, it is dropping at the rate of approximately 0.0004 lbs per square inch/foot.

9. Suppose the amount of phosphorus 32 at time t is given by
$$Q(t) = Q_0 e^{-kt}$$
where Q_0 is the amount present initially and k is the decay constant. Since this element has a half–life of 14.2 days, we have
$$\tfrac{1}{2}Q_0 = Q_0 e^{-14.2k}, \quad e^{-14.2k} = \tfrac{1}{2}, \quad -14.2k = \ln\tfrac{1}{2}, \quad k = \frac{\frac{1}{2}}{14.2} \approx 0.0488.$$
Therefore, the amount of phosphorus 32 present at any time t is given by
$$Q(t) = 100 e^{-0.0488t}$$
The amount left after 7.1 days is given by
$$Q(7.1) = 100 e^{-0.0488(7.1)} = 100 e^{-0.3465}$$
$$= 70.617, \text{ or } 70.617 \text{ grams.}$$
The rate at which the phosphorus 32 is decaying when $t = 7.1$ is given by
$$Q'(t) = \frac{d}{dt}[100 e^{-0.0488t}] = 100(-0.0488)e^{-0.0488t} = -4.88 e^{-0.0488t}.$$
Therefore, $Q'(7.1) = -4.88 e^{-0.0488(7.1)} \approx -3.451.$
That is, it is changing at the rate of 3.451 gms/day.

11. We solve the equation
$$0.2Q_0 = Q_0 e^{-0.00012t}$$
obtaining $\quad t = \dfrac{\ln 0.2}{-0.00012} \approx 13{,}412, \text{ or approximately } 13{,}412 \text{ years.}$

13. The graph of $Q(t)$ follows.

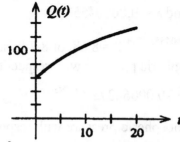

a. $Q(0) = 120(1 - e^0) + 60 = 60$, or 60 w.p.m.
b. $Q(10) = 120(1 - e^{-0.5}) + 60 = 107.22$, or approximately 107 w.p.m.
c. $Q(20) = 120(1 - e^{-1}) + 60 = 135.65$, or approximately 136 w.p.m.

15. The graph of $D(t)$ follows.

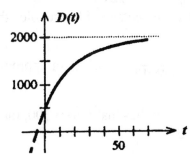

a. After one month, the demand is $D(1) = 2000 - 1500e^{-0.05} \approx 573$.
After twelve months, the demand is $D(12) = 2000 - 1500e^{-0.6} \approx 1177$.
After twenty-four months the demand is
$$D(24) = 2000 - 1500e^{-1.2} \approx 1548.$$
After sixty months, the demand is
$$D(60) = 2000 - 1500e^{-3} \approx 1925.$$
b. $\qquad \lim_{t \to \infty} D(t) = \lim_{t \to \infty} 2000 - 1500e^{-0.05t} = 2000$

and we conclude that the demand is expected to stabilize at 2000 computers per month.

c. $D'(t) = -1500e^{-0.05t}(-0.05) = 75e^{-0.05t}$.
Therefore, the rate of growth after ten months is given by
$$D'(10) = 75e^{-0.5} \approx 45.49,$$
or approximately 46 computers per month.

17. a. $Q(1) = \dfrac{1000}{1 + 199e^{-0.8}} \approx 11.06$, or 11 children.

b. $Q(10) = \dfrac{1000}{1 + 199e^{-8}} \approx 937.4$, or 937 children.

c. $\lim_{t \to \infty} \dfrac{1000}{1 + 199e^{-0.8t}} = 1000$, or 1000 children.

19. $P(t) = \dfrac{68}{1 + 21.67e^{-0.62t}}$.
The percentage of households that owned VCRs at the beginning of 1985 is given by

$$P(0) = \frac{68}{1+21.67e^{-0.62(0)}} = \frac{68}{22.67} \approx 3,$$

or approximately 3 percent. The percentage of households that owned VCRs at the beginning of 1995 is given by

$$P(10) = \frac{68}{1+21.67e^{-0.62(10)}} \approx 65.14, \quad \text{or approximately 65.14 percent.}$$

21. The first of the given conditions implies that $f(0) = 300$, that is,

$$300 = \frac{3000}{1+Be^0} = \frac{3000}{1+B}.$$

So $1+B = 10$, or $B = 9$. Therefore,

$$f(t) = \frac{3000}{1+9e^{-kt}}.$$

Next, the condition $f(2) = 600$ gives the equation

$$600 = \frac{3000}{1+9e^{-2k}}, \quad 1+9e^{-2k} = 5, \quad e^{-2k} = \frac{4}{9}, \quad \text{or } k = -\frac{1}{2}\ln\left(\frac{4}{9}\right).$$

Therefore, $f(t) = \dfrac{3000}{1+9e^{(1/2)t \cdot \ln(4/9)}} = \dfrac{3000}{1+9\left(\frac{4}{9}\right)^{t/2}}.$

The number of students who had heard about the policy four hours later is given by

$$f(4) = \frac{3000}{1+9\left(\frac{4}{9}\right)^2} = 1080, \quad \text{or 1080 students.}$$

To find the rate at which the rumor was spreading at any time time, we compute

$$f'(t) = \frac{d}{dt}\left[3000(1+9e^{-0.405465t})^{-1}\right]$$

$$= (3000)(-1)(1+9e^{-0.405465})^{-2}\frac{d}{dt}(9e^{-0.405465t})$$

$$= -3000(9)(-0.405465)e^{-0.405465t}(1+9e^{-0.405465t})^{-2}$$

$$= \frac{10947.555\, e^{-0.405465t}}{(1+9e^{-0.405465t})^2}$$

In particular, the rate at which the rumor was spreading 4 hours after the cermoney is given by $f'(4) = \dfrac{10947.555e^{-0.405465(4)}}{(1+9e^{-0.405465(4)})^2} \approx 280.25737.$

So , the rumor is spreading at the rate of 280 students per hour.

23. a. $\lim\limits_{t \to \infty}\left\{ \dfrac{r}{k} - \left[\left(\dfrac{r}{k} \right) - C_0 \right] e^{-kt} \right\} = \dfrac{r}{k},$ and this shows that in the long run the

concentration of the glucose solution approaches r/k.

b.

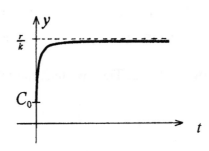

CHAPTER 5 REVIEW EXERCISES, page 437

1.   a-b

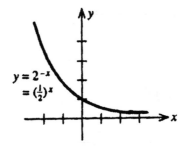

Since $y = \left(\dfrac{1}{2} \right)^x = \dfrac{1}{2^x} = 2^{-x}$, it has the same graph as that of $y = 2^{-x}$.

3. $16^{-3/4} = 0.125$ is equivalent to $-\dfrac{3}{4} = \log_{16} 0.125.$

5.
$$\ln (x - 1) + \ln 4 = \ln (2x + 4) - \ln 2$$
$$\ln (x - 1) - \ln (2x + 4) = -\ln 2 - \ln 4 = -(\ln 2 + \ln 4)$$
$$\ln \left(\frac{x - 1}{2x + 4} \right) = -\ln 8 = \ln \tfrac{1}{8}\,.$$
$$\left(\frac{x - 1}{2x + 4} \right) = \frac{1}{8}$$

$$8x - 8 = 2x + 4$$
$$6x = 12, \text{ or } x = 2.$$

CHECK: l.h.s. $\ln(2-1) + \ln 4 = \ln 4$

r.h.s $\ln(4+4) - \ln 2 = \ln 8 - \ln 2 = \ln \frac{8}{2} = \ln 4$.

7. $\ln 3.6 = \ln \frac{36}{10} = \ln 36 - \ln 10 = \ln 6^2 - \ln 2 \cdot 5 = 2\ln 6 - \ln 2 - \ln 5$
    $= 2(\ln 2 + \ln 3) - \ln 2 - \ln 5 = 2(x + y) - x - z = x + 2y - z$.

9. We first sketch the graph of $y = 2^x - 3$. Then we take the reflection of this graph with respect to the line $y = x$.

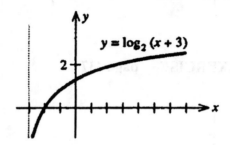

11. $f(x) = xe^{2x}; f'(x) = e^{2x} + xe^{2x}(2) = (1 + 2x)e^{2x}$.

13. $g(t) = \sqrt{t}e^{-2t}; g'(t) = \frac{1}{2}t^{-1/2}e^{-2t} + \sqrt{t}e^{-2t}(-2) = \dfrac{1 - 4t}{2\sqrt{t}e^{2t}}$.

15. $y = \dfrac{e^{2x}}{1 + e^{-2x}}; y' = \dfrac{(1+e^{-2x})e^{2x}(2) - e^{2x} \cdot e^{-2x}(-2)}{(1+e^{-2x})^2} = \dfrac{2(e^{2x}+2)}{(1+e^{-2x})^2}$.

17. $f(x) = xe^{-x^2}; f'(x) = e^{-x^2} + xe^{-x^2}(-2x) = (1 - 2x^2)e^{-x^2}$.

19. $f(x) = x^2e^x + e^x$;
    $f'(x) = 2xe^x + x^2e^x + e^x = (x^2 + 2x + 1)e^x = (x + 1)^2 e^x$.

21. $f(x) = \ln(e^{x^2} + 1)$; $f'(x) = \dfrac{e^{x^2}(2x)}{e^{x^2}+1} = \dfrac{2xe^{x^2}}{e^{x^2}+1}$.

23. $f(x) = \dfrac{\ln x}{x+1}$. $f'(x) = \dfrac{(x+1)\left(\dfrac{1}{x}\right) - \ln x}{(x+1)^2} = \dfrac{1+\dfrac{1}{x}-\ln x}{(x+1)^2} = \dfrac{x - x\ln x + 1}{x(x+1)^2}$.

25. $y = \ln(e^{4x}+3)$; $y' = \dfrac{e^{4x}(4)}{e^{4x}+3} = \dfrac{4e^{4x}}{e^{4x}+3}$.

27. $f(x) = \dfrac{\ln x}{1+e^x}$;

$$f'(x) = \dfrac{(1+e^x)\dfrac{d}{dx}\ln x - \ln x\dfrac{d}{dx}(1+e^x)}{(1+e^x)^2} = \dfrac{(1+e^x)\left(\dfrac{1}{x}\right) - (\ln x)e^x}{(1+e^x)^2}$$

$$= \dfrac{1+e^x - xe^x \ln x}{x(1+e^x)^2} = \dfrac{1+e^x(1 - x\ln x)}{x(1+e^x)^2}.$$

29. $y = \ln(3x+1)$; $y' = \dfrac{3}{3x+1}$;

$$y'' = 3\dfrac{d}{dx}(3x+1)^{-1} = -3(3x+1)^{-2}(3) = -\dfrac{9}{(3x+1)^2}.$$

31. $h'(x) = g'(f(x))f'(x)$. But $g'(x) = 1 - \dfrac{1}{x^2}$ and $f'(x) = e^x$.

So $f(0) = e^0 = 1$ and $f'(0) = e^0 = 1$. Therefore,
$h'(0) = g'(f(0))f'(0) = g'(1)f'(0) = 0 \cdot 1 = 0$.

33. $y = (2x^3 + 1)(x^2 + 2)^3$. $\ln y = \ln(2x^3 + 1) + 3\ln(x^2 + 2)$.

$$\dfrac{y'}{y} = \dfrac{6x^2}{2x^3+1} + \dfrac{3(2x)}{x^2+2} = \dfrac{6x^2(x^2+2) + 6x(2x^3+1)}{(2x^3+1)(x^2+2)}$$

$$= \frac{6x^4 + 12x^2 + 12x^4 + 6x}{(2x^3 + 1)(x^2 + 2)} = \frac{18x^4 + 12x^2 + 6x}{(2x^3 + 1)(x^2 + 2)}.$$

Therefore, $y' = 6x(3x^3 + 2x + 1)(x^2 + 2)^2$.

35. $y = e^{-2x}$. $y' = -2e^{-2x}$ and this gives the slope of the tangent line to the graph of $y = e^{-2x}$ at any point (x, y). In particular, the slope of the tangent line at $(1, e^{-2})$ is $y'(1) = -2e^{-2}$. The required equation is $y - e^{-2} = -2e^{-2}(x - 1)$ or

$$y = \frac{1}{e^2}(-2x + 3).$$

37. $f(x) = xe^{-2x}$.

We first gather the following information on f.

1. The domain of f is $(-\infty, \infty)$.
2. Setting $x = 0$ gives 0 as the y-intercept.
3. $\lim\limits_{x \to -\infty} xe^{-2x} = -\infty$ and $\lim\limits_{x \to \infty} xe^{-2x} = 0$.
4. The results of (3) show that $y = 0$ is a horizontal asymptote.
5. $f'(x) = e^{-2x} + xe^{-2x}(-2) = (1 - 2x)e^{-2x}$. Observe that $f'(x) = 0$ if $x = 1/2$, a critical point of f. The sign diagram of f'

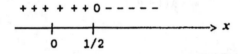

shows that f is increasing on $(-\infty, \frac{1}{2})$ and decreasing on $(\frac{1}{2}, \infty)$.

6. The results of (5) show that $(\frac{1}{2}, \frac{1}{2}e^{-1})$ is a relative maximum.

7. $f''(x) = -2e^{-2x} + (1 - 2x)e^{-2x}(-2) = 4(x - 1)e^{-2x}$ and is equal to zero if $x = 1$. The sign diagram of f''

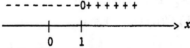

shows that the graph of f is concave downward on $(-\infty, 1)$ and concave upward on $(1, \infty)$.

The graph of f follows.

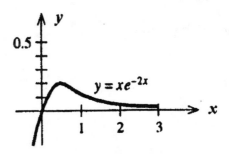

39. $f(t) = te^{-t}$. $f'(t) = e^{-t} + t(-e^{-t}) = e^{-t}(1 - t)$. Setting $f'(t) = 0$ gives $t = 1$ as the only critical point of f. From the sign diagram of f'

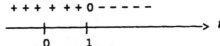

we see that $f(1) = e^{-1} = 1/e$ is the absolute maximum value of f.

41. We want to find r where r satisfies the equation $8.2 = 4.5\,e^{r(5)}$. We have

$$e^{5r} = \frac{8.2}{4.5} \quad \text{or} \quad r = \frac{1}{5}\ln\left(\frac{8.2}{4.5}\right) \approx 0.12$$

and so the annual rate of return is 12 percent per year.

43. a. $Q(t) = 2000e^{kt}$. Now $Q(120) = 18{,}000$ gives $2000e^{120k} = 18{,}000$, $e^{120k} = 9$, or $120k = \ln 9$. So $k = \frac{1}{120}\ln 9 \approx 0.01831$ and $Q(t) = 2000e^{0.01831t}$.

b. $Q(4) = 2000e^{0.01831(240)} \approx 161{,}992$, or approximately 162,000.

45.

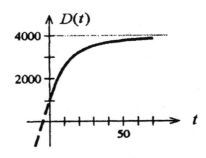

a. $D(1) = 4000 - 3000\,e^{-0.06} = 1175$, $D(12) = 4000 - 3000\,e^{-0.72} = 2540$, and

$$D(24) = 4000 - 3000\, e^{-1.44} = 3289.$$

b. $\lim\limits_{t \to \infty} D(t) = \lim\limits_{t \to \infty} (4000 - 3000e^{-0.06t}) = 4000.$

CHAPTER 6

EXERCISES 6.1, page 450

1. $F(x) = \frac{1}{3}x^3 + 2x^2 - x + 2$; $F'(x) = x^2 + 4x - 1 = f(x)$.

3. $F(x) = (2x^2 - 1)^{1/2}$; $F'(x) = \frac{1}{2}(2x^2 - 1)^{-1/2}(4x) = 2x(2x^2 - 1)^{-1/2} = f(x)$.

5. a. $G'(x) = \dfrac{d}{dx}(2x) = 2 = f(x)$ 　　　　　　b. $F(x) = G(x) + C = 2x + C$

    c.

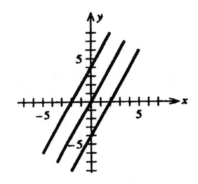

7. a. $G'(x) = \dfrac{d}{dx}(\frac{1}{3}x^3) = x^2 = f(x)$ 　　b. $F(x) = G(x) + C = \frac{1}{3}x^3 + C$

    c.

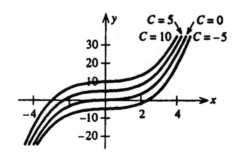

6 Integration

9. $\int 6\,dx = 6x + C.$

11. $\int x^3 dx = \frac{1}{4}x^4 + C$

13. $\int x^{-4} dx = -\frac{1}{3}x^{-3} + C$

15. $\int x^{2/3} dx = \frac{3}{5}x^{5/3} + C$

17. $\int x^{-5/4} dx = -4x^{-1/4} + C$

19. $\int \frac{2}{x^2}\,dx = 2\int x^{-2} dx = 2(-1x^{-1}) + C = -\frac{2}{x} + C$

21. $\int \pi\sqrt{t}\,dt = \pi\int t^{1/2}dt = \pi(\frac{2}{3}t^{3/2}) + C = \frac{2\pi}{3}t^{3/2} + C$

23. $\int (3-2x)\,dx = \int 3\,dx - 2\int x\,dx = 3x - x^2 + C$

25. $\int (x^2 + x + x^{-3})\,dx = \int x^2\,dx + \int x\,dx + \int x^{-3}\,dx = \frac{1}{3}x^3 + \frac{1}{2}x^2 - \frac{1}{2}x^{-2} + C$

27. $\int 4e^x\,dx = 4e^x + C$

29. $\int (1 + x + e^x)\,dx = x + \frac{1}{2}x^2 + e^x + C$

31. $\int (4x^3 - \frac{2}{x^2} - 1)\,dx = \int (4x^3 - 2x^{-2} - 1)\,dx = x^4 + 2x^{-1} - x + C = x^4 + \frac{2}{x} - x + C$

33. $\int (x^{5/2} + 2x^{3/2} - x)\,dx = \frac{2}{7}x^{7/2} + \frac{4}{5}x^{5/2} - \frac{1}{2}x^2 + C$

35. $\int (x^{1/2} + 3x^{-1/2})\,dx = \frac{2}{3}x^{3/2} + 6x^{1/2} + C$

37. $\int \left(\dfrac{u^3+2u^2-u}{3u}\right) du = \dfrac{1}{3}\int (u^2+2u-1)\,du = \dfrac{1}{9}u^3+\dfrac{1}{3}u^2-\dfrac{1}{3}u+C$

39. $\int (2t+1)(t-2)\,dt = \int (2t^2-3t-2)\,dt = \tfrac{2}{3}t^3-\tfrac{3}{2}t^2-2t+C$

41. $\int \dfrac{1}{x^2}(x^4-2x^2+1)\,dx = \int (x^2-2+x^{-2})\,dx = \dfrac{1}{3}x^3-2x-x^{-1}+C$

$\qquad\qquad = \dfrac{1}{3}x^3-2x-\dfrac{1}{x}+C$

43. $\int \dfrac{ds}{(s+1)^{-2}} = \int (s+1)^2\,ds = \int (s^2+2s+1)\,ds = \tfrac{1}{3}s^3+s^2+s+C$

45. $\int (e^t+t^e)\,dt = e^t + \dfrac{1}{e+1}t^{e+1}+C$

47. $\int \left(\dfrac{x^3+x^2-x+1}{x^2}\right) dx = \int \left(x+1-\dfrac{1}{x}+\dfrac{1}{x^2}\right) dx = \dfrac{1}{2}x^2+x-\ln|x|-x^{-1}+C$

49. $\int \left(\dfrac{(x^{1/2}-1)^2}{x^2}\right) dx = \int \left(\dfrac{x-2x^{1/2}+1}{x^2}\right) dx = \int (x^{-1}-2x^{-3/2}+x^{-2})\,dx$

$\qquad\qquad = \ln|x|+4x^{-1/2}-x^{-1}+C = \ln|x|+\dfrac{4}{\sqrt{x}}-\dfrac{1}{x}+C$

51. $\int f'(x)\,dx = \int (2x+1)\,dx = x^2+x+C.$ The condition $f(1)=3$ gives
$f(1) = 1+1+C = 3,$ or $C = 1.$ Therefore, $f(x) = x^2+x+1.$

53. $f'(x) = 3x^2+4x-1;\ f(x) = x^3+2x^2-x+C.$ Using the given initial condition,
we have $\qquad f(2) = 8+2(4)-2+C = 9,$ so $16-2+C = 9,$ or $C = -5.$ Therefore,
$f(x) = x^3+2x^2-x-5.$

55. $f(x) = \int f'(x)\,dx = \int \left(1 + \frac{1}{x^2}\right)dx = \int (1 + x^{-2})\,dx = x - \frac{1}{x} + C.$

Using the given initial condition, we have $f(1) = 1 - 1 + C = 2$, or $C = 2$.

Therefore, $f(x) = x - \frac{1}{x} + 2.$

57. $f(x) = \int \frac{x+1}{x}\,dx = \int \left(1 + \frac{1}{x}\right)dx = x + \ln|x| + C.$ Using the initial condition, we

have $f(1) = 1 + \ln 1 + C = 1 + C = 1$, or $C = 0$. So $f(x) = x + \ln|x|.$

59. $f(x) = \int f'(x)\,dx = \int \frac{1}{2}x^{-1/2}\,dx = \frac{1}{2}(2x^{1/2}) + C = x^{1/2} + C; \ f(2) = \sqrt{2} + C = \sqrt{2}$

implies $C = 0$. So $f(x) = \sqrt{x}.$

61. $f'(x) = e^x + x; \ f(x) = e^x + \frac{1}{2}x^2 + C; f(0) = e^0 + \frac{1}{2}(0) + C = 1 + C$

So $3 = 1 + C$ or $2 = C$. Therefore, $f(x) = e^x + \frac{1}{2}x^2 + 2.$

63. The position of the car is

$s(t) = \int f(t)\,dt = \int 2\sqrt{t}\,dt = \int 2t^{1/2}\,dt = 2(\frac{2}{3}t^{3/2}) + C = \frac{4}{3}t^{3/2} + C.$

$s(0) = 0$ implies $s(0) = C = 0$. So $s(t) = \frac{4}{3}t^{3/2}.$

65. $C(x) = \int C'(x)\,dx = \int (0.000009x^2 - 0.009x + 8)\,dx$

$= 0.000003x^3 - 0.0045x^2 + 8x + k.$

$C(0) = k = 120$ and so $C(x) = 0.000003x^3 - 0.0045x^2 + 8x + 120.$

$C(500) = 0.000003(500)^3 - 0.0045(500)^2 + 8(500) + 120,$ or $\$3370.$

67. $P'(x) = -0.004x + 20, \ P(x) = -0.002x^2 + 20x + C$. Since $C = -16{,}000$, we find
that $P(x) = -0.002x^2 + 20x - 16{,}000$. The company realizes a maximum profit
when $P'(x) = 0$, that is, when $x = 5000$ units. Next,

$P(5000) = -0.002(5000)^2 + 20(5000) - 16{,}000 = 34{,}000.$

Thus, a maximum profit of $\$34{,}000$ is realized at a production level of 5000 units.

69. $N(t) = \int N'(t)\,dt = \int (-3t^2 + 12t + 45)\,dt = -t^3 + 6t^2 + 45t + C$. But $N(0) = C = 0$ and so $N(t) = -t^3 + 6t^2 + 45t$. The number is $N(4) = -4^3 + 6(4)^2 + 45(4) = 212$.

71. The number of new subscribers at any time is
$$N(t) = \int (100 + 210t^{3/4})\,dt = 100t + 120t^{7/4} + C.$$

The given condition implies that $N(0) = 5000$. Using this condition, we find $C = 5000$. Therefore, $N(t) = 100t + 120t^{7/4} + 5000$. The number of subscribers 16 months from now is
$$N(16) = 100(16) + 120(16)^{7/4} + 5000, \text{ or } 21{,}960.$$

73. The rate of change of the population at any time t is $P'(t) = 4500t^{1/2} + 1000$. Therefore, $P(t) = 3000t^{3/2} + 1000t + C$. But $P(0) = 30{,}000$ and this implies that
$$P(t) = 3000t^{3/2} + 1000t + 30{,}000.$$
Finally, the projected population 9 years after the construction has begun is
$$P(9) = 3000(9)^{3/2} + 1000(9) + 30{,}000 = 120{,}000.$$

75. $S'(W) = 0.131773W^{-0.575}; \quad S = \int 0.131773W^{-0.575}\,dW = 0.310054W^{0.425} + C$

$S(70) = 0.310054(70)^{0.425} + C = 1.8867 + C = 1.886277$.
Therefore $C = -0.000007 \approx 0$. $\quad S(75) = 0.310054(75)^{0.425} \approx 1.9424$.

77. $v(r) = \int v'(r)\,dr = \int -kr\,dr = -\tfrac{1}{2}kr^2 + C$.

But $v(R) = 0$ and so $v(R) = -\tfrac{1}{2}kR^2 + C = 0$, or $C = \tfrac{1}{2}kR^2$. Therefore,
$v(R) = -\tfrac{1}{2}kr^2 + \tfrac{1}{2}kR^2 = \tfrac{1}{2}k(R^2 - r^2)$.

79. Denote the constant deceleration by k (ft/sec^2). Then $f''(t) = -k$, so
$f'(t) = v(t) = -kt + C_1$. Next, the given condition implies that $v(0) = 88$. This gives
$C_1 = 88$, or $f'(t) = -kt + 88$.
$$s = f(t) = \int f'(t)\,dt = \int (-kt + 88)\,dt = -\tfrac{1}{2}kt^2 + 88t + C_2.$$

Also, $f(0) = 0$ gives $s = f(t) = -\tfrac{1}{2}kt^2 + 88t$. Since the car is brought to rest in 9 seconds, we have $v(9) = -9k + 88 = 0$, or $k = \tfrac{88}{9}$, or $9\tfrac{7}{9}$. So the deceleration is $9\tfrac{7}{9}$ ft/sec^2. The distance covered is

$$s = f(9) = -\tfrac{1}{2}\left(\tfrac{88}{9}\right)(81) + 88(9) = 396.$$

So the stopping distance is 396 ft.

81. Suppose the acceleration is k ft/sec^2. The distance covered is $s = f(t)$ and satisfies $f''(t) = k$. So $f'(t) = v(t) = \int k\, dt = kt + C_1$. Next, $v(0) = 0$ gives $v(t) = kt$, and

$s = f(t) = \int kt\, dt = \tfrac{1}{2}kt^2 + C_2$. Also, $f(0) = 0$ gives $s = \tfrac{1}{2}kt^2$. If it travelled 800 ft,

we have $800 = \dfrac{1}{2}kt^2$, or $t = \dfrac{40}{\sqrt{k}}$. Its speed at this time is

$$v(t) = kt = k\left(\frac{40}{\sqrt{k}}\right) = 40\sqrt{k}.$$

We want the speed to be at least 240 ft/sec. So we require $40\sqrt{k} > 240$, or $k > 36$. In other words, the minimum acceleration must be 36 ft/sec^2.

83. True. See proof in Section 6.1 in the text.

85. True. Use the Sum Rule followed by the Constant Multiple Rule.

EXERCISES 6.2, page 463

1. Put $u = 4x + 3$ so that $du = 4\, dx$, or $dx = \tfrac{1}{4}du$. Then

$$\int 4(4x+3)^4\, dx = \int u^4\, du = \tfrac{1}{5}u^5 + C = \tfrac{1}{5}(4x+3)^5 + C.$$

3. Let $u = x^3 - 2x$ so that $du = (3x^2 - 2)\, dx$. Then

$$\int (x^3 - 2x)^2 (3x^2 - 2)\, dx = \int u^2\, du = \tfrac{1}{3}u^3 + C = \tfrac{1}{3}(x^3 - 2x)^3 + C.$$

5. Let $u = 2x^2 + 3$ so that $du = 4x\, dx$. Then

$$\int \frac{4x}{(2x^2+3)^3}\, dx = \int \frac{1}{u^3}\, du = \int u^{-3}\, du = -\tfrac{1}{2}u^{-2} + C = -\frac{1}{2(2x^2+3)^2} + C.$$

7. Put $u = t^3 + 2$ so that $du = 3t^2\, dt$ or $t^2\, dt = \tfrac{1}{3}du$. Then

$$\int 3t^2 \sqrt{t^3 + 2}\, dt = \int u^{1/2}\, du = \tfrac{2}{3}u^{3/2} + C = \tfrac{2}{3}(t^3 + 2)^{3/2} + C$$

9. Let $u = x^2 - 1$ so that $du = 2x\,dx$ and $x\,dx = \frac{1}{2}du$. Then,

$$\int (x^2 - 1)^9\, x\,dx = \int \tfrac{1}{2}u^9\,du = \tfrac{1}{20}u^{10} + C = \tfrac{1}{20}(x^2 - 1)^{10} + C.$$

11. Let $u = 1 - x^5$ so that $du = -5x^4\,dx$ or $x^4\,dx = -\frac{1}{5}du$. Then

$$\int \frac{x^4}{1-x^5}\,dx = -\frac{1}{5}\int \frac{du}{u} = -\frac{1}{5}\ln|u| + C = -\frac{1}{5}\ln\left|1-x^5\right| + C.$$

13. Let $u = x - 2$ so that $du = dx$. Then

$$\int \frac{2}{x-2}\,dx = 2\int \frac{du}{u} = 2\ln|u| + C = \ln u^2 + C = \ln(x-2)^2 + C$$

15. Let $u = 0.3x^2 - 0.4x + 2$. Then $du = (0.6x - 0.4)\,dx = 2(0.3x - 0.2)\,dx$.

$$\int \frac{0.3x - 0.2}{0.3x^2 - 0.4x + 2}\,dx = \int \frac{1}{2u}\,du = \frac{1}{2}\ln|u| + C = \frac{1}{2}\ln(0.3x^2 - 0.4x + 2) + C.$$

17. Let $u = 3x^2 - 1$ so that $du = 6x\,dx$, or $x\,dx = \frac{1}{6}du$. Then

$$\int \frac{x}{3x^2 - 1}\,dx = \frac{1}{6}\int \frac{du}{u} = \frac{1}{6}\ln|u| + C = \frac{1}{6}\ln\left|3x^2 - 1\right| + C.$$

19. Let $u = -2x$ so that $du = -2\,dx$ or $dx = -\frac{1}{2}du$. Then

$$\int e^{-2x}\,dx = -\frac{1}{2}\int e^u\,du = -\frac{1}{2}e^u + C = -\frac{1}{2}e^{-2x} + C.$$

21. Let $u = 2 - x$ so that $du = -\,dx$ or $dx = -\,du$. Then

$$\int e^{2-x}\,dx = -\int e^u\,du = -e^u + C = -e^{2-x} + C.$$

23. Let $u = -x^2$, then $du = -2x\,dx$ or $x\,dx = -\frac{1}{2}du$.

$$\int xe^{-x^2}\,dx = \int -\tfrac{1}{2}e^u\,du = -\tfrac{1}{2}e^u + C = -\tfrac{1}{2}e^{-x^2} + C.$$

25. $\displaystyle\int (e^x - e^{-x})\,dx = \int e^x\,dx - \int e^{-x}\,dx = e^x - \int e^{-x}\,dx.$

To evaluate the second integral on the right, let $u = -x$ so that $du = -dx$ or $dx = -du.$ Therefore,

$\displaystyle\int (e^x - e^{-x})\,dx = e^x + \int e^u\,du = e^x + e^u + C = e^x + e^{-x} + C.$

27. Let $u = 1 + e^x$ so that $du = e^x\,dx.$ Then

$\displaystyle\int \frac{e^x}{1+e^x}\,dx = \int \frac{du}{u} = \ln|u| + C = \ln(1+e^x) + C.$

29. Let $u = \sqrt{x} = x^{1/2}.$ Then $du = \frac{1}{2}x^{-1/2}\,dx$ or $2\,du = x^{-1/2}\,dx.$

$\displaystyle\int \frac{e^{\sqrt{x}}}{\sqrt{x}}\,dx = \int 2e^u\,du = 2e^u + C = 2e^{\sqrt{x}} + C.$

31. Let $u = e^{3x} + x^3$ so that $du = (3e^{3x} + 3x^2)\,dx = 3(e^{3x} + x^2)\,dx$ or $(e^{3x} + x^2)\,dx = \frac{1}{3}\,du.$ Then

$\displaystyle\int \frac{e^{3x} + x^2}{(e^{3x} + x^3)^3}\,dx = \frac{1}{3}\int \frac{du}{u^3} = \frac{1}{3}\int u^{-3}\,du = -\frac{1}{6}u^{-2} + C = -\frac{1}{6(e^{3x} + x^3)^2} + C.$

33. Let $u = e^{2x} + 1,$ so that $du = 2e^{2x}\,dx,$ or $\frac{1}{2}\,du = e^{2x}\,dx.$

$\displaystyle\int e^{2x}(e^{2x} + 1)^3\,dx = \int \frac{1}{2}u^3\,du = \frac{1}{8}u^4 + C = \frac{1}{8}(e^{2x} + 1)^4 + C.$

35. Let $u = \ln 5x$ so that $du = \dfrac{1}{x}\,dx.$ Then

$\displaystyle\int \frac{\ln 5x}{x}\,dx = \int u\,du = \frac{1}{2}u^2 + C = \frac{1}{2}(\ln 5x)^2 + C.$

37. Let $u = \ln x$ so that $du = \frac{1}{x}\,dx.$ Then

$$\int \frac{1}{x \ln x} dx = \int \frac{du}{u} = \ln|u| + C = \ln|\ln x| + C.$$

39. Let $u = \ln x$ so that $du = \frac{1}{x} dx$. Then

$$\int \frac{\sqrt{\ln x}}{x} dx = \int \sqrt{u} \, du = \frac{2}{3} u^{3/2} + C = \frac{2}{3} (\ln x)^{3/2} + C$$

41. $$\int \left(xe^{x^2} - \frac{x}{x^2 + 2} \right) dx = \int xe^{x^2} - \int \frac{x}{x^2 + 2} dx.$$

To evaluate the first integral, let $u = x^2$ so that $du = 2x \, dx$, or $x \, dx = \frac{1}{2} du$. Then

$$\int xe^{x^2} dx = \frac{1}{2} \int e^u \, du + C_1 = \frac{1}{2} e^u + C_1 = \frac{1}{2} e^{x^2} + C_1.$$

To evaluate the second integral, let $u = x^2 + 2$ so that $du = 2x \, dx$, or $x \, dx = \frac{1}{2} du$. Then

$$\int \frac{x}{x^2 + 2} dx = \frac{1}{2} \int \frac{du}{u} = \frac{1}{2} \ln|u| + C_2 = \frac{1}{2} \ln(x^2 + 2) + C_2.$$

Therefore, $$\int \left(xe^{x^2} - \frac{x}{x^2 + 2} \right) dx = \frac{1}{2} e^{x^2} - \frac{1}{2} \ln(x^2 + 2) + C.$$

43. Let $u = \sqrt{x} - 1$ so that $du = \frac{1}{2} x^{-1/2} dx = \frac{1}{2\sqrt{x}} dx$ or $dx = 2\sqrt{x} \, du$.

Also, we have $\sqrt{x} = u + 1$, so that $x = (u + 1)^2 = u^2 + 2u + 1$ and $dx = 2(u + 1) \, du$.
So

$$\int \frac{x+1}{\sqrt{x}-1} dx = \int \frac{u^2 + 2u + 2}{u} \cdot 2(u+1) \, du = 2 \int \frac{(u^3 + 3u^2 + 4u + 2)}{u} du$$

$$= 2 \int \left(u^2 + 3u + 4 + \frac{2}{u} \right) du = 2 \left(\frac{1}{3} u^3 + \frac{3}{2} u^2 + 4u + 2 \ln|u| \right) + C$$

$$= 2 \left[\frac{1}{3} (\sqrt{x} - 1)^3 + \frac{3}{2} (\sqrt{x} - 1)^2 + 4(\sqrt{x} - 1) + 2 \ln|\sqrt{x} - 1| \right] + C.$$

6 Integration

45. Let $u = x - 1$ so that $du = dx$. Also, $x = u + 1$ and so

$$\int x(x-1)^5 \, dx = \int (u+1)u^5 \, du = \int (u^6 + u^5) \, du$$

$$= \frac{1}{7}u^7 + \frac{1}{6}u^6 + C = \frac{1}{7}(x-1)^7 + \frac{1}{6}(x-1)^6 + C$$

$$= \frac{(6x+1)(x-1)^6}{42} + C.$$

47. Let $u = 1 + \sqrt{x}$ so that $du = \frac{1}{2}x^{-1/2} \, dx$ and $dx = 2\sqrt{x} = 2(u-1) \, du$

$$\int \frac{1-\sqrt{x}}{1+\sqrt{x}} \, dx = \int \left(\frac{1-(u-1)}{u}\right) \cdot 2(u-1) \, du = 2 \int \frac{(2-u)(u-1)}{u} \, du$$

$$= 2 \int \frac{-u^2 + 3u - 2}{u} \, du = 2 \int \left(-u + 3 - \frac{2}{u}\right) du = -u^2 + 6u - 4\ln|u| + C$$

$$= -(1+\sqrt{x})^2 + 6(1+\sqrt{x}) - 4\ln(1+\sqrt{x}) + C$$

$$= -1 - 2\sqrt{x} - x + 6 + 6\sqrt{x} - 4\ln(1+\sqrt{x}) + C$$

$$= -x + 4\sqrt{x} + 5 - 4\ln(1+\sqrt{x}) + C.$$

49. $I = \int v^2(1-v)^6 \, dv$. Let $u = 1 - v$, then $du = -dv$. Also, $1 - u = v$, and
$(1-u)^2 = v^2$. Therefore,

$$I = \int -(1 - 2u + u^2)u^6 \, du = \int -(u^6 - 2u^7 + u^8) \, du = -\left(\frac{u^7}{7} - \frac{2u^8}{8} + \frac{u^9}{9}\right) + C$$

$$= -u^7\left(\frac{1}{7} - \frac{1}{4}u + \frac{1}{9}u^2\right) + C = -\frac{1}{252}(1-v)^7[36 - 63(1-v) + 28(1 - 2v + v^2)]$$

$$= -\frac{1}{252}(1-v)^7[36 - 63 + 63v + 28 - 56v + 28v^2]$$

$$= -\frac{1}{252}(1-v)^7(28v^2 + 7v + 1) + C.$$

51. $f(x) = \int f'(x) \, dx = 5\int (2x-1)^4 \, dx$. Let $u = 2x - 1$ so that $du = 2x-1$ so that
$du = 2 \, dx$, or $dx = \frac{1}{2} \, du$. Then

$$f(x) = \frac{5}{2}\int u^4\, du = \frac{1}{2}u^5 + C = \frac{1}{2}(2x-1)^5 + C.$$

Next, $f(1) = 3$ implies $\frac{1}{2} + C = 3$ or $C = \frac{5}{2}$. Therefore,

$$f(x) = \frac{1}{2}(2x-1)^5 + \frac{5}{2}.$$

53. $f(x) = \int -2xe^{-x^2+1}\, dx$. Let $u = -x^2 + 1$ so that $du = -2x\, dx$. Then

$f(x) = \int e^u\, du = e^u + C = e^{-x^2+1} + C$. The condition $f(1) = 0$ implies

$f(1) = 1 + C = 0$, or $C = -1$. Therefore, $f(x) = e^{-x^2+1} - 1$.

55. $N'(t) = 2000(1 + 0.2t)^{-3/2}$. Let $u = 1 + 0.2t$. Then $du = 0.2\, dt$ and $5\, du = dt$. Therefore, $N(t) = (5)(2000)$

$$\int u^{-3/2}\, du = -20,000u^{-1/2} + C = -20,000(1 + 0.2t)^{-1/2} + C.$$

Next, $N(0) = -20,000(1)^{-1/2} + C = 1000$. Therefore, $C = 21,000$ and

$$N(t) = -\frac{20,000}{\sqrt{1 + 0.2t}} + 21,000. \text{ In particular, } N(5) = -\frac{20,000}{\sqrt{2}} + 21,000 \approx 6,858.$$

57. $p(x) = \int -\frac{250x}{(16 + x^2)^{3/2}}\, dx = -250 \int \frac{x}{(16 + x^2)^{3/2}}\, dx.$

Let $u = 16 + x^2$ so that $du = 2x\, dx$ and $x\, dx = \frac{1}{2} du$.

Then $p(x) = -\frac{250}{2} \int u^{-3/2}\, du = (-125)(-2)u^{-1/2} + C = \frac{250}{\sqrt{16 + x^2}} + C.$

$p(3) = \frac{250}{\sqrt{16 + 9}} + C = 50$ implies $C = 0$ and $p(x) = \frac{250}{\sqrt{16 + x^2}}.$

59. Let $u = 2t + 4$, so that $du = 2\, dt$. Then

$$r(t) = \int \frac{30}{\sqrt{2t + 4}}\, dt = 30 \int \frac{1}{2}u^{-1/2}\, du = 30u^{1/2} + C = 30\sqrt{2t + 4} + C.$$

$r(0) = 60 + C = 0$, and $C = -60$. Therefore, $r(t) = 30\left(\sqrt{2t + 4} - 2\right).$ Then

$r(16) = 30\left(\sqrt{36} - 2\right) = 120\,\text{ft}.$ Therefore, the polluted area is
$$\pi r^2 = \pi(120)^2 = 14{,}400\pi, \qquad \text{or } 14{,}400\pi \ \text{sq ft.}$$

61. Let $u = 2.449e^{-0.3277t}$ so that $du = -0.73565373e^{-0.3277t}\,dt$ and $e^{-0.3277t}\,dt = -1.359335\,du.$

Then $h(t) = \displaystyle\int \frac{52.8706e^{-0.3277t}}{(1+2.449e^{-0.3277t})^2}\,dt = (52.8706)(-1.359335)\int \frac{du}{u^2}$

$= 71.86887u^{-1} + C = \dfrac{71.86887}{1+2.449e^{-0.3277t}} + C.$

$h(0) = \dfrac{71.86887}{1+2.449} + C = 19.4, \ \text{ and } \ C = -1.43760.$

Therefore, $h(t) = \dfrac{71.86887}{1+2.449e^{-0.3277t}} - 1.43760,$

and $h(8) = \dfrac{71.86887}{1+2.449e^{-0.3277(8)}} - 1.43760 \approx 59.6, \ \text{ or } \ 59.6 \text{ inches.}$

63. The number of speakers sold at the end of t years is
$$f(t) = \int f'(t)\,dt = \int 2000(3 - 2e^{-t})\,dt = 2000(3t + 2e^{-t}) + C$$

But 2000 pairs of speakers were sold in the first year, and this implies that $f(1) = 2000.$ So $2000 = 2000(3 + 2e^{-1}) + C$ and $C = -5472.$ Therefore,
$$f(t) = 2000(3t + 2e^{-t}) - 5472.$$
The number of pairs sold in the first five years is
$$f(5) = 2000(15 + 2e^{-5}) - 5472 = 24{,}555, \text{ or } 24{,}555 \text{ pairs.}$$

65. $x(t) = \displaystyle\int x'(t)\,dt = \int \frac{1}{V}(ac - bx_0)e^{-bt/V}\,dt = \frac{1}{V}(ac - bx_0)\int e^{-bt/V}\,dt.$

Let $u = -bt/V$, so that $du = -b/V\,dt$ and $dt = -V/b\,du.$ Then
$$x(t) = \frac{1}{V}(ac - bx_0)\int -\frac{V}{b}e^{u}\,du = \left(-\frac{ac}{b} + x_0\right)e^{u} = \left(-\frac{ac}{b} + x_0\right)e^{-bt/V} + C.$$

Since $x(0) = \left(-\dfrac{ac}{b} + x_0\right) + C = x_0, \ C = \dfrac{ac}{b},$ and

$$x(t) = \frac{ac}{b} + \left(x_0 - \frac{ac}{b}\right)e^{-bt/V}.$$

1. $\frac{1}{3}(1.9 + 1.5 + 1.8 + 2.4 + 2.7 + 2.5) = \frac{12.8}{3} \approx 4.27$.

3. a. $A = \frac{1}{2}(2)(6) = 6$ sq units.

   b. $\Delta x = \frac{2}{4} = \frac{1}{2}$; $x_1 = 0$, $x_2 = \frac{1}{2}$, $x_3 = 1$, $x_4 = \frac{3}{2}$.

   $A \approx \frac{1}{2}[3(0) + 3(\frac{1}{2}) + 3(1) + 3(\frac{3}{2})] = \frac{9}{2}$

   $= 4.5$ sq units.

   c. $\Delta x = \frac{2}{8} = \frac{1}{4}$. $x_1 = 0, \dots, x_8 = \frac{7}{4}$.

   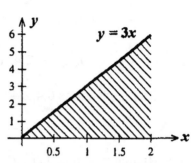

   $A \approx \frac{1}{4}\left[3(0) + 3(\frac{1}{4}) + 3(\frac{1}{2}) + 3(\frac{3}{4}) + 3(1) + 3(\frac{5}{4}) + 3(\frac{3}{2}) + 3(\frac{7}{4})\right]$

   $= \frac{21}{4} = 5.25$ sq units.

   d. Yes.

5. a. $A = 4$ sq units

   b. $\Delta x = \frac{2}{5} = 0.4$; $x_1 = 0$, $x_2 = 0.4$, $x_3 = 0.8$, $x_4 = 1.2$

   $x_5 = 1.6$,

   $A \approx 0.4\{[4 - 2(0)] + [4 - 2(0.4)] + [4 - 2(0.8)]$

   $+ [4 - 2(1.2)] + [4 - 2(1.6)]\}$

   $= 4.8$ sq units

   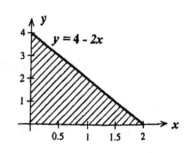

   c. $\Delta x = \frac{2}{10} = 0.2$, $x_1 = 0$, $x_2 = 0.2$, $x_3 = 0.4$, ..., $x_{10} = 1.8$.

   $A \approx 0.2\{[4 - 2(0)] + [4 - 2(0.2)] + [4 - 2(0.4)]$

   $+ \cdots + [4 - 2(1.8)]\} = 4.4$ sq units

   d. Yes.

7. a. $\Delta x = \dfrac{4-2}{2} = 1$; $x_1 = 2.5$, $x_2 = 3.5$;   The Riemann sum is $[(2.5)^2 + (3.5)^2] = 18.5$.

   b. $\Delta x = \dfrac{4-2}{5} = 0.4$; $x_1 = 2.2$, $x_2 = 2.6$, $x_3 = 3.0$, $x_4 = 3.4$, $x_5 = 3.8$

   The Riemann sum is $0.4[2.2^2 + 2.6^2 + 3.0^2 + 3.4^2 + 3.8^2] = 18.64$.

c. $\Delta x = \dfrac{4-2}{10} = 0.2$; $x_1 = 2.1$, $x_2 = 2.3$, $x_2 = 2.5$, ..., $x_{10} = 3.9$

The Riemann sum is $0.2[2.1^2 + 2.3^2 + 2.5^2 + \cdots + 3.9^2] = 18.66$.
The area seems to be $18\frac{2}{3}$ sq units.

9.   a. $\Delta x = \dfrac{4-2}{2} = 1$; $x_1 = 3$, $x_2 = 4$. The Riemann sum is $(1)[3^2 + 4^2] = 25$.

    b. $\Delta x = \dfrac{4-2}{5} = 0.4$; $x_1 = 2.4$, $x_2 = 2.8$, $x_3 = 3.2$, $x_4 = 3.6$, $x_5 = 4$.

The Riemann sum is $0.4[2.4^2 + 2.8^2 + \cdots + 4^2] = 21.12$.

    c. $\Delta x = \dfrac{4-2}{10} = 0.2$; $x_1 = 2.2$, $x_2 = 2.4$, $x_3 = 2.6$, ..., $x_{10} = 4$.

The Riemann sum is $0.2[2.2^2 + 2.4^2 + 2.6^2 + \cdots + 4^2] = 19.88$.
d. 19.9 sq units

11.   a. $\Delta x = \dfrac{1}{2}$, $x_1 = 0$, $x_2 = \dfrac{1}{2}$. The Riemann sum is

$f(x_1)\Delta x + f(x_2)\Delta x = \left[(0)^3 + (\frac{1}{2})^3\right]\frac{1}{2} = \frac{1}{16} = 0.0625$.

    b. $\Delta x = \dfrac{1}{5}$, $x_1 = 0$, $x_2 = \dfrac{1}{5}$, $x_3 = \dfrac{2}{5}$, $x_4 = \dfrac{3}{5}$, $x_5 = \dfrac{4}{5}$. The Riemann sum

    is $f(x_1)\Delta x + f(x_2)\Delta x + \cdots f(x_5)\Delta x = \left[(\frac{1}{5})^3 + (\frac{2}{5})^3 + \cdots + (\frac{4}{5})^3\right]\frac{1}{5} = \frac{100}{625} = 0.16$.

    c. $\Delta x = \dfrac{1}{10}$; $x_1 = 0$, $x_2 = \dfrac{1}{10}$, $x_3 = \dfrac{2}{10}$, ..., $x_{10} = \dfrac{9}{10}$.

The Riemann sum is

$f(x_1)\Delta x + f(x_2)\Delta x + \cdots + f(x_{10})\Delta x = \left[(\frac{1}{10})^3 + (\frac{2}{10})^3 + \cdots + (\frac{9}{10})^3\right]\frac{1}{10}$

$$= \tfrac{2025}{10,000} = 0.2025 \approx 0.2 \text{ sq units.}$$

The Riemann sum seems to approach 0.2.

13. $\Delta x = \dfrac{2-0}{5} = \dfrac{2}{5}$; $x_1 = \dfrac{1}{5}$, $x_2 = \dfrac{3}{5}$, $x_3 = \dfrac{5}{5}$, $x_4 = \dfrac{7}{5}$, $x_5 = \dfrac{9}{5}$.

$A \approx \left\{\left[(\frac{1}{5})^2 + 1\right] + \left[(\frac{3}{5})^2 + 1\right] + \left[(\frac{5}{5})^2 + 1\right] + \left[(\frac{7}{5})^2 + 1\right] + \left[(\frac{9}{5})^2 + 1\right](\frac{2}{5})\right\}$

$= \tfrac{580}{125} = 4.64$ sq units.

15. $\Delta x = \dfrac{3-1}{4} = \dfrac{1}{2};\ x_1 = \dfrac{3}{2},\ x_2 = \dfrac{4}{2},\ x_3 = \dfrac{5}{2},\ x_4 = 3.$

$A \approx \left[\dfrac{1}{\frac{3}{2}} + \dfrac{1}{\frac{4}{2}} + \dfrac{1}{\frac{5}{2}} + \dfrac{1}{3}\right]\dfrac{1}{2} \approx 0.95$ sq units.

17. $A = 20[f(10) + f(30) + f(50) + f(70) + f(90)]$

$\qquad = 20(80 + 100 + 110 + 100 + 80) = 9400$ sq ft.

EXERCISES 6.4, page 486

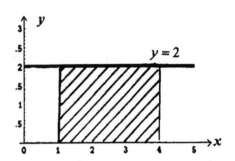

1. $A = \displaystyle\int_1^4 2\,dx = 2x\Big|_1^4 = 2(4-1) = 6,$ or 6 square units. The region is a rectangle whose area is $3 \cdot 2$, or 6, square units.

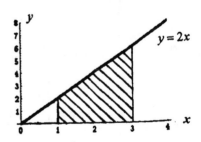

3. $A = \displaystyle\int_1^3 2x\,dx = x^2\Big|_1^3 = 9 - 1 = 8,$ or 8 sq units. The region is a parallelogram of area $(1/2)(3 - 1)(2 + 6) = 8$ sq units.

5. $A = \displaystyle\int_{-1}^2 (2x+3)\,dx = x^2 + 3x\Big|_{-1}^2 = (4+6) - (1-3) = 12,$ or 12 sq. units.

7. $A = \displaystyle\int_{-1}^2 (-x^2 + 4)\,dx = -\dfrac{1}{3}x^3 + 4x\Big|_{-1}^2 = \left(-\dfrac{8}{3}+8\right) - \left(\dfrac{1}{3}-4\right) = 9,$ or 9 sq units.

9. $A = \int_1^2 \frac{1}{x} dx = \ln|x|\Big|_1^2 = \ln 2 - \ln 1 = \ln 2$, or $\ln 2$ sq units.

11. $A = \int_1^9 \sqrt{x}\, dx = \frac{2}{3} x^{3/2}\Big|_1^9 = \frac{2}{3}(27-1) = \frac{52}{3}$, or $17\frac{1}{3}$ sq units.

13. $A = \int_{-8}^{-1}(1-x^{1/3})\, dx = x - \frac{3}{4}x^{4/3}\Big|_{-8}^{-1} = (-1-\frac{3}{4})-(-8-12) = 18\frac{1}{4}$, or $18\frac{1}{4}$ sq units.

15. $A = \int_0^2 e^x dx = e^x\Big|_0^2 = (e^2-1)$, or approximately 6.39 sq units.

17. $\int_2^4 3\, dx = 3x\Big|_2^4 = 3(4-2) = 6.$

19. $\int_1^3 (2x+3)\, dx = x^2 + 3x\Big|_1^3 = (9+9)-(1+3) = 14.$

21. $\int_{-1}^3 2x^2\, dx = \frac{2}{3}x^3\Big|_{-1}^3 = \frac{2}{3}(27)-\frac{2}{3}(-1) = \frac{56}{3}.$

23. $\int_{-2}^2 (x^2-1)\, dx = \frac{1}{3}x^3 - x\Big|_{-2}^2 = \left(\frac{8}{3}-2\right)-\left(-\frac{8}{3}+2\right) = \frac{4}{3}.$

25. $\int_1^8 4x^{1/3}\, dx = (4)(\frac{3}{4})x^{4/3}\Big|_1^8 = 3(16-1) = 45.$

27. $\int_0^1 (x^3 - 2x^2 + 1)\, dx = \frac{1}{4}x^4 - \frac{2}{3}x^3 + x\Big|_0^1 = \frac{1}{4}-\frac{2}{3}+1 = \frac{7}{12}$

29. $\int_2^4 \frac{1}{x} dx = \ln|x|\Big|_2^4 = \ln 4 - \ln 2 = \ln(\frac{4}{2}) = \ln 2.$

31. $\int_0^4 x(x^2-1)\, dx = \int_0^4 (x^3 - x)\, dx = \frac{1}{4}x^4 - \frac{1}{2}x^2\Big|_0^4 = 64-8 = 56.$

33. $\int_1^3 (t^2 - t)^2 \, dt = \int_1^3 t^4 - 2t^3 + t^2) \, dt = \frac{1}{5}t^5 - \frac{1}{2}t^4 + \frac{1}{3}t^3 \Big|_1^3$

$$= \left(\frac{243}{5} - \frac{81}{2} + \frac{27}{3} \right) - \left(\frac{1}{5} - \frac{1}{2} + \frac{1}{3} \right) = \frac{512}{30} = \frac{256}{15}.$$

35. $\int_{-3}^{-1} x^{-2} \, dx = -\frac{1}{x} \Big|_{-3}^{-1} = 1 - \frac{1}{3} = \frac{2}{3}.$

37. $\int_1^4 \left(\sqrt{x} - \frac{1}{\sqrt{x}} \right) dx = \int_1^4 (x^{1/2} - x^{-1/2}) \, dx = \frac{2}{3}x^{3/2} - 2x^{1/2} \Big|_1^4$

$$= \left(\frac{16}{3} - 4 \right) - \left(\frac{2}{3} - 2 \right) = \frac{8}{3}.$$

39. $\int_1^4 \frac{3x^3 - 2x^{2.} + 4}{x^2} \, dx = \int_1^4 (3x - 2 + 4x^{-2}) \, dx = \frac{3}{2}x^2 - 2x - \frac{4}{x} \Big|_1^4$

$$= (24 - 8 - 1) - (\tfrac{3}{2} - 2 - 40 = \tfrac{39}{2}.$$

41. *a.* $C(300) - C(0) = \int_0^{300} (0.0003x^2 - 0.12x + 20) \, dx = 0.0001x^3 - 0.06x^2 + 20x \Big|_0^{300}$

$$= 0.0001(300)^3 - 0.06(300)^2 + 20(300) = 3300.$$

Therefore $C(300) = 3300 + C(0) = 3300 + 800 = 4100$, or $4100.

b. $\int_{200}^{300} C'(x) \, dx = (0.0001x^3 - 0.06x^2 + 20x) \Big|_{200}^{300}$

$$= [0.0001(300)^3 - 0.06(300)^2 + 20(300)]$$
$$-[0.0001(200)^3 - 0.06(200)^2 + 20(200)]$$
$$= 900 \text{ or } \$900.$$

43. *a.* The profit is $\int_0^{200} (-0.0003x^2 + 0.02x + 20) \, dx + P(0)$

$$= -0.0001x^3 + 0.01x^2 + 20x \Big|_0^{200} + P(0)$$
$$= 3600 + P(0) = 3600 - 800, \text{ or } \$2800.$$

243

b. $\int_{200}^{220} P'(x)\,dx = P(220) - P(200) = -0.0001x^3 + 0.01x^2 + 20x\Big|_{200}^{220}$

$\qquad\qquad = 219.20, \text{ or } \$219.20.$

45. The distance is

$$\int_0^{20} v(t)\,dt = \int_0^{20} (-t^2 + 20t + 440)\,dt = -\tfrac{1}{3}t^3 + 10t^2 + 440t\Big|_0^{20}$$

$$\approx 10{,}133\tfrac{1}{3}\text{ ft.}$$

47. The number will be

$$\int (0.00933t^3 + 0.019t^2 - 0.10833t + 1.3467)\,dt$$

$$= 0.0023325t^4 + 0.0063333t^3 - 0.054165t^2 + 1.3467\,t\Big|_0^{10} = 37.7,$$

or approximately 37.7 million Americans.

49. False. The integrand $f(x) = \dfrac{1}{x^3}$ is discontinuous at $x = 0$.

51. False. It is given by $\int_0^{5000} R'(x)\,dx = R(5000) - R(0)$, where $R(x)$ is the total revenue.

USING TECHNOLOGY EXERCISES 6.4, page 489

1. 6.1787   3. 0.7873   5. −0.5888   7. 2.7044

9. 3.9973   11. 37.7 million   13. 333,209   15. 963,213

EXERCISES 6.5, page 493

1. Let $u = x^2 - 1$ so that $du = 2x\,dx$ or $x\,dx = \tfrac{1}{2}\,du$. Also, if $x = 0$, then $u = -1$ and if $x = 2$, then $u = 3$. So

$$\int_0^2 x(x^2 - 1)^3\,dx = \tfrac{1}{2}\int_{-1}^3 u^3\,du = \tfrac{1}{8}u^4\Big|_{-1}^3 = \tfrac{1}{8}(81) - \tfrac{1}{8}(1) = 10.$$

3. Let $u = 5x^2 + 4$ so that $du = 10x\,dx$ or $x\,dx = \tfrac{1}{10}\,du$. Also, if

$x = 0$, then $u = 4$, and if $x = 1$, then $u = 9$. So

$$\int_0^1 x\sqrt{5x^2+4}\,dx = \frac{1}{10}\int_4^9 u^{1/2}\,du = \frac{1}{15}u^{3/2}\Big|_4^9 = \frac{1}{15}(27) - \frac{1}{15}(8) = \frac{19}{15}.$$

5. Let $u = x^3 + 1$ so that $du = 3x^2\,dx$ or $x^2\,dx = \frac{1}{3}\,du$. Also, if $x = 0$, then $u = 1$, and if $x = 2$, then $u = 9$. So,

$$\int_0^2 x^2(x^3+1)^{3/2}\,dx = \frac{1}{3}\int_1^9 u^{3/2}\,du = \frac{2}{15}u^{5/2}\Big|_1^9 = \frac{2}{15}(243) - \frac{2}{15}(1) = \frac{484}{15}.$$

7. Let $u = 2x + 1$ so that $du = 2\,dx$ or $dx = \frac{1}{2}\,du$. Also, if $x = 0$, then $u = 1$ and if $x = 1$ then $u = 3$. So

$$\int_0^1 \frac{1}{\sqrt{2x+1}}\,dx = \frac{1}{2}\int_1^3 \frac{1}{\sqrt{u}}\,du = \frac{1}{2}\int_1^3 u^{-1/2}\,du = u^{1/2}\Big|_1^3 = \sqrt{3} - 1.$$

9. $\int_1^2 (2x-1)^4\,dx$. Put $u = 2x-1$ so that $du = 2\,dx$ or $dx = \frac{1}{2}\,du$.

Then $\int_1^2 (2x-1)^4\,dx = \frac{1}{2}\int_1^3 u^4\,du = \frac{1}{10}u^5\Big|_1^3 = \frac{1}{10}(243-1) = \frac{121}{5} = 24\frac{1}{5}.$

11. Let $u = x^3 + 1$ so that $du = 3x^2\,dx$ or $x^2\,dx = \frac{1}{3}\,du$. Also, if $x = -1$, then $u = 0$ and if $x = 1$, then $u = 2$. So

$$\int_{-1}^1 x^2(x^3+1)^4\,dx = \frac{1}{3}\int_0^2 u^4\,du = \frac{1}{15}u^5\Big|_0^2 = \frac{32}{15}.$$

13. Let $u = x - 1$ so that $du = dx$. Then if $x = 1$, $u = 0$, and if $x = 5$, then $u = 4$.

$$\int_1^5 x\sqrt{x-1}\,dx = \int_0^4 (u+1)u^{1/2}\,du = \int_0^4 (u^{3/2} + u^{1/2})\,du$$
$$= \frac{2}{5}u^{5/2} + \frac{2}{3}u^{3/2}\Big|_0^4 = \frac{2}{5}(32) + \frac{2}{3}(8) = 18\frac{2}{15}.$$

15. Let $u = x^2$ so that $du = 2x\,dx$ or $x\,dx = \frac{1}{2}\,du$. If $x = 0$, $u = 0$ and if $x = 2$, $u = 4$. So

$$\int_0^2 xe^{x^2} dx = \tfrac{1}{2}\int_0^4 e^u \, du = \tfrac{1}{2}e^u\Big|_0^4 = \tfrac{1}{2}(e^4 - 1).$$

17. $\displaystyle\int_0^1 (e^{2x} + x^2 + 1)\, dx = \tfrac{1}{2}e^{2x} + \tfrac{1}{3}x^3 + x\Big|_0^1 = (\tfrac{1}{2}e^2 + \tfrac{1}{3} + 1) - \tfrac{1}{2}$

$$= \tfrac{1}{2}e^2 + \tfrac{5}{6}.$$

19. Put $u = x^2 + 1$ so that $du = 2x\, dx$ or $x\, dx = \tfrac{1}{2}\, du$. Then

$$\int_{-1}^1 xe^{x^2+1} dx = \frac{1}{2}\int_2^2 e^u \, du = \frac{1}{2}e^u\Big|_2^2 = 0$$

(Since the upper and lower limits are equal.)

21. Let $u = x - 2$ so that $du = dx$. If $x = 3$, $u = 1$ and if $x = 6$, $u = 4$. So

$$\int_3^6 \frac{2}{x-2}\, dx = 2\int_1^4 \frac{du}{u} = 2\ln|u|\Big|_1^4 = 2\ln 4.$$

23. Let $u = x^3 + 3x^2 - 1$ so that $du = (3x^2 + 6x)dx = 3(x^2 + 2x)dx$. If $x = 1$, $u = 3$, and if $x = 2$, $u = 19$. So

$$\int_1^2 \frac{x^2+2x}{x^3+3x^2-1}\, dx = \frac{1}{3}\int_3^{19} \frac{du}{u} = \frac{1}{3}\ln u\Big|_3^{19} = \frac{1}{3}(\ln 19 - \ln 3).$$

25. $\displaystyle\int_1^2 \left(4e^{2u} - \frac{1}{u}\right) du = 2e^{2u} - \ln u\Big|_1^2 = (2e^4 - \ln 2) - (2e^2 - 0) = 2e^4 - 2e^2 - \ln 2.$

27. $\displaystyle\int (2e^{-4x} - x^{-2})\, dx = -\frac{1}{2}e^{-4x} + \frac{1}{x}\Big|_1^2 = (-\frac{1}{2}e^{-8} + \frac{1}{2}) - (-\frac{1}{2}e^{-4} + 1)$

$$= -\frac{1}{2}e^{-8} + \frac{1}{2}e^{-4} - \frac{1}{2} = \frac{1}{2}(e^{-4} - e^{-8} - 1).$$

29. $\displaystyle\text{AV} = \frac{1}{2}\int_0^2 (2x+3)\, dx = \frac{1}{2}(x^2 + 3x)\Big|_0^2 = \frac{1}{2}(10) = 5.$

31. $AV = \frac{1}{2}\int_{1}^{3}(2x^2 - 3)\,dx = \frac{1}{2}(\frac{2}{3}x^3 - 3x)\big|_{1}^{3} = \frac{1}{2}(9 + \frac{7}{3}) = \frac{17}{3}.$

33. $AV = \frac{1}{3}\int_{-1}^{2}(x^2 + 2x - 3)\,dx = \frac{1}{3}(\frac{1}{3}x^3 + x^2 - 3x)\big|_{-1}^{2}$

$= \frac{1}{3}[(\frac{8}{3} + 4 - 6) - (-\frac{1}{3} + 1 + 3)] = \frac{1}{3}(\frac{8}{3} - 2 + \frac{1}{3} - 4) = -1.$

35. $AV = \frac{1}{4}\int_{0}^{4}(2x + 1)^{1/2}\,dx = (\frac{1}{4})(\frac{1}{2})(\frac{2}{3})(2x + 1)^{3/2}\big|_{0}^{4} = \frac{1}{12}(27 - 1) = \frac{13}{6}$

37. $AV = \frac{1}{2}\int_{0}^{2}xe^{x^2}\,dx = \frac{1}{4}e^{x^2}\bigg|_{0}^{2} = \frac{1}{4}(e^4 - 1).$

39. The amount produced was
$$\int_{0}^{20}3.5e^{0.05t}\,dt = \frac{3.5}{0.05}e^u\bigg|_{0}^{20} \qquad \text{(Use the substitution } u = 0.05t.)$$
$$= 70(e - 1) \approx 120.3,$$
or 120.3 billion metric tons.

41. The amount is $\int_{1}^{2}t(\frac{1}{2}t^2 + 1)^{1/2}\,dt.$ Let $u = \frac{1}{2}t^2 + 1$, so that $du = t\,dt.$ Therefore,
$$\int_{1}^{2}t(\frac{1}{2}t^2 + 1)^{1/2}\,dt = \int_{3/2}^{3}u^{1/2}\,du = \frac{2}{3}u^{3/2}\bigg|_{3/2}^{3} = \frac{2}{3}[(3)^{3/2} - (\frac{3}{2})^{3/2}]$$
$$\approx 2.24 \text{ million dollars.}$$

43. The tractor will depreciate
$$\int_{0}^{5}13388.61e^{-0.22314t}\,dt = \frac{13388.61}{-0.22314}e^{-0.22314t}\bigg|_{0}^{5}$$

$$= -60,000.94e^{-0.22314t}\bigg|_{0}^{5} = -60,000.94(-0.672314)$$
$$= 40,339.47, \quad \text{or } \$40,339.$$

45. The average temperature is $\frac{1}{12}\int_0^{12}(-0.05t^3+0.4t^2+3.8t+5.6)\,dt$

$$=\frac{1}{12}\left(-\frac{0.05}{4}t^4+\frac{0.4}{3}t^3+1.9t^2+5.6t\right)\Big|_0^{12}=26°\,F.$$

47. The average number is

$$\frac{1}{5}\int_0^5\left(-\frac{40,000}{\sqrt{1+0.2t}}+50,000\right)dt=-8000\int_0^5(1+0.2t)^{-1/2}\,dt+10,000\int_0^5 dt$$

Integrating the first integral by substitution with $u=1+0.2t$, so that $du=0.2\,dt$ or $dt=5\,du$, we find that the average value is

$$-8000\int_1^2 5u^{-1/2}\,du+\int_0^5 10,000\,dt=-40,000(2u^{1/2})\Big|_1^2+10,000t\Big|_0^5$$

$$=-40,000(2\sqrt{2}-2)+50,000=16,863$$

or 16,863 subscribers.

49. The average concentration of the drug is

$$\frac{1}{4}\int_0^4\frac{0.2t}{t^2+1}\,dt=\frac{0.2}{4}\int_0^4\frac{t}{t^2+1}\,dt=\frac{0.2}{(4)(2)}\ln(t^2+1)\Big|_0^4$$

$$=0.025\ln 17\approx 0.071,$$

or 0.071 milligrams per cubic centimeter.

51. The average velocity of the blood is

$$\frac{1}{R}\int_0^R k(R^2-r^2)\,dr=\frac{k}{R}\int_0^R(R^2-r^2)\,dr=\frac{k}{r}(R^2r-\tfrac{1}{3}r^3)\Big|_0^R$$

$$=\frac{k}{R}(R^3-\tfrac{1}{3}R^3)=\frac{k}{R}\cdot\frac{2}{3}R^3=\frac{2k}{3}R^2\ \text{cm/sec.}$$

53. $\displaystyle\int_a^b f(x)\,dx=F(x)\Big|_a^b=F(b)-F(a)=-[F(a)-F(b)]$

$$=-F(x)\Big|_b^a=-\int_b^a f(x)\,dx$$

55. $\displaystyle\int_a^b cf(x)\,dx=xF(x)\Big|_a^b=c[F(b)-F(a)]=c\int_a^b f(x)\,dx.$

57. $\int_0^1 (1+x-e^x)\,dx = x + \frac{1}{2}x^2 - e^x \Big|_0^1 = (1+\frac{1}{2}-e)+1 = \frac{5}{2}-e.$

$\int_0^1 dx + \int_0^1 x\,dx - \int_0^1 e^x\,dx = x\Big|_0^1 + \frac{1}{2}x^2\Big|_0^1 - e^x\Big|_0^1$

$= (1-0) + (\frac{1}{2}-0) - (e-1) = \frac{5}{2}-e.$

59. $\int_0^3 (1+x^3)\,dx = x + \frac{1}{4}x^4\Big|_0^3 = 3 + \frac{81}{4} = \frac{93}{4}.$

$\int_0^1 (1+x^3)\,dx + \int_1^2 (1+x^3)\,dx + \int_2^3 (1+x^3)\,dx$

$= (x+\frac{1}{4}x^4)\Big|_0^1 + (x+\frac{1}{4}x^4)\Big|_1^2 + (x+\frac{1}{4}x^4)\Big|_2^3$

$= (1+\frac{1}{4}) + (2+4) - (1+\frac{1}{4}) + (3+\frac{81}{4}) - (2+4) = \frac{93}{4}.$

61. $\int_3^0 f(x)\,dx = -\int_0^3 f(x)\,dx$ (Property 2)

    $= -4.$

63. a. $\int_{-1}^2 [2f(x)+g(x)]\,dx = 2\int_{-1}^2 f(x)\,dx + \int_{-1}^2 g(x)\,dx$

         $= 2(-2) + 3 = -1.$

  b. $\int_{-1}^2 [g(x)-f(x)]\,dx = \int_{-1}^2 g(x)\,dx - \int_{-1}^2 f(x)\,dx$

         $= 3 - (-2) = 5.$

  c. $\int_{-1}^2 [2f(x)-3g(x)]\,dx = 2\int_{-1}^2 f(x)\,dx - 3\int_{-1}^2 g(x)\,dx$

         $= 2(-2) - 3(3) = -13.$

65. True. this follows from Property 1 of the definite integral.

67. False. Only a constant can be "moved out" of the integral sign.

69. True. This follows from Properties 3 and 4 of the definite integral.

1. 7.71667   3. 17.5649   5. 10,140

EXERCISES 6.6, page 511

1. $-\int_0^6 (x^3 - 6x^2)\,dx = -\frac{1}{4}x^4 + 2x^3 \Big|_0^6 = -\frac{1}{4}(6^4) + 2(6^3) = 108$ sq units.

3. $A = -\int_{-1}^0 x\sqrt{1-x^2}\,dx + \int_0^1 x\sqrt{1-x^2}\,dx = 2\int_0^1 x(1-x^2)^{1/2}\,dx$ (by symmetry). Let
   $u = 1 - x^2$ so that $du = -2x\,dx$ or $x\,dx = -\frac{1}{2}\,du$. Also, if $x = 0$, then $u = 1$ and
   if $x = 1$, $u = 0$. So $A = (2)(-\frac{1}{2})\int_0^1 u^{1/2}\,du = -\frac{2}{3}u^{3/2}\Big|_1^0 = \frac{2}{3}$, or $\frac{2}{3}$ sq units.

5. $A = -\int_0^4 (x - 2\sqrt{x})\,dx = \int_0^4 (-x + 2x^{1/2})\,dx = -\frac{1}{2}x^2 + \frac{4}{3}x^{3/2}\Big|_0^4$
   $= 8 + \frac{32}{3} = \frac{8}{3}$ sq units.

7. The required area is given by
   $$\int_{-1}^0 (x^2 - x^{1/3})\,dx + \int_0^1 (x^{1/3} - x^2)\,dx = \frac{1}{3}x^3 - \frac{3}{4}x^{4/3}\Big|_{-1}^0 + \frac{3}{4}x^{4/3} - \frac{1}{3}x^3\Big|_0^1$$
   $$= -(-\frac{1}{3} - \frac{3}{4}) + (\frac{3}{4} - \frac{1}{3}) = 1\frac{1}{2} \quad \text{sq units.}$$

9. The required area is given by
   $-\int_{-1}^2 -x^2\,dx = \frac{1}{3}x^3\Big|_{-1}^2 = \frac{8}{3} + \frac{1}{3} = 3$ sq units.

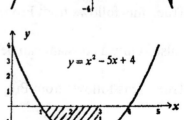

11. $y = x^2 - 5x + 4 = (x-4)(x-1) = 0$
    if $x = 1$ or 4. These give the x-intercepts.
    $A = -\int_1^3 (x^2 - 5x + 4)\,dx = -\frac{1}{3}x^3 + \frac{5}{2}x^2 - 4x\Big|_1^3$
    $= (-9 + \frac{45}{2} - 12) - (-\frac{1}{3} + \frac{5}{2} - 4) = \frac{10}{3} = 3\frac{1}{3}.$

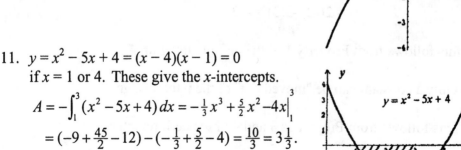

13. The required area is given by

$$-\int_0^9 -(1+\sqrt{x})\,dx = x + \tfrac{2}{3}x^{3/2}\Big|_0^9 = 9+18 = 27.$$

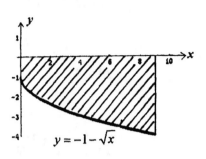

$$y = -1 - \sqrt{x}$$

15. $$-\int_{-2}^4 -e^{(1/2)x}\,dx = 2e^{(1/2)x}\Big|_{-2}^4$$

$$= 2(e^2 - e^{-1})\text{ sq units.}$$

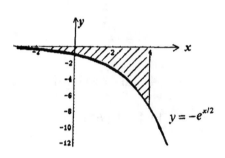

$$y = -e^{x/2}$$

17. $$A = \int_1^3 [(x^2+3)-1]\,dx$$

$$= \int_1^3 (x^2+2)\,dx = \tfrac{1}{3}x^3 + 2x\Big|_1^3$$

$$= (9+6) - (\tfrac{1}{3}+2) = \tfrac{38}{3}.$$

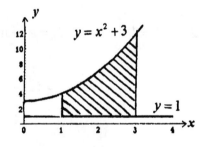

$$y = x^2 + 3$$
$$y = 1$$

19. $$A = \int_0^2 (-x^2 + 2x + 3 + x - 3)\,dx$$

$$= \int_0^2 (-x^2 + 3x)\,dx$$

$$= -\tfrac{1}{3}x^3 + \tfrac{3}{2}x^2\Big|_0^2 = -\tfrac{1}{3}(8) + \tfrac{3}{2}(4)$$

$$= 6 - \tfrac{8}{3} = \tfrac{10}{3}\text{ sq units}$$

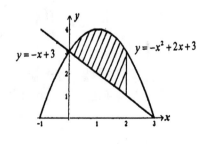

$$y = -x+3$$
$$y = -x^2 + 2x + 3$$

21. $A = \int_{-1}^{2} [(x^2 + 1) - \tfrac{1}{3} x^3] \, dx$

$= \int_{-1}^{2} (-\tfrac{1}{3} x^3 + x^2 + 1) \, dx$

$= -\tfrac{1}{12} x^4 + \tfrac{1}{3} x^3 + x \Big|_{-1}^{2}$

$= (-\tfrac{4}{3} + \tfrac{8}{3} + 2) - (-\tfrac{1}{12} - \tfrac{1}{3} - 1) = 4\tfrac{3}{4}$ sq units.

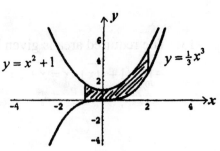

23. $A = \int_{1}^{4} \left[(2x - 1) - \dfrac{1}{x} \right] dx = \int_{1}^{4} \left(2x - 1 - \dfrac{1}{x} \right) dx$

$= (x^2 - x - \ln x) \Big|_{1}^{4}$

$= (16 - 4 - \ln 4) - (1 - 1 - \ln 1)$

$= 12 - \ln 4 \approx 10.6$ sq units.

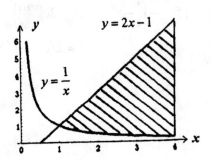

25. $A = \int_{1}^{2} \left(e^x - \dfrac{1}{x} \right) dx = e^x - \ln x \Big|_{1}^{2}$

$= (e^2 - \ln 2) - e = (e^2 - e - \ln 2)$ sq units.

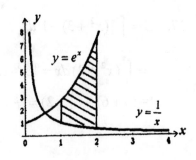

27.
$A = -\int_{-1}^{0} x \, dx + \int_{0}^{2} x \, dx$

$= -\tfrac{1}{2} x^2 \Big|_{-1}^{0} + \tfrac{1}{2} x^2 \Big|_{0}^{2}$
$= \tfrac{1}{2} + 2 = 2\tfrac{1}{2}$ sq units.

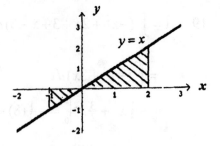

29. The x–intercepts are found by solving
$x^2 - 4x + 3 = (x-3)(x-1) = 0$ giving $x = 1$
or 3. The region is shown in the figure.

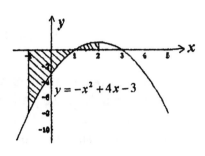

$$A = -\int_{-1}^{1}[(-x^2 + 4x - 3)\,dx + \int_{1}^{2}(-x^2 + 4x - 3)\,dx$$

$$= \tfrac{1}{3}x^3 - 2x^2 + 3x\big|_{-1}^{1} + (-\tfrac{1}{3}x^3 + 2x^2 - 3x)\big|_{1}^{2}$$

$$= (\tfrac{1}{3} - 2 + 3) - (-\tfrac{1}{3} - 2 - 3)$$

$$+(-\tfrac{8}{3} + 8 - 6) - (-\tfrac{1}{3} + 2 - 3) = \tfrac{22}{3} \text{ sq units.}$$

31. The region is shown in the figure at the right.

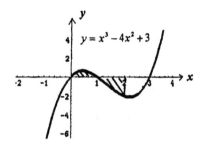

$$A = \int_{0}^{1}(x^3 - 4x^2 + 3x)\,dx - \int_{1}^{2}(x^3 - 4x^2 + 3x)\,dx$$

$$= (\tfrac{1}{4}x^4 - \tfrac{4}{3}x^3 + \tfrac{3}{2}x^2)\big|_{0}^{1}$$

$$-(\tfrac{1}{4}x^4 - \tfrac{4}{3}x^3 + \tfrac{3}{2}x^2)\big|_{1}^{2} = \tfrac{3}{2}\text{ sq units.}$$

33. The region is shown in the figure at the right.

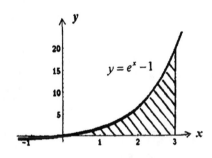

$$A = -\int_{-1}^{0}(e^x - 1)\,dx + \int_{0}^{3}(e^x - 1)\,dx$$

$$= (-e^x + x)\big|_{-1}^{0} + (e^x - x)\big|_{0}^{3}$$

$$= -1 - (-e^{-1} - 1) + (e^3 - 3) - 1$$

$$= e^3 - 4 + \tfrac{1}{e} \approx 16.5 \quad \text{sq units.}$$

35. To find the points of intersection of the two curves, we solve the equation
$x^2 - 4 = x + 2$
$x^2 - x - 6 = (x-3)(x+2) = 0$, obtaining x
$= -2$ or $x = 3$.
The region is shown in the figure at the right

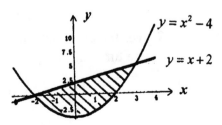

$$A = \int_{-2}^{3}[(x+2) - (x^2 - 4)]\,dx$$

$$= \int_{-2}^{3} (-x^2 + x + 6) \, dx = (-\tfrac{1}{3}x^3 + \tfrac{1}{2}x^2 + 6x)\big|_{-2}^{3}$$

$$= (-9 + \tfrac{9}{2} + 18) - (\tfrac{8}{3} + 2 - 12) = \tfrac{125}{6} \text{ sq units.}$$

37. To find the points of intersection of the two curves, we solve the equation $x^3 = x^2$ or $x^3 - x^2 = x^2(x - 1) = 0$ giving $x = 0$ or 1. The region is shown in the figure.

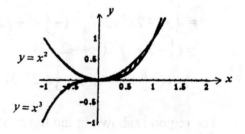

$$A = -\int_{0}^{1} (x^2 - x^3) \, dx$$

$$= (\tfrac{1}{3}x^3 - \tfrac{1}{4}x^4)\big|_{0}^{1} = \tfrac{1}{3} - \tfrac{1}{4} = \tfrac{1}{12} \text{ sq units}$$

39. The graphs intersect at the points where $\sqrt{x} = x^2$ or $x = x^4$ and $x(x^3 - 1) = 0$; that is, when $x = 0$ and $x = 1$. The required area is

$$A = \int_{0}^{1} (x^{1/2} - x^2) \, dx = \tfrac{2}{3}x^{3/2} - \tfrac{1}{3}x^3\big|_{0}^{1} = \tfrac{2}{3} - \tfrac{1}{3} = \tfrac{1}{3}.$$

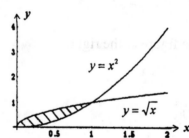

41. S gives the additional revenue that the company would realize if it used a different advertising agency.

$$S = \int_{0}^{b} [g(x) - f(x)] \, dx.$$

43. a. S gives the difference in the amount of smoke removed by the two brands over the same time interval $[a, b]$.

   b. $\qquad S = \int_{a}^{b} [f(t) - g(t)] \, dt$

45. $\displaystyle \int_{T_1}^{T} [g(t) - f(t)] \, dt - \int_{0}^{T_1} [f(t) - g(t)] \, dt$

47. The additional amount of coal that will be produced is

$$\int_0^{20} (3.5e^{0.05t} - 3.5e^{0.01t})\,dt = 3.5\int_0^{20} (e^{0.05t} - e^{0.01t})\,dt$$

$$= 3.5(20e^{0.05t} - 100e^{0.01t})\Big|_0^{20} = 3.5[20e - 100e^{0.2}) - (20 - 100)]$$

$$= 42.8 \text{ billion metric tons.}$$

49. If the campaign is mounted, there will be

$$\int_0^5 (60e^{0.02t} + t^2 - 60)\,dt = 3000e^{0.02t} + \tfrac{t^3}{3} - 60t\Big|_0^5$$

$$= 3315.5 + \frac{125}{3} - 300 - 3000 \approx 57.179,$$

or 57,179 fewer people. (Remember t is measured in thousands.)

51. False. The area is given by $\int_0^2 [g(x) - f(x)]\,dx$ since $g(x) \geq f(x)$ on $[0, 2]$.

USING TECHNOLOGY EXERCISES 6.6, page 517

1. a.

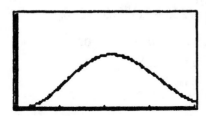

b. 1074.2857 sq units

3. a.

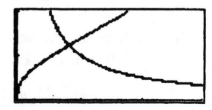

b. 0.9961 sq units

5. a.

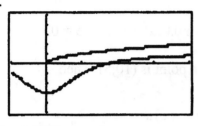

b. 5.4603 sq units

7. a.

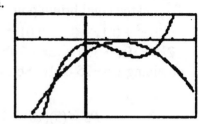

b. 25.8549 sq units

9.   a.

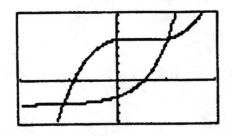

11.   a.

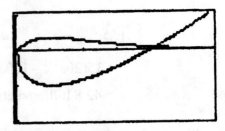

b. 10.5144 sq units   b. 3.5799 sq units

EXERCISES 6.7, page 531

1.   When $p = 4$, $-0.01x^2 - 0.1x + 6 = 4$ or $x^2 + 10x - 200 = 0$, $(x - 10)(x + 20) = 0$ and $x = 10$ or -20. We reject the root $x = -20$. The consumers' surplus is

$$CS = \int_0^{10} (-0.01x^2 - 0.1x + 6)\, dx - (4)(10)$$

$$= -\frac{0.01}{3}x^3 - 0.05x^2 + 6x\Big|_0^{10} - 40 \approx 11.667, \text{ or } \$11,667.$$

3.   Setting $p = 10$, we have $\sqrt{225 - 5x} = 10$, $225 - 5x = 100$, or $x = 25$.

Then $CS = \int_0^{25} \sqrt{225 - 5x}\, dx - (10)(25) = \int_0^{25} (225 - 5x)^{1/2}\, dx - 250$.

To evaluate the integral, let $u = 225 - 5x$ so that $du = -5\, dx$ or $dx = -\frac{1}{5}\, du$. If $x = 0$, $u = 225$ and if $x = 25$, $u = 100$. So

$$CS = -\frac{1}{5}\int_{225}^{100} u^{1/2}\, du - 250 = -\frac{2}{15}u^{3/2}\Big|_{225}^{100} - 250$$

$$= -\frac{2}{15}(1000 - 3375) - 250 = 66.667, \text{ or } \$6,667.$$

5.   To find the equilibrium point, we solve

$$0.01x^2 + 0.1x + 3 = -0.01x^2 - 0.2x + 8, \text{ or } 0.02x^2 + 0.3x - 5 = 0,$$
$$2x^2 + 30x - 500 = (2x - 20)(x + 25) = 0$$

obtaining $x = -25$ or 10. So the equilibrium point is $(10,5)$. Then

$$PS = (5)(10) - \int_0^{10} (0.01x^2 + 0.1x + 3)\, dx$$

$$= 50 - (\frac{0.01}{3}x^3 + 0.05x^2 + 3x)\Big|_0^{10} = 50 - \frac{10}{3} - 5 - 30 = \frac{35}{3},$$

or approximately \$11,667.

7. To find the market equilibrium, we solve
$$-0.2x^2 + 80 = 0.1x^2 + x + 40, \quad 0.3x^2 + x - 40 = 0,$$
$$3x^2 + 10x - 400 = 0, \ (3x + 40)(x - 10) = 0$$
giving $x = -\frac{40}{3}$ or $x = 10$. We reject the negative root. The corresponding equilibrium price is \$60. The consumers' surplus is
$$CS = \int_0^{10}(-0.2x^2 + 80)dx - (60)(10) = -\frac{0.2}{3}x^3 + 80x\Big|_0^{10} - 600 = 133\tfrac{1}{3},$$
or \$13,333. The producers' surplus is
$$PS = 600 - \int_0^{10}(0.1x^2 + x + 40)dx = 600 - [\tfrac{0.1}{3}x^3 + \tfrac{1}{2}x^2 + 40x]\Big|_0^{10}$$
$$= 116\tfrac{2}{3}, \text{ or } \$11,667.$$

9. Here $P = 200{,}000$, $r = 0.08$, and $T = 5$. So
$$PV = \int_0^5 200{,}000e^{-0.08t}dt = -\frac{200{,}000}{0.08}e^{-0.08t}\Big|_0^5 = -2{,}500{,}000(e^{-0.4} - 1)$$
$$\approx 824{,}199.85, \text{ or } \$824{,}200.$$

11. Here $P = 250$, $m = 12$, $T = 20$, and $r = 0.08$. So
$$A = \frac{mP}{r}(e^{rT} - 1) = \frac{12(250)}{0.08}(e^{1.6} - 1) \approx 148{,}238.70$$
or approximately \$148,239.

13. Here $P = 150$, $m = 12$, $T = 15$, and $r = 0.08$. So
$$A = \frac{12(150)}{0.08}(e^{1.2} - 1) \approx 52{,}202.60, \text{ or approximately } \$52{,}203.$$

15. Here $P = 2000$, $m = 1$, $T = 15.75$, and $r = 0.1$. So
$$A = \frac{1(2000)}{0.1}(e^{1.575} - 1) \approx 76{,}615, \text{ or approximately } \$76{,}615.$$

17. Here $P = 1200$, $m = 12$, $T = 15$, and $r = 0.1$. So

$$PV = \frac{12(1200)}{0.1}(1 - e^{-1.5}) \approx 111{,}869, \text{ or approximately } \$111{,}869.$$

19. We want the present value of an annuity with $P = 300$, $m = 12$,

$T = 10$, and $r = 0.12$. So

$$PV = \frac{12(300)}{0.12}(1 - e^{-1.2}) \approx 20{,}964, \text{ or approximately } \$20{,}964.$$

21. a.

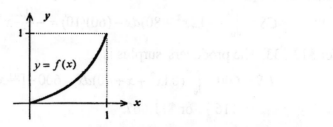

b. $f(0.4) = \frac{15}{16}(0.16) + \frac{1}{16}(0.4) \approx 0.175$; $f(0.9) = \frac{15}{16}(0.81) + \frac{1}{16}(0.9) \approx 0.816$.

So, the lowest 40 percent of the people receive 17.5 percent of the total income and the lowest 90 percent of the people receive 81.6 percent of the income.

23. a.

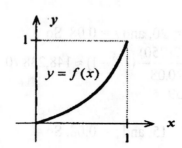

b. $f(0.3) = \frac{14}{15}(0.09) + \frac{1}{15}(0.3) = 0.104$

$f(0.7) = \frac{14}{15}(0.49) + \frac{1}{15}(0.7) \approx 0.504$.

1. Consumer's surplus: $18,000,000; producer's surplus: $11,700,000.

3. Consumer's surplus: $33,120; producer's surplus: $2,880.

EXERCISES 6.8, page 541

1. $V = \pi \int_a^b [f(x)]^2\, dx = \pi \int_0^1 (3x)^2\, dx = 9\pi \int_0^1 x^2\, dx = 3\pi x^3 \Big|_0^1 = 3\pi$ cu units.

3. $V = \pi \int_a^b [f(x)]^2\, dx = \pi \int_1^4 (\sqrt{x})^2\, dx = \pi \int_1^4 x\, dx = \dfrac{\pi}{2}(16-1) = \dfrac{15\pi}{2}$ cu units.

5. $V = \pi \int_a^b [f(x)]^2\, dx = \pi \int_0^1 (\sqrt{1+x^2})^2\, dx = \pi \int_0^1 (1+x^2)\, dx = \pi(x + \tfrac{1}{3}x^3)\Big|_0^1$
   $= \pi(1 + \tfrac{1}{3}) = \tfrac{4\pi}{3}$ cu units.

7. $V = \pi \int_{-1}^{1} [f(x)]^2\, dx = \pi \int_{-1}^{1} (1-x^2)^2\, dx = \pi \int_{-1}^{1} (1 - 2x^2 + x^4)\, dx = \pi(x - \tfrac{2}{3}x^3 + \tfrac{1}{5}x^5)\Big|_{-1}^{1}$
   $= \pi[(1 - \tfrac{2}{3} + \tfrac{1}{5}) - (-1 + \tfrac{2}{3} - \tfrac{1}{5})] = \tfrac{16\pi}{15}$ cu units.

9. $V = \pi \int_a^b [f(x)]^2\, dx = \pi \int_0^1 (e^x)^2\, dx = \pi \int_0^1 e^{2x}\, dx = \tfrac{1}{2}\pi e^{2x}\Big|_0^1 = \tfrac{1}{2}\pi(e^2 - 1)$ cu units.

11. $V = \pi \int_a^b \{[f(x)]^2 - [g(x)]^2\}\, dx = \pi \int_0^1 [(x)^2 - (x^2)^2]\, dx = \pi \int_0^1 (x^2 - x^4)\, dx$
   $= \pi(\tfrac{1}{3}x^3 - \tfrac{1}{5}x^5)\Big|_0^1 = \pi(\tfrac{1}{3} - \tfrac{1}{5}) = \tfrac{2\pi}{15}$ cu units.

13. $V = \pi \int_a^b \{[f(x)]^2 - [g(x)]^2\}\, dx = \pi \int_{-1}^{1} [(4-x^2)^2 - 3^2]\, dx$
   $= \pi \int_{-1}^{1} (16 - 8x^2 + x^4 - 9)\, dx = \pi \int_{-1}^{1} (7 - 8x^2 + x^4)\, dx$
   $= \pi(7x - \tfrac{8}{3}x^3 + \tfrac{1}{5}x^5)\Big|_{-1}^{1} = \pi[(7 - \tfrac{8}{3} + \tfrac{1}{5}) - (-7 + \tfrac{8}{3} - \tfrac{1}{5})] = \tfrac{136\pi}{15}$ cu units.

15. $V = \pi \int_a^b \{[f(x)]^2 - [g(x)]^2\}dx = \pi \int_0^{2\sqrt{2}} [(\sqrt{16-x^2})^2 - (x)^2]dx$

$$= \pi \int_0^{2\sqrt{2}} (16-x^2 - x^2)\,dx = \pi \int_0^{2\sqrt{2}} (16-2x^2)\,dx = \pi(16x - \tfrac{2}{3}x^3)\Big|_0^{2\sqrt{2}}$$

$$= \pi[32\sqrt{2} - \tfrac{2}{3}(2\sqrt{2})^3] = \pi\left[32\sqrt{2} - \tfrac{32\sqrt{2}}{3}\right] = \tfrac{64\sqrt{2}\pi}{3} \text{ cu units.}$$

17. $V = \pi \int_a^b \{[f(x)]^2 - [g(x)]^2\}dx = \pi \int_0^1 \left[(e^x)^2 - (e^{-x})^2\right]dx = \pi \int_0^1 (e^{2x} - e^{-2x})dx$

$$= \pi \int_0^1 (e^{2x} - e^{-2x})dx = \pi(\tfrac{1}{2}e^{2x} + \tfrac{1}{2}e^{-2x})\Big|_0^1 = \pi[(\tfrac{1}{2}e^2 + \tfrac{1}{2}e^{-2}) - (\tfrac{1}{2} + \tfrac{1}{2})]$$

$$= \tfrac{\pi}{2}(e^2 - 2 + e^{-2}) \text{ cu units.}$$

19. The region is shown in the figure at the right. The points of intersection of the curves are found by solving the simultaneous system of equations

$y = x$ and $y = \sqrt{x}$, giving

$$x = \sqrt{x}$$

$$x^2 = x$$

$$x^2 - x = 0$$

$$x(x-1) = 0, \text{ or } x = 0 \text{ and } x = 1.$$

$$V = \pi \int_a^b \{[f(x)]^2 - [g(x)]^2\,dx\} = \pi \int_0^1 \left[(\sqrt{x})^2 - x^2\right]dx$$

$$\pi \int_0^1 (x - x^2)\,dx = \pi(\tfrac{1}{2}x^2 - \tfrac{1}{3}x^3)\Big|_0^1 = \pi(\tfrac{1}{2} - \tfrac{1}{3}) = \tfrac{\pi}{6} \text{ cu units.}$$

21. The region is shown in the figure. The points of intersection of the curves are found by solving the simultaneous system of equations

$$y = x^2 \text{ and } y = \tfrac{1}{2}x + 3$$

$$2x^2 - x - 6 = 0$$

$$(2x+3)(x-2) = 0,$$

$$\text{or } x = -\tfrac{3}{2} \text{ and } x = 2.$$

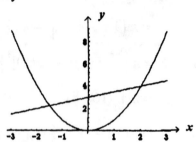

$$V = \pi \int_a^b \left\{ [f(x)]^2 - [g(x)]^2 \, dx \right\} = \pi \int_{-3/2}^2 \left[\left(\tfrac{1}{2}x + 3 \right)^2 - (x)^2 \right] dx$$

$$= \pi \int_{-3/2}^2 \left(\tfrac{1}{4}x^2 + 3x + 9 - x^4 \right) dx = \pi \left(\tfrac{1}{12}x^3 + \tfrac{3}{2}x^2 + 9x - \tfrac{1}{5}x^5 \right) \Big|_{-3/2}^2$$

$$= \pi \left[\left(\tfrac{2}{3} + 6 + 18 - \tfrac{32}{5} \right) - \left(-\tfrac{27}{96} + \tfrac{27}{8} - \tfrac{27}{2} + \tfrac{243}{160} \right) \right]$$

$$= \tfrac{6517\pi}{240} \text{ cu units.}$$

23. The region is shown in the figure. To find
where the graphs intersect, we solve $y = x^2$ and $y = 4 - x^2$, simultaneously,
obtaining

$$x^2 = 4 - x^2, \ 2x^2 = 4, \ x^2 = 2$$

giving $x = \pm\sqrt{2}$. The required volume is

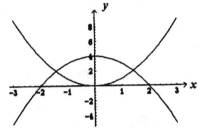

$$V = \pi \int_a^b \left\{ [f(x)]^2 - [g(x)]^2 \right\} dx$$

$$= \pi \int_{-\sqrt{2}}^{\sqrt{2}} \left[(4 - x^2)^2 - (x^2)^2 \right] dx$$

$$= 2\pi \int_0^{\sqrt{2}} \left[16 - 8x^2 + x^4 - x^4 \right] dx$$

$$= 2\pi \left[16x - \frac{8}{3}x^3 \Big|_0^{\sqrt{2}} \right]$$

$$= 16\pi \left[16\sqrt{2} - \frac{8}{3}(2\sqrt{2}) \right] = \frac{64\sqrt{2}\pi}{3} \text{ cu units.}$$

25. The region is shown in the figure. To find the
points of intersection of $y = 2x$ and $y = \dfrac{1}{x}$, we solve

$$2x = \frac{1}{x}$$

$$2x^2 = 1$$

or $x = \pm\dfrac{1}{\sqrt{2}} = \pm\dfrac{\sqrt{2}}{2}$.

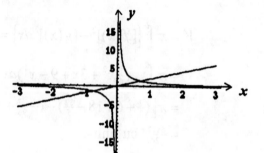

To find the points of intersection of

$y = x$ and $y = \dfrac{1}{x}$, we solve

$$x = \frac{1}{x}$$

giving $x = \pm 1$. By symmetry

$$V = 2\pi \int_0^{\sqrt{2}/2} \left[(2x)^2 - (x)^2\right]dx + 2\pi \int_{\sqrt{2}/2}^1 \left[\left(\frac{1}{x}\right)^2 - (x)^2\right]dx$$

$$= 2\pi \int_0^{\sqrt{2}/2} \left[4x^2 - x^2\right]dx + 2\pi \int_{\sqrt{2}/2}^1 \left[\frac{1}{x^2} - x^2\right]dx$$

$$= 2\pi \int_0^{\sqrt{2}/2} 3x^2\, dx + 2\pi \int_{\sqrt{2}/2}^1 (x^{-2} - x^2)dx = 2\pi x^3 \Big|_0^{\sqrt{2}/2} + 2\pi\left(-\frac{1}{x} - \frac{1}{3}x^3\right)\Big|_{\sqrt{2}/2}^1$$

$$= 2\pi\left(\frac{2\sqrt{2}}{8}\right) + 2\pi\left[(-1-\tfrac{1}{3}) - \left(-\frac{2}{\sqrt{2}} - \frac{\sqrt{2}}{12}\right)\right] = \frac{8(\sqrt{2}-1)\pi}{3} \quad \text{cu units.}$$

27. $V = \pi \int_a^b [f(x)]^2\, dx = \pi \int_{-r}^r \left[\sqrt{r^2 - x^2}\right]^2 dx = \pi \int_{-r}^r (r^2 - x^2)\, dx$

$$= 2\pi \int_0^r (r^2 - x^2)dx = 2\pi(r^2 x - \tfrac{1}{3}x^3)\Big|_0^r = 2\pi(r^3 - \tfrac{1}{3}x^3) = \tfrac{4}{3}\pi r^3 \quad \text{cu units.}$$

29. $V = \pi \int_{-10}^0 x^2\, dy = \pi \int_{-10}^0 [f(y)]^2\, dy.$

Solving the given equation for x in terms of y, we have

$$\frac{y}{10} = \left(\frac{x}{100}\right)^2 - 1$$

$$\left(\frac{x}{100}\right)^2 = 1 + \frac{y}{10}, \quad \text{or} \quad x = 100\sqrt{1 + \frac{y}{10}}.$$

Therefore,

$$V = \pi \int_{-10}^{0} 10{,}000\left(1 + \tfrac{y}{10}\right) dy = 10{,}000\pi (y + \tfrac{1}{20} y^2)\Big|_{-10}^{0}$$

$$= -10{,}000\pi(-10 + \tfrac{100}{20}) = 50{,}000\pi \text{ cu ft.}$$

CHAPTER 6 REVIEW EXERCISES, page 536

1. $\int (x^3 + 2x^2 - x)\, dx = \tfrac{1}{4}x^4 + \tfrac{2}{3}x^3 - \tfrac{1}{2}x^2 + C.$

3. $\int \left(x^4 - 2x^3 + \dfrac{1}{x^2}\right) dx = \dfrac{x^5}{5} - \dfrac{1}{2}x^4 - \dfrac{1}{x} + C$

5. $\int x(2x^2 + x^{1/2})\, dx = \int (2x^3 + x^{3/2})\, dx = \tfrac{1}{2}x^4 + \tfrac{2}{5}x^{5/2} + C.$

7. $\int (x^2 - x + \tfrac{2}{x} + 5)\, dx = \int x^2\, dx - \int x\, dx + 2\int \tfrac{dx}{x} + 5\int dx$

$$= \tfrac{1}{3}x^3 - \tfrac{1}{2}x^2 + 2\ln|x| + 5x + C.$$

9. Let $u = 3x^2 - 2x + 1$ so that $du = (6x - 2)\, dx = 2(3x - 1)\, dx$ or $(3x - 1)\, dx = \tfrac{1}{2} du.$

So $\int (3x - 1)(3x^2 - 2x + 1)^{1/3}\, dx = \tfrac{1}{2}\int u^{1/3}\, du = \tfrac{3}{8}u^{4/3} + C = \tfrac{3}{8}(3x^2 - 2x + 1)^{4/3} + C.$

11. Let $u = x^2 - 2x + 5$ so that $du = 2(x - 1)\, dx$ or $(x - 1)\, dx = \tfrac{1}{2} du.$

$$\int \frac{x - 1}{x^2 - 2x + 5}\, dx = \frac{1}{2}\int \frac{du}{u} = \frac{1}{2}\ln|u| + C = \frac{1}{2}\ln(x^2 - 2x + 5) + C.$$

13. Put $u = x^2 + x + 1$ so that $du = (2x + 1)\, dx = 2(x + \tfrac{1}{2})\, dx$ and $(x + \tfrac{1}{2})\, dx = \tfrac{1}{2} du.$

$$\int (x + \tfrac{1}{2})e^{x^2 + x + 1}\, dx = \tfrac{1}{2}\int e^u\, du = \tfrac{1}{2}e^u + C = \tfrac{1}{2}e^{x^2 + x + 1} + C.$$

15. Let $u = \ln x$ so that $du = \tfrac{1}{x}\, dx.$ Then

$$\int \frac{(\ln x)^5}{x}\,dx = \int u^5\,du = \frac{1}{6}u^6 + C = \frac{1}{6}(\ln x)^6 + C.$$

17. Let $u = x^2 + 1$ so that $du = 2x\,dx$ or $x\,dx = \frac{1}{2}\,du$. Then

$$\int x^3(x^2+1)^{10}\,dx = \frac{1}{2}\int (u-1)u^{10}\,du \qquad\qquad (x^2 = u - 1)$$

$$= \frac{1}{2}\int (u^{11} - u^{10})\,du = \frac{1}{2}(\frac{1}{12}u^{12} - \frac{1}{11}u^{11}) + C$$

$$= \frac{1}{264}u^{11}(11u - 12) + C = \frac{1}{264}(x^2+1)^{11}(11x^2 - 1) + C.$$

19. Put $u = x - 2$ so that $du = dx$. Then $x = u + 2$ and

$$\int \frac{x}{\sqrt{x-2}}\,dx = \int \frac{u+2}{\sqrt{u}}\,du = \int (u^{1/2} + 2u^{-1/2})\,du = \int u^{1/2}\,du + 2\int u^{-1/2}\,du$$

$$= \frac{2}{3}u^{3/2} + 4u^{1/2} + C = \frac{2}{3}u^{1/2}(u+6) + C = \frac{2}{3}\sqrt{x-2}(x-2+6) + C$$

$$= \frac{2}{3}(x+4)\sqrt{x-2} + C.$$

21. $\int_0^1 (2x^3 - 3x^2 + 1)\,dx = \frac{1}{2}x^4 - x^3 + x\Big|_0^1 = \frac{1}{2} - 1 + 1 = \frac{1}{2}.$

23. $\int_1^4 (x^{1/2} + x^{-3/2})\,dx = \frac{2}{3}x^{3/2} - 2x^{-1/2}\Big|_1^4 = \frac{2}{3}x^{3/2} - \frac{2}{\sqrt{x}}\Big|_1^4 = (\frac{16}{3} - 1) - (\frac{2}{3} - 2) = \frac{17}{3}.$

25. Put $u = x^3 - 3x^2 + 1$ so that $du = (3x^2 - 6x)\,dx = 3(x^2 - 2x)\,dx$ or $(x^2 - 2x)\,dx = \frac{1}{3}\,du$. Then if $x = -1$, $u = -3$, and if $x = 0$, $u = 1$,

$$\int_{-1}^0 12(x^2 - 2x)(x^3 - 3x^2 + 1)^3\,dx = (12)(\frac{1}{3})\int_{-3}^1 u^3\,du = 4(\frac{1}{4})u^4\Big|_{-3}^1$$

$$= 1 - 81 = -80.$$

27. Let $u = x^2 + 1$ so that $du = 2x\,dx$ or $x\,dx = \frac{1}{2}\,du$. Then, if $x = 0$, $u = 1$, and if $x = 2$, $u = 5$, so

$$\int_0^2 \frac{x}{x^2+1}\,dx = \frac{1}{2}\int_1^5 \frac{du}{u} = \frac{1}{2}\ln u\big|_1^5 = \frac{1}{2}\ln 5.$$

29. Let $u = 1 + 2x^2$ so that $du = 4x\,dx$ or $x\,dx = \frac{1}{4}\,du$. If $x = 0$, then $u = 1$ and if $x = 2$, then $u = 9$.

$$\int_0^2 \frac{4x}{\sqrt{1+2x^2}}\,dx = \int_1^9 \frac{du}{u^{1/2}} = 2u^{1/2}\big|_1^9 = 2(3-1) = 4.$$

31. Let $u = 1 + e^{-x}$ so that $du = -e^{-x}\,dx$ and $e^{-x}\,dx = -\,du$. Then

$$\int_{-1}^0 \frac{e^{-x}}{(1+e^{-x})^2}\,dx = -\int_{1+e}^2 \frac{du}{u^2} = \frac{1}{u}\bigg|_{1+e}^2 = \frac{1}{2} - \frac{1}{1+e} = \frac{e-1}{2(1+e)}.$$

33. $f(x) = \int f'(x)\,dx = \int (3x^2 - 4x + 1)\,dx = 3\int x^2\,dx - 4\int x\,dx + \int dx$

$$= x^3 - 2x^2 + x + C.$$

The given condition implies that $f(1) = 1$ or $1 - 2 + 1 + C = 1$ and $C = 1$. Therefore, the required function is

$$f(x) = x^3 - 2x^2 + x + 1.$$

35. $f(x) = \int f'(x)\,dx = \int (1 - e^{-x})\,dx = x + e^{-x} + C$, $f(0) = 2$ implies $0 + 1 + C = 2$ or $C = 1$. So $f(x) = x + e^{-x} + 1$.

37. $\Delta x = \frac{2-1}{5} = \frac{1}{5}$; $x_1 = \frac{6}{5}$, $x_2 = \frac{7}{5}$, $x_3 = \frac{8}{5}$, $x_4 = \frac{9}{5}$, $x_5 = \frac{10}{5}$. The Riemann sum is

$$f(x_1)\Delta x + \cdots + f(x_5)\Delta x = \left\{\left[-2(\tfrac{6}{5})^2 + 1\right] + \left[-2(\tfrac{7}{5})^2 + 1\right] + \cdots + \left[-2(\tfrac{10}{5})^2 + 1\right]\right\}(\tfrac{1}{5})$$

$$= \tfrac{1}{5}(-1.88 - 2.92 - 4.12 - 5.48 - 7) = -4.28.$$

39. a. $R(x) = \int R'(x)\,dx = \int (-0.03x + 60)\,dx = -0.015x^2 + 60x + C$.

$R(0) = 0$ implies that $C = 0$. So, $R(x) = -0.015x^2 + 60x$.

b. From $R(x) = px$, we have $-0.015x^2 + 60x = px$ or $p = -0.015x + 60$.

41. The total number of systems that Vista may expect to sell t months from the time they are put on the market is given by $f(t) = 3000t - 50,000(1 - e^{-0.04t})$.

The number is $\int_0^{12} (3000 - 2000e^{-0.04t})\, dt = \left(3000t - \dfrac{2000}{-0.04} e^{-0.04t} \right)\Big|_0^{12}$

$$= 3000(12) + 50,000e^{-0.48} - 50,000$$

$$= 16,939.$$

43. $C(x) = \int C'(x)\, dx = \int (0.00003x^2 - 0.03x + 10)\, dx$

$$= 0.00001x^3 - 0.015x^2 + 10x + k.$$

But $C(0) = 600$ and this implies that $k = 600$. Therefore,
$$C(x) = 0.00001x^3 - 0.015x^2 + 10x + 600.$$
The total cost incurred in producing the first 500 corn poppers is
$$C(500) = 0.00001(500)^3 - 0.015(500)^2 + 10(500) + 600$$
$$= 3,100, \text{ or } \$3,100.$$

45. $A = \int_{-1}^{2} (3x^2 + 2x + 1)\, dx = x^3 + x^2 + x\big|_{-1}^{2} = [2^3 + 2^2 + 2] - [(-1)^3 + 1 - 1]$

$$= 14 - (-1) = 15 \text{ sq units.}$$

47. $A = \int_1^3 \dfrac{1}{x^2}\, dx = \int_1^3 x^{-2}\, dx = -\dfrac{1}{x}\Big|_1^3 = -\dfrac{1}{3} + 1 = \dfrac{2}{3}.$

49.

$A = \int_a^b [f(x) - g(x)]\, dx$

$= \int_0^2 (e^x - x)\, dx$

$= \left(e^x - \dfrac{1}{2}x^2 \right)\Big|_0^2$

$= (e^2 - 2) - (1 - 0) = e^2 - 3$ sq units.

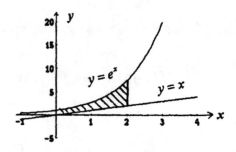

51. $A = \int_0^1 (x^3 - 3x^2 + 2x)\,dx - \int_1^2 (x^3 - 3x^2 + 2x)\,dx$

$\quad = \frac{x^4}{4} - x^3 + x^2\Big|_0^1 - \left(\frac{x^4}{4} - x^3 + x^2\right)\Big|_1^2$

$\quad = \frac{1}{4} - 1 + 1 - [(4 - 8 + 4) - (\frac{1}{4} - 1 + 1)]$

$\quad = \frac{1}{4} + \frac{1}{4} = \frac{1}{2}$ sq units.

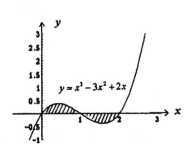

53.

$A = \frac{1}{3}\int_0^3 \frac{x}{\sqrt{x^2 + 16}}\,dx = \frac{1}{3} \cdot \frac{1}{2} \cdot 2(x^2 + 16)^{1/2}\Big|_0^3$

$\quad = \frac{1}{3}(x^2 + 16)^{1/2}\Big|_0^3 = \frac{1}{3}(5 - 4) = \frac{1}{3}$ sq units.

55. To find the equilibrium point, we solve $0.1x^2 + 2x + 20 = -0.1x^2 - x + 40$
$0.2x^2 + 3x - 20 = 0$, $x^2 + 15x - 100 = 0$, $(x + 20)(x - 5) = 0$, or $x = 5$.
Therefore, $p = -0.1(25) - 5 + 40 = 32.5$.

$CS = \int_0^5 (-0.1x^2 - x + 40)\,dx - (5)(32.5) = -\frac{0.1}{3}x^3 - \frac{1}{2}x^2 + 40x\Big|_0^5 - 162.5$

$\quad = 20.833$, or \$2083.

$PS = (5)(32.5) - \int_0^5 (0.1x^2 + 2x + 20)\,dx = 162.5 - \frac{0.1}{3}x^3 + x^2 + 20x\Big)\Big|_0^5$

$\quad = 33.333$, or \$3,333.

57. Use Equation (18) with $P = 925$, $m = 12$, $T = 30$, and $r = 0.12$, obtaining

$$PV = \frac{mP}{r}(1 - e^{-rT}) = \frac{(12)(925)}{(0.12)}(1 - e^{-0.12(30)}) = 89{,}972.56,$$

and we conclude that the present value of the purchase price of the house is \$89,972.56 + \$9000 , or \$98,972.56.

59. a.

b. $f(0.3) = \frac{17}{18}(0.3)^2 + \frac{1}{18}(0.3) \approx 0.1$

so that 30 percent of the people receive 10 percent of the total income.

$f(0.6) = \frac{17}{18}(0.6)^2 + \frac{1}{18}(0.6) \approx 0.37$

so that 60 percent of the people receive 37 percent of the total revenue.

c. The coefficient of inequality for this curve is

$$L = 2\int_0^1 [x - \frac{17}{18}x^2 - \frac{1}{18}x]\,dx = \frac{17}{9}\int_0^1 (x - x^2)\,dx = \frac{17}{9}\left(\frac{1}{2}x^2 - \frac{1}{3}x^3\right)\Big|_0^1 = \frac{17}{54} \approx 0.315.$$

61. $V = \pi\int_1^3 \frac{dx}{x^2} = -\frac{\pi}{x}\Big|_1^3 = \pi(-\frac{1}{3}+1) = \frac{2\pi}{3}$ cu units.

CHAPTER 7

EXERCISES 7.1, page 547

1. $I = \int xe^{2x}\,dx$. Let $u = x$ and $dv = e^{2x}\,dx$. Then $du = dx$ and $v = \frac{1}{2}e^{2x}$. Therefore,

$$I = uv - \int v\,du = \frac{1}{2}xe^{2x} - \int \frac{1}{2}e^{2x}\,dx = \frac{1}{2}xe^{2x} - \frac{1}{4}e^{2x} = \frac{1}{4}e^{2x}(2x-1) + C.$$

3. Let $u = x$ and $dv = e^{x/4}\,dx$. Then $du = dx$ and $v = 4e^{x/4}$.

$$\int xe^{x/4}\,dx = uv - \int v\,du = 4xe^{x/4} - 4\int e^{x/4}\,dx = 4xe^{x/4} - 16e^{x/4} + C$$

$$= 4(x-4)e^{x/4} + C.$$

5. $\int (e^x - x)^2\,dx = \int (e^{2x} - 2xe^x + x^2)\,dx = \int e^{2x}\,dx - 2\int xe^x\,dx + \int x^2\,dx.$

Using the result $\int xe^x\,dx = (x-1)e^x + k$, from Example 1, we see that

$$\int (e^x - x)^2\,dx = \frac{1}{2}e^{2x} - 2(x-1)e^x + \frac{1}{3}x^3 + C.$$

7. $I = \int (x+1)e^x\,dx$. Let $u = x+1$, $dv = e^x\,dx$. Then $du = dx$ and $v = e^x$. Therefore,

$$I = (x+1)e^x - \int e^x\,dx = (x+1)e^x - e^x + C = xe^x + C.$$

9. Let $u = x$ and $dv = (x+1)^{-3/2}\,dx$. Then $du = dx$ and $v = -2(x+1)^{-1/2}$.

$$\int x(x+1)^{-3/2}\,dx = uv - \int v\,du = -2x(x+1)^{-1/2} + 2\int (x+1)^{-1/2}\,dx$$

$$= -2x(x+1)^{-1/2} + 4(x+1)^{1/2} + C$$

$$= 2(x+1)^{-1/2}[-x + 2(x+1)] + C = \frac{2(x+2)}{\sqrt{x+1}} + C.$$

11. $I = \int x(x-5)^{1/2}\,dx$. Let $u = x$ and $dv = (x-5)^{1/2}\,dx$. Then $du = dx$ and

$v = \frac{2}{3}(x-5)^{3/2}$. Therefore,

$$I = \tfrac{2}{3}x(x-5)^{3/2} - \int \tfrac{2}{3}(x-5)^{3/2}\,dx = \tfrac{2}{3}x(x-5)^{3/2} - \tfrac{2}{3}\cdot\tfrac{2}{5}(x-5)^{5/2} + C$$

$$= \tfrac{2}{3}(x-5)^{3/2}[x - \tfrac{2}{5}(x-5)] + C = \tfrac{2}{15}(x-5)^{3/2}(5x-2x+10) + C$$

$$= \tfrac{2}{15}(x-5)^{3/2}(3x+10) + C.$$

13. $I = \displaystyle\int x\,\ln 2x\,dx.$ Let $u = \ln 2x$ and $dv = x\,dx$. Then $du = \tfrac{1}{x}\,dx$ and $v = \tfrac{1}{2}x^2$.

Therefore,

$$I = \tfrac{1}{2}x^2 \ln 2x - \int \tfrac{1}{2}x\,dx = \tfrac{1}{2}x^2 \ln 2x - \tfrac{1}{4}x^2 + C = \tfrac{1}{4}x^2(2\ln 2x - 1) + C.$$

15. Let $u = \ln x$ and $dv = x^3\,dx$, then $du = \tfrac{1}{x}\,dx$, and $v = \tfrac{1}{4}x^4$.

$$\int x^3 \ln x\,dx = \tfrac{1}{4}x^4 \ln x - \tfrac{1}{4}\int x^3\,dx = \tfrac{1}{4}x^4 \ln x - \tfrac{1}{16}x^4 + C$$

$$= \tfrac{1}{16}x^4(4\ln x - 1) + C.$$

17. Let $u = \ln x^{1/2}$ and $dv = x^{1/2}\,dx$. Then $du = \tfrac{1}{2x}\,dx$ and $v = \tfrac{2}{3}x^{3/2}$,

and $\displaystyle\int \sqrt{x}\,\ln \sqrt{x}\,dx = uv - \int v\,du = \tfrac{2}{3}x^{3/2}\ln x^{1/2} - \tfrac{1}{3}\int x^{1/2}\,dx$

$$= \tfrac{2}{3}x^{3/2}\ln x^{1/2} - \tfrac{2}{9}x^{3/2} + C = \tfrac{2}{9}x\sqrt{x}(3\ln\sqrt{x} - 1) + C.$$

19. Let $u = \ln x$ and $dv = x^{-2}\,dx$. Then $du = \tfrac{1}{x}\,dx$ and $v = -x^{-1}$,

$$\int \frac{\ln x}{x^2}\,dx = uv - \int v\,du = -\frac{\ln x}{x} + \int x^{-2}\,dx = -\frac{\ln x}{x} - \frac{1}{x} + C$$

$$= -\frac{1}{x}(\ln x + 1) + C.$$

21. Let $u = \ln x$ and $dv = dx$. Then $du = \tfrac{1}{x}\,dx$ and $v = x$ and

$$\int \ln x\,dx = uv - \int v\,du = x\ln x - \int dx = x\ln x - x + C = x(\ln x - 1) + C.$$

23. Let $u = x^2$ and $dv = e^{-x}\,dx$. Then $du = 2x\,dx$ and $v = -e^{-x}$, and

$$\int x^2 e^{-x}\,dx = uv - \int v\,du = -x^2 e^{-x} + 2\int xe^{-x}\,dx.$$

We can integrate by parts again, or, using the result of Problem 2, we find

$$\int x^2 e^{-x}\, dx = -x^2 e^{-x} + 2[-(x+1)e^{-x}] + C = -x^2 e^{-x} - 2(x+1)e^{-x} + C$$

$$= -(x^2 + 2x + 2)e^{-x} + C.$$

25. $I = \int x(\ln x)^2\, dx$. Let $u = (\ln x)^2$ and $dv = x\, dx$, so that

$$du = 2(\ln x)\left(\frac{1}{x}\right) = \frac{2\ln x}{x} \text{ and } v = \tfrac{1}{2}x^2. \text{ Then}$$

$$I = \tfrac{1}{2}x^2(\ln x)^2 - \int x \ln x\, dx.$$

Next, we evaluate $\int x \ln x\, dx$, by letting $u = \ln x$ and $dv = x\, dx$, so that $du = \tfrac{1}{x}\, dx$

and $v = \tfrac{1}{2}x^2$. Then

$$\int x \ln x\, dx = \tfrac{1}{2}x^2(\ln x) - \tfrac{1}{2}\int x\, dx = \tfrac{1}{2}x^2 \ln x - \tfrac{1}{4}x^2 + C.$$

Therefore, $\int x(\ln x)^2\, dx = \tfrac{1}{2}x^2(\ln x)^2 - \tfrac{1}{2}x^2 \ln x + \tfrac{1}{4}x^2 + C$

$$= \tfrac{1}{4}x^2[2(\ln x)^2 - 2\ln x + 1] + C.$$

27. $\int_0^{\ln 2} xe^x\, dx = (x-1)e^x\Big|_0^{\ln 2}$ (Using the results of Example 1.)

$$= (\ln 2 - 1)e^{\ln 2} - (-e^0) = 2(\ln 2 - 1) + 1 \quad (\text{Recall } e^{\ln 2} = 2.)$$
$$= 2\ln 2 - 1.$$

29. We first integrate $I = \int \ln x\, dx$. Integrating by parts with $u = \ln x$ and $dv = dx$ so

that $du = \tfrac{1}{x}\, dx$ and $v = x$, we find

$$I = x \ln x - \int dx = x \ln x - x + C = x(\ln x - 1) + C.$$

Therefore, $\int_1^4 \ln x\, dx = x(\ln x - 1)\Big|_1^4 = 4(\ln 4 - 1) - 1(\ln 1 - 1) = 4\ln 4 - 3.$

31. Let $u = x$ and $dv = e^{2x}\, dx$. Then $u = dx$ and $v = \tfrac{1}{2}e^{2x}$ and

$$\int_0^2 xe^{2x}\,dx = \tfrac{1}{2}xe^{2x}\Big|_0^2 - \tfrac{1}{2}\int_0^2 e^{2x}\,dx = e^4 - \tfrac{1}{4}e^{2x}\Big|_0^2$$

$$= e^4 - \tfrac{1}{4}e^4 + \tfrac{1}{4} = \tfrac{1}{4}(3e^4 + 1).$$

33. Let $u = x$ and $dv = e^{-2x}\,dx$, so that $du = dx$ and $v = -\tfrac{1}{2}e^{-2x}$.

$$f(x) = \int xe^{-2x}\,dx = -\tfrac{1}{2}xe^{-2x} - \tfrac{1}{4}e^{-2x} + C$$

$$f(0) = -\tfrac{1}{4} + C = 3 \quad \text{and} \quad C = \tfrac{13}{4}.$$

Therefore, $y = -\tfrac{1}{2}xe^{-2x} - \tfrac{1}{4}e^{-2x} + \tfrac{13}{4}$.

35. The required area is given by $\displaystyle\int_1^5 \ln x\,dx$.

We first find $\int \ln x\,dx$. Using the technique of integration by parts with $u = \ln x$ and $dv = dx$ so that $du = \tfrac{1}{x}\,dx$ and $v = x$, we have

$$\int \ln x\,dx = x\ln x - \int dx = x\ln x - x = x(\ln x - 1) + C.$$

Therefore,

$$\int_1^5 \ln x\,dx = x(\ln x - 1)\Big|_1^5 = 5(\ln 5 - 1) - 1(\ln 1 - 1) = 5\ \ln 5 - 4$$

and the required area is (5 ln 5 - 4) sq units.

37. The distance covered is given by $\displaystyle\int_0^{10} 100te^{-0.2t}\,dt = 100\int_0^{10} te^{-0.2t}\,dt$.

We integrate by parts, letting $u = t$ and $dv = e^{-0.2t}\,dt$ so that $du = dt$ and $v = -\dfrac{1}{0.2}e^{-0.2t} = -5e^{-0.2t}$. Therefore,

$$100\int_0^{10} te^{-0.2t}\,dt = 100\left[-5te^{-0.2t}\Big|_0^{10}\right] + 5\int_0^{10} e^{-0.2t}\,dt$$

$$= 100[-5te^{-0.2t} - 25e^{-0.2t}]\Big|_0^{10} = -500e^{-0.2t}(t+5)\Big|_0^{10}$$

$$= -500e^{-2}(15) + 500(5) = 1485, \text{ or } 1485 \text{ feet.}$$

39. The average concentration is $C = \dfrac{1}{12}\displaystyle\int_0^{12} 3te^{-t/3}\,dt = \dfrac{1}{4}\int_0^{12} te^{-t/3}\,dt$.

Let $u = t$ and $dv = e^{-t/3}\, dt$. So $du = dt$ and $v = -3e^{-t/3}$. Then

$$C = \frac{1}{4}\left[-3te^{-t/3}\Big|_0^{12} + 3\int_0^{12} e^{-t/3}\, dt\right] = \frac{1}{4}\left\{-36e^{-4} - \left[9e^{-t/3}\Big|_0^{12}\right]\right\}$$

$$= \tfrac{1}{4}(-36e^{-4} - 9e^{-4} + 9) \approx 2.04 \quad \text{mg/ml.}$$

41. $N = 2\int te^{-0.1t}\, dt$. Let $u = t$ and $dv = e^{-0.1t}$, so that $du = dt$ and $v = -10e^{-0.1t}$. Then

$v = -10e^{-0.1t}$. Then

$$N(t) = 2[-10te^{-0.1t} + 10\int e - 0.1t\, dt] = 2(-10te^{-0.1t} - 100e^{-0.1t}) + C$$

$$= -20e^{-0.1t}(t + 10) + 200. \qquad\qquad [N(0) = 0]$$

43. The present value of the franchise is

$$PV = \int_0^T P(t)e^{-rt}\, dt = \int_0^{15}(50{,}000 + 3000t)e^{-0.1t}\, dt$$

$$= 50{,}000\int_0^{15} e^{-0.1t}\, dt + 3000\int_0^{15} te^{-0.1t}\, dt.$$

The first integral is

$$50{,}000\int_0^{15} e^{-0.1t}\, dt = \frac{50{,}000}{-0.1}e^{-0.1t}\Big|_0^{15} = -500{,}000(e^{-1.5} - 1) \approx 388{,}435.$$

The second integral is evaluated by integration by parts . Let $u = t$ and $dv = e^{-0.1t}\, dt$ so that $du = dt$ and $v = -10e^{-0.1t}$.

$$3000\int_0^{15} te^{-0.1t}\, dt = 3000[-10te^{-0.1t}\Big|_0^{15} + 10\int_0^{15} e^{-0.1t}\, dt]$$

$$= 3000\left[-150e^{-1.5} - 100e^{-0.1t}\Big|_0^{15}\right]$$

$$= 3000(-150\, e^{-1.5} - 100e^{-1.5} + 100) \approx 132{,}652.$$

Therefore, $PV \approx 388{,}435 + 132{,}652 = 521{,}087$ or \$521,087.

45. True. This is just the integration by parts formula

EXERCISES 7.2, page 563

1. First we note that

$$\int \frac{2x}{2 + 3x}\, dx = 2\int \frac{x}{2 + 3x}\, dx.$$

Next, we use Formula 1 with $a = 2$, $b = 3$, and $u = x$. Then

$$\int \frac{2x}{2+3x} dx = \frac{2}{9}[2 + 3x - 2\ln|2 + 3x|] + C.$$

3. $\displaystyle \int \frac{3x^2}{2+4x} dx = \frac{3}{2} \int \frac{x^2}{1+2x} dx.$

Use Formula 2 with $a = 1$ and $b = 2$ obtaining

$$\int \frac{3x^2}{2+4x} dx = \frac{3}{32}[(1+2x)^2 - 4(1+2x) + 2\ln|1+2x|] + C.$$

5. $\displaystyle \int x^2\sqrt{9+4x^2}\, dx = \int x^2\sqrt{4(\tfrac{9}{4})+x^2)}\, dx = 2\int x^2\sqrt{(\tfrac{3}{2})^2 + x^2}\, dx.$

Use Formula (8) with $a = 3/2$, we find that

$$\int x^2\sqrt{9+4x^2}\, dx == 2[\tfrac{x}{8}(\tfrac{9}{4}+2x^2)\sqrt{\tfrac{9}{4}+x^2} - \tfrac{81}{128}\ln\left|x + \sqrt{\tfrac{9}{4}+x^2}\right| + C.$$

7. Use Formula 6 with $a = 1$, $b = 4$, and $u = x$, then

$$\int \frac{dx}{x\sqrt{1+4x}} = \ln\left|\frac{\sqrt{1+4x}-1}{\sqrt{1+4x}+1}\right| + C.$$

9. Use Formula 9 with $a = 3$ and $u = 2x$. Then $du = 2\, dx$ and

$$\int_0^2 \frac{dx}{\sqrt{9+4x^2}} = \frac{1}{2}\int_0^4 \frac{du}{\sqrt{3^2+u^2}} = \frac{1}{2}\ln\left|u + \sqrt{9+u^2}\right|\Big|_0^4$$

$$= \frac{1}{2}(\ln 9 - \ln 3) = \frac{1}{2}\ln 3.$$

Note that the limits of integration have been changed from $x = 0$ to $x = 2$ and from $u = 0$ to $u = 4$.

11. Using Formula 22 with $a = 3$, we see that $\displaystyle \int \frac{dx}{(9-x^2)^{3/2}} = \frac{x}{9\sqrt{9-x^2}} + C.$

13. $\displaystyle \int x^2\sqrt{x^2-4}\, dx.$

Use Formula 14 with $a = 2$ and $u = x$, obtaining

$$\int x^2\sqrt{x^2-4}\, dx = \tfrac{x}{8}(2x^2-4)\sqrt{x^2-4} - 2\ln\left|x + \sqrt{x^2-4}\right| + C.$$

15. Using Formula 19 with $a = 2$ and $u = x$, we have

$$\int \frac{\sqrt{4-x^2}}{x} dx = \sqrt{4-x^2} - 2\ln\left|\frac{2+\sqrt{4-x^2}}{x}\right| + C.$$

17. $\int xe^{2x}\, dx.$

Use Formula 23 with $u = x$ and $a = 2$, obtaining

$$\int xe^{2x}\, dx = \frac{1}{4}(2x-1)e^{2x} + C.$$

19. $\int \dfrac{dx}{(x+1)\ln(x+1)}.$

Let $u = x + 1$ so that $du = dx$. Then $\displaystyle\int \frac{dx}{(x+1)\ln(x+1)} = \int \frac{du}{u\ln u}.$

Use Formula 28 with $u = x$, obtaining

$$\int \frac{du}{u\ln u} = \ln|\ln u| + C.$$

Therefore, $\displaystyle\int \frac{dx}{(x+1)\ln(x+1)} = \ln|\ln(x+1)| + C$

21. $\int \dfrac{e^{2x}}{(1+3e^x)^2}\, dx.$

Put $u = e^x$ then $du = e^x dx$. Then we use Formula 3 with $a = 1, b = 3$. Then

$$I = \int \frac{u}{(1+3u)^2}\, du = \frac{1}{9}\left[\frac{1}{1+3u} + \ln|1+3u|\right] + C$$

$$= \frac{1}{9}\left[\frac{1}{1+3e^x} + \ln(1+3e^x)\right] + C$$

23. $\int \dfrac{3e^x}{1+e^{x/2}}\,dx = 3\int \dfrac{e^{x/2}}{e^{-x/2}+1}\,dx$.

Let $v = e^{x/2}$ so that $dv = \frac{1}{2}e^{x/2}dx$ or $e^{x/2}\,dx = 2\,dv$. Then

$$\int \dfrac{3e^x}{1+e^{x/2}}\,dx = 6\int \dfrac{dv}{\frac{1}{v}+1} = 6\int \dfrac{v}{v+1}\,dv.$$

Use Formula 1 with $a = 1$, $b = 1$, and $u = v$, obtaining

$$6\int \dfrac{v}{v+1}\,dv = 6[1+v-\ln|1+v|]+C. \text{ So } \int \dfrac{3e^x}{1+e^{x/2}}\,dx = 6[1+e^{x/2}-\ln(1+e^{x/2})]+C.$$

This answer may be written in the form $6[e^{x/2}-\ln(1+e^{x/2})]+C$ since C is an arbitrary constant.

25. $\int \dfrac{\ln x}{x(2+3\ln x)}\,dx.$ Let $v = \ln x$ so that $dv = \dfrac{1}{x}dx.$ Then

$$\int \dfrac{\ln x}{x(2+3\ln x)}\,dx = \int \dfrac{v}{2+3v}\,dv.$$

Use Formula 1 with $a = 2$, $b = 3$, and $u = v$ to obtain

$$\int \dfrac{v}{2+3v}\,dv = \frac{1}{9}[2+3\ln x - 2\ln|2+3\ln x|]+C . \text{ So}$$

$$\int \dfrac{\ln x}{x(2+3\ln x)}\,dx = \frac{1}{9}[2+3\ln x - 2\ln|2+3\ln x|]+C.$$

27. Using Formula 24 with $a = 1$, $n = 2$, and $u = x$. Then

$$\int_0^1 x^2 e^x\,dx = x^2 e^x\Big|_0^1 - 2\int_0^1 xe^x\,dx = x^2 e^x - 2(xe^x - e^x)\Big|_0^1$$

$$= x^2 e^x - 2xe^x + 2e^x\Big|_0^1 = e - 2e + 2e - 2 = e - 2.$$

29. $\int x^2 \ln x\,dx.$

Use Formula 27 with $n = 2$ and $u = x$, obtaining

$$\int x^2 \ln x\,dx = \dfrac{x^3}{9}(3\ln x - 1)+C.$$

31. $\int (\ln x)^3\,dx.$

Use Formula 29 with $n = 3$ to write

$$\int (\ln x)^3 \, dx = x(\ln x)^3 - 3\int (\ln x)^2 \, dx.$$

Using Formula 29 again with $n = 2$, we obtain

$$\int (\ln x)^3 \, dx = x(\ln x)^3 - 3[x(\ln x)^2 - 2\int \ln x \, dx].$$

Using Formula 29 one more time with $n = 1$ gives

$$\int (\ln x)^3 \, dx = x(\ln x)^3 - 3x(\ln x)^2 + 6(x \ln x - x) + C$$

$$= x(\ln x)^3 - 3x(\ln x)^2 + 6x \ln x - 6x + C.$$

33. Letting $p = 50$ gives

$$50 = \frac{250}{\sqrt{16 + x^2}}$$

from which we deduce that $\sqrt{16 + x^2} = 5$, $16 + x^2 = 25$, and $x = 3$.
Using Formula 9 with $u = 3$, we see that

$$CS = \int_0^3 \frac{250}{\sqrt{16 + x^2}} \, dx = 50(3) = 250 \int_0^3 \frac{1}{\sqrt{16 + x^2}} \, dx - 150$$

$$= 250 \ln \left| x + \sqrt{16 + x^2} \right| \Big|_0^3 - 150 = 250[\ln 8 - \ln 4] - 150$$

$$= 23.286795, \text{ or approximately } \$2{,}329.$$

35. The number of visitors admitted to the amusement park by noon is found by evaluating the integral

$$\int_0^3 \frac{60}{(2 + t^2)^{3/2}} \, dt = 60 \int_0^3 \frac{dt}{(2 + t)^{3/2}}.$$

Using Formula 12 with $a = \sqrt{2}$ and $u = t$, we find

$$60 \int_0^3 \frac{dt}{(2 + t^2)^{3/2}} = 60 \left[\frac{t}{2\sqrt{2 + t^3}} \right]_0^3 = 60 \left[\frac{3}{2\sqrt{11} - 0} \right] = \frac{90}{\sqrt{11}} = 27.136,$$

or 27,136.

37. In the first 10 days:

$$\frac{1}{10}\int_0^{10}\frac{1000}{1+24e^{-0.02t}}\,dt = 100\int_0^{10}\frac{1}{1+24e^{-0.02t}}\,dt = 100\left[t+\frac{1}{0.02}\ln(1+24e^{-0.02t})\right]_0^{10}$$

(Use Formula 25 with $a = 0.02$ and $b = 24$.)

$$= 100[10+50\ln 20.64953807 - 50\ln 25] = 44.0856,$$

or approximately 44 fruitflies.

In the first 20 days:

$$\frac{1}{20}\int_0^{20}\frac{1000}{1+24e^{-0.02t}}\,dt = 50\int_0^{10}\frac{1}{1+24e^{-0.02t}}\,dt$$

$$= 500[t+\ln(1+24e^{-0.02t})]_0^{20}$$

$$= 50[20+50\ln 17.0876822 - 50\ln 25] = 48.71$$

or approximately 49 fruitflies.

39. $\dfrac{1}{5}\displaystyle\int_0^5\frac{100,000}{2(1+1.5e^{-0.2t})}\,dt = 10,000\int_0^5\frac{1}{1+1.5e^{-0.2t}}\,dt$

$$= 10,000[t+5\ln(1+1.5e^{-0.2t})]\Big|_0^5$$

(Use Formula 25 with $a = -0.2$ and $b = 1.5$.)

$$= 10,000[5+5\ln 1.551819162 - 5\ln 2.5] \approx 26157,$$

or approximately 26,157 people.

41. $\displaystyle\int_0^5 20,000te^{0.15t}\,dt = 20,000\int_0^5 te^{0.15t}\,dt$

$$= 20,000\left[\frac{1}{(0.15)^2}(0.15t-1)e^{0.15t}\right]_0^5$$

(Use Formula 23 with $a = 0.15$.)

$$=888,888.8889[-0.25e^{0.75}+1] = \$418,444.$$

EXERCISES 7.3, page 578

1. $\Delta x = \frac{2}{6} = \frac{1}{3}, x_0 = 0, x_1 = \frac{1}{3}, x_2 = \frac{2}{3}, x_3 = 1, x_4 = \frac{4}{3}, x_5 = \frac{5}{3}, x_6 = 2.$

    Trapezoidal Rule:

    $$\int_0^2 x^2\, dx \approx \tfrac{1}{6}\left[0 + 2(\tfrac{1}{3})^2 + 2(\tfrac{2}{3})^2 + 2(1)^2 + 2(\tfrac{4}{3})^2 + 2(\tfrac{5}{3})^2 + 2^2\right]$$

    $$\approx \tfrac{1}{6}\ (0.22222 + 0.88889 + 2 + 3.55556 + 5.55556 + 4)$$

    $$\approx 2.7037.$$

    Simpson's Rule:

    $$\int x^2\, dx = \tfrac{1}{9}[0 + 4(\tfrac{1}{3})^2 + 2(\tfrac{2}{3})^2 + 4(1)^2 + 2(\tfrac{4}{3})^2 + 4(\tfrac{5}{3})^2 + 2^2]$$

    $$\approx \tfrac{1}{9}\ (0.44444 + 0.88889 + 4 + 3.55556 + 11.11111 + 4) \approx 2.6667.$$

    Exact Value: $\displaystyle\int_0^2 x^2\, dx = \tfrac{1}{3}x^3\Big|_0^2 = \tfrac{8}{3} = 2\tfrac{2}{3}.$

3. $\Delta x = \dfrac{b-a}{n} = \dfrac{1-0}{4} = \tfrac{1}{4}; x_0 = 0, x_1 = \tfrac{1}{4}, x_2 = \tfrac{1}{2}, x_3 = \tfrac{3}{4}, x_4 = 1.$

    Trapezoidal Rule:

    $$\int_0^1 x^3\, dx \approx \tfrac{1}{8}\left[0 + 2(\tfrac{1}{4})^3 + 2(\tfrac{1}{2})^3 + 2(\tfrac{3}{4})^3 + 1^3\right] \approx \tfrac{1}{8}(0 + 0.3125 + 0.25 + 0.8)$$

    $$\approx 0.265625.$$

    Simpson's Rule:

    $$\int_0^1 x^3\, dx \approx \tfrac{1}{12}\left[0 + 4(\tfrac{1}{4})^3 + 2(\tfrac{1}{2})^3 + 4(\tfrac{3}{4})^3 + 1\right] \approx \tfrac{1}{12}[0 + 0.625 + 0.25 + 1.6875 + 1]$$

    $$\approx 0.25.$$

    Exact Value:

    $$\int_0^1 x^3\, dx = \tfrac{1}{4}x^4\Big|_0^1 = \tfrac{1}{4} - 0 = \tfrac{1}{4}.$$

5. a. Here $a = 1$, $b = 2$, and $n = 4$; so $\Delta x = \frac{2-1}{4} = \tfrac{1}{4} = 0.25$, and $x_0 = 1$, $x_1 = 1.25$, $x_2 = 1.5$, $x_3 = 1.75$, $x_4 = 2$.

    Trapezoidal Rule:

    $$\int_1^2 \frac{1}{x}\, dx \approx \frac{0.25}{2}\left[1 + 2\left(\frac{1}{1.25}\right) + 2\left(\frac{1}{1.5}\right) + 2\left(\frac{1}{1.75}\right) + \frac{1}{2}\right] \approx 0.697.$$

Simpson's Rule:

$$\int_1^2 \frac{1}{x}\,dx \approx \frac{0.25}{3}\left[1+4\left(\frac{1}{1.25}\right)+2\left(\frac{1}{1.5}\right)+4\left(\frac{1}{1.75}\right)+\frac{1}{2}\right] \approx 0.6933.$$

$$\int_1^2 \frac{1}{x}\,dx = \ln x\big|_1^2 = \ln 2 - \ln 1 \approx 0.6931.$$

7. $\Delta x = \frac{1}{4}$, $x_0 = 1$, $x_1 = \frac{5}{4}$, $x_2 = \frac{3}{2}$, $x_3 = \frac{7}{4}$, $x_4 = 2$.

Trapezoidal Rule:

$$\int_1^2 \frac{1}{x^2}\,dx \approx \tfrac{1}{8}\left[1+2\left(\tfrac{4}{5}\right)^2+2\left(\tfrac{2}{3}\right)^2+2\left(\tfrac{4}{7}\right)^2+\left(\tfrac{1}{2}\right)^2\right] \approx 0.5090.$$

Simpson's Rule:

$$\int_1^2 \frac{1}{x^2}\,dx \approx \tfrac{1}{12}\left[1+4\left(\tfrac{4}{5}\right)^2+2\left(\tfrac{2}{3}\right)^2+4\left(\tfrac{4}{7}\right)^2+\left(\tfrac{1}{2}\right)^2\right] \approx 0.5004.$$

Exact Value:

$$\int_1^2 \frac{1}{x^2}\,dx = -\frac{1}{x}\bigg|_1^2 = -\frac{1}{2}+1 = \frac{1}{2}.$$

9. $\Delta x = \frac{b-a}{n} = \frac{4-0}{8} = \frac{1}{2}$; $x_0 = 0, x_1 = \frac{1}{2}, x_2 = \frac{2}{2}, x_3 = \frac{3}{2}, \ldots, x_8 = \frac{8}{2}$.

Trapezoidal Rule:

$$\int_0^4 \sqrt{x}\,dx \approx \frac{\frac{1}{2}}{2}\left[0+2\sqrt{0.5}+2\sqrt{1}+2\sqrt{1.5}+\cdots+2\sqrt{3.5}+\sqrt{4}\right] \approx 5.26504.$$

Simpson's Rule:

$$\int_0^4 \sqrt{x}\,dx \approx \frac{\frac{1}{2}}{3}\left[0+4\sqrt{0.5}+2\sqrt{1}+4\sqrt{1.5}+\cdots+4\sqrt{3.5}+\sqrt{4}\right] \approx 5.30463.$$

The actual value is

$$\int_0^4 \sqrt{x}\,dx \approx \frac{2}{3}x^{3/2}\bigg|_0^4 = \frac{2}{3}(8) = \frac{16}{3} \approx 5.333333.$$

11. $\Delta x = \frac{1-0}{6} = \frac{1}{6}$; $x_0 = 0, x_1 = \frac{1}{6}, x_2 = \frac{2}{6}, \ldots, x_6 = \frac{6}{6}$.

Trapezoidal Rule:

$$\int_0^1 e^{-x}\,dx \approx \tfrac{\frac{1}{6}}{2}[1+2e^{-1/6}+2e^{-2/6}+\cdots+2e^{-5/6}+e^{-1}] \approx 0.633583.$$

Simpson's Rule:

$$\int_0^1 e^{-x}\, dx \approx \tfrac{1}{6}[1 + 4e^{-1/6} + 2e^{-2/6} + \cdots + 4e^{-5/6} + e^{-1}] \approx 0.632123.$$

The actual value is

$$\int_0^1 e^{-x}\, dx = -e^{-x}\Big|_0^1 = -e^{-1} + 1 \approx 0.632121.$$

13. $\Delta x = \tfrac{1}{4};\ x_0 = 0, x_1 = \tfrac{5}{4}, x_2 = \tfrac{3}{2}, x_3 = \tfrac{7}{4}, x_4 = 2.$

Trapezoidal Rule:

$$\int_1^2 \ln x\, dx \approx \tfrac{1}{8}[\ln 1 + 2\ln\tfrac{5}{4} + 2\ln\tfrac{3}{2} + 2\ln\tfrac{7}{4} + \ln 2] \approx 0.38370.$$

Simpson's Rule:

$$\int_1^2 \ln x\, dx \approx \tfrac{1}{12}[\ln 1 + 4\ln\tfrac{5}{4} + 2\ln\tfrac{3}{2} + 4\ln\tfrac{7}{4} + \ln 2] \approx 0.38626.$$

Exact Value:

$$\int_1^2 \ln x\, dx \approx x(\ln x - 1)\Big|_1^2 = 2(\ln 2 - 1) + 1 = 2\ln 2 - 1.$$

15. $\Delta x = \tfrac{1-0}{4} = \tfrac{1}{4};\ x_0 = 0, x_1 = \tfrac{1}{4}, x_2 = \tfrac{2}{4}, x_3 = \tfrac{3}{4}, x_4 = \tfrac{4}{4}.$

Trapezoidal Rule:

$$\int_0^1 \sqrt{1+x^3}\, dx \approx \tfrac{1}{2}\left[\sqrt{1} + 2\sqrt{1+(\tfrac{1}{4})^3} + \cdots + 2\sqrt{1+(\tfrac{3}{4})^3} + \sqrt{2}\right] \approx 1.1170.$$

Simpson's Rule:

$$\int_0^1 \sqrt{1+x^3}\, dx \approx \tfrac{1}{3}\left[\sqrt{1} + 4\sqrt{1+(\tfrac{1}{4})^3} + 2\sqrt{1+(\tfrac{2}{4})^3} \cdots + 4\sqrt{1+(\tfrac{3}{4})^3} + \sqrt{2}\right] \approx 1.1114.$$

17. $\Delta x = \tfrac{2-0}{4} = \tfrac{1}{2};\ x_0 = 0, x_1 = \tfrac{1}{2}, x_2 = \tfrac{2}{2}, x_3 = \tfrac{3}{2}, x_4 = \tfrac{4}{2}.$

Trapezoidal Rule:

$$\int_0^2 \frac{1}{\sqrt{x^3+1}}\, dx = \frac{\tfrac{1}{2}}{2}\left[1 + \frac{2}{\sqrt{(\tfrac{1}{2})^3+1}} + \frac{2}{\sqrt{(1)^3+1}} + \frac{2}{\sqrt{(\tfrac{3}{2})^3+1}} + \frac{1}{\sqrt{(2)^3+1}}\right]$$

$$\approx 1.3973$$

Simpson's Rule:

$$\int_0^2 \frac{1}{\sqrt{x^3+1}}\, dx = \frac{\tfrac{1}{2}}{3}\left[1 + \frac{4}{\sqrt{(\tfrac{1}{2})^3+1}} + \frac{2}{\sqrt{(1)^3+1}} + \frac{4}{\sqrt{(\tfrac{3}{2})^3+1}} + \frac{1}{\sqrt{(2)^3+1}}\right]$$

$$\approx 1.4052$$

19. $\Delta x = \frac{2}{4} = \frac{1}{2}; x_0 = 0, x_1 = \frac{1}{2}, x_2 = 1, x_3 = \frac{3}{2}, x_4 = 2.$

Trapezoidal Rule:

$$\int_0^2 e^{-x^2}\, dx = \tfrac{1}{4}[e^{-0} + 2e^{-(1/2)^2} + 2e^{-1} + 2e^{-(3/2)^2} + e^{-4}] \approx 0.8806.$$

Simpson's Rule:

$$\int_0^2 e^{-x^2}\, dx = \tfrac{1}{6}[e^{-0} + 4e^{-(1/2)^2} + 2e^{-1} + 4e^{-(3/2)^2} + e^{-4}] \approx 0.8818.$$

21. $\Delta x = \frac{2-1}{4} = \frac{1}{4}; x_0 = 1, x_1 = \frac{5}{4}, x_2 = \frac{6}{4}, x_3 = \frac{7}{4}, x_4 = \frac{8}{4}.$

Trapezoidal Rule:

$$\int_1^2 x^{-1/2} e^x\, dx = \tfrac{1}{4}\left[e + \frac{2e^{5/4}}{\sqrt{\frac{5}{4}}} + \cdots + \frac{2e^{7/4}}{\sqrt{\frac{7}{4}}} + \frac{e^2}{\sqrt{2}} \right] \approx 3.7757.$$

Simpson's Rule:

$$\int_1^2 x^{-1/2} e^x\, dx = \tfrac{1}{4}\left[e + \frac{4e^{5/4}}{\sqrt{\frac{5}{4}}} + \cdots + \frac{4e^{7/4}}{\sqrt{\frac{7}{4}}} + \frac{e^2}{\sqrt{2}} \right] \approx 3.7625.$$

23. a. Here $a = -1$, $b = 2$, $n = 10$, and $f(x) = x^5$. $f'(x) = 5x^4$ and $f''(x) = 20x^3$.
Because $f'''(x) = 60x^2 > 0$ on $(-1,0) \cup (0,2)$, we see that $f''(x)$ is increasing on $(-1,0) \cup (0,2)$. So, we take

$$M = f''(2) = 20(2^3) = 160.$$

Using (7), we see that the maximum error incurred is

$$\frac{M(b-a)^3}{12n^2} = \frac{160[2 - (-1)]^3}{12(100)} = 3.6.$$

b. We compute $f''' = 60x^2$ and $f^{(iv)}(x) = 120x$. $f^{(iv)}(x)$ is clearly increasing on $(-1,2)$, so we can take $M = f^{(iv)}(2) = 240$. Therefore, using (8), we see that an error bound is

$$\frac{M(b-a)^3}{180n^4} = \frac{240(3)^5}{180(10^4)} \approx 0.0324.$$

25. a. Here $a = 1$, $b = 3$, $n = 10$, and $f(x) = \dfrac{1}{x}$. We find

$$f'(x) = -\frac{1}{x^2}, \quad f'''(x) = \frac{2}{x^3}.$$

Since $f'''(x) = -\frac{6}{x^4} < 0$ on $(1,3)$, we see that $f''(x)$ is decreasing there. We may take $M = f''(1) = 2$. Using (7), we find an error bound is

$$\frac{M(b-a)^3}{12n^2} = \frac{2(3-1)^3}{12(100)} \approx 0.013.$$

b. $f'''(x) = -\frac{6}{x^4}$ and $f^{(iv)}(x) = \frac{24}{x^5}$. $f^{(iv)}(x)$ is decreasing on $(1,3)$, so we can

take $M = f^{(iv)}(1) = 24$. Using (8), we find an error bound is $\dfrac{24(3-1)^5}{180(10^4)} \approx 0.00043.$

27. a. Here $a = 0$, $b = 2$, $n = 8$, and $f(x) = (1+x)^{-1/2}$. We find

$$f'(x) = -\tfrac{1}{2}(1+x)^{-3/2}, \quad f''(x) = \tfrac{3}{4}(1+x)^{-5/2}.$$

Since f'' is positive and decreasing on $(0,2)$, we see that $\left|f''(x)\right| \le \tfrac{3}{4}$.

So the maximum error is $\dfrac{\tfrac{3}{4}(2-0)^3}{12(8)^2} = 0.0078125.$

b. $f''' = -\tfrac{15x}{8}(1+x)^{-7/2}$ and $f^{(4)}(x) = \dfrac{105}{16}(1+x)^{-9/2}$. Since $f^{(4)}$ is positive and

decreasing on $(0,2)$, we find $\left|f^{(4)}(x)\right| \le \tfrac{105}{16}.$

Therefore, the maximum error is $\dfrac{\tfrac{105}{16}(2-0)^5}{180(8)^4} = 0.000285.$

29. The distance covered is given by

$$d = \int_0^2 V(t)\,dt = \tfrac{\tfrac{1}{4}}{2}\left[V(0) + 2V(\tfrac{1}{4}) + \cdots + 2V(\tfrac{7}{4}) + V(2)\right]$$
$$= \tfrac{1}{8}[19.5 + 2(24.3) + 2(34.2) + 2(40.5) + 2(38.4) + 2(26.2)$$
$$+ 2(18) + 2(16) + 8] \approx 52.84, \text{ or } 52.84 \text{ miles.}$$

31. $\dfrac{1}{13}\displaystyle\int_0^{13} f(t)\,dt = (\tfrac{1}{13})(\tfrac{1}{2})\{13.2 + 2[14.8 + 16.6 + 17.2 + 18.7 + 19.3 + 22.6 + 24.2 + 25$
$$+ 24.6 + 25.6 + 26.4 + 26.6] + 26.6\}$$
$$\approx 21.65.$$

33. Solving the equation $25 = \dfrac{50}{0.01x^2 + 1}$, we see that $0.01x^2 + 1 = 2$, $0.01x^2 = 1$, and

$x = 10$. Therefore, $CS = \displaystyle\int_0^{10} \dfrac{50}{0.01x^2 + 1}\, dx - (25)(10)$ and $\Delta x = \tfrac{10}{8} = 1.25$, $x_0 = 0$,

$x_1 = 1.25$, $x_2 = 2.50$, ..., $x_8 = 10$.

a. $CS = \dfrac{1.25}{2}\left\{50 + 2\left[\dfrac{50}{0.01(1.25)^2 + 1}\right] + \cdots + \left[\dfrac{50}{0.01(10)^2 + 1}\right]\right\} - 250$

$\approx 142{,}373.56$, or \$142,373.56.

b. $CS = \dfrac{1.25}{3}\left\{50 + 4\left[\dfrac{50}{0.01(1.25)^2 + 1}\right] + \cdots + \left[\dfrac{50}{0.01(10)^2 + 1}\right]\right\} - 250$

$\approx 142{,}698.12$, or \$142,698.12.

35. $\Delta t = \tfrac{5}{10} = 0.5$, $t_0 = 0$, $t_1 = 0.5$, $t_2 = 1$, ..., $t_{10} = 5$.

$PSI = \dfrac{1}{5}\displaystyle\int_0^5\left[\dfrac{136}{1 + 0.25(t - 4.5)^2} + 28\right]dt = 27.2\int_0^5\dfrac{1}{1 + 0.25(t - 4.5)^2} + 28\,dt$

$\approx \dfrac{(0.5)(27.2)}{2}\left[\dfrac{1}{1 + 0.25(-4.5)^2} + \dfrac{2}{1 + 0.25(5 - 4.5)^2} + \cdots + \dfrac{1}{1 + 0.25(5 - 4.5)^2}\right] + 28$

≈ 103.9

37. $\Delta x = \tfrac{21 - 19}{10} = 0.2$; $x_0 = 19$, $x_1 = 19.2$ $x_2 = 19.4$, ..., $x_{10} = 21$

$P = \dfrac{100}{2.6\sqrt{2\pi}}\displaystyle\int_{19}^{21} e^{-0.5[(x-20)/2.6]^2}\, dx$

$\approx \dfrac{100}{2.6\sqrt{2\pi}}\left(\dfrac{0.2}{3}\right)\left[e^{-0.5[(19-20)/2.6]^2} + 4e^{-0.5[19.2-20)/2.6]^2} + \cdots + e^{-0.5[(21-20)/2.6]^2}\right]$

≈ 29.94, or 30 percent.

39. $R = \dfrac{60D}{\displaystyle\int_0^T C(t)\, dt} = \dfrac{480}{\displaystyle\int_0^{24} C(t)\, dt}$. Now,

$\displaystyle\int_0^{24} C(t)\, dt \approx \tfrac{24}{12} \cdot \tfrac{1}{3}[0 + 4(0) + 2(2.8) + 4(6.1) + 2(9.7) + 4(7.6) + 2(4.8)$

$\qquad\qquad + 4(3.7) + 2(1.9) + 4(0.8) + 2(0.3) + 4(0.1) + 0]$

$$\approx 74.8$$

and $R = \frac{480}{74.8} \approx 6.42$, or 6.42 liters/min.

41. False. The number n in Simpson's Rule must be even.

43. True. Using Formula 8, we see that the error incurred in the approximation is zero since, in this situation $f^{(4)}(x) = 0$ for all x in $[a, b]$.

EXERCISES 7.4, page 590

1. The required area is given by

$$\int_3^\infty \frac{2}{x^2}\, dx = \lim_{b \to \infty} \int_3^b \frac{2}{x^2}\, dx = \lim_{b \to \infty} \left(-\frac{2}{x} \right)\Big|_3^b = \lim_{b \to \infty} \left(-\frac{2}{b} + \frac{2}{3} \right) = \frac{2}{3}.$$

3. $$A = \int_3^\infty \frac{1}{(x-2)^2}\, dx = \lim_{b \to \infty} \int_3^b (x-2)^{-2}\, dx = \lim_{b \to \infty} -\frac{1}{x-2}\Big|_3^b = \lim_{b \to \infty} \left(-\frac{1}{b-2} + 1 \right) = 1.$$

5. $$A = \int_1^\infty \frac{1}{x^{3/2}}\, dx = \lim_{b \to \infty} \int_1^b x^{-3/2}\, dx = \lim_{b \to \infty} -\frac{2}{\sqrt{x}}\Big|_1^b = \lim_{b \to \infty} \left(-\frac{2}{\sqrt{b}} + 2 \right) = 2.$$

7. $$A = \int_0^\infty \frac{1}{(x+1)^{5/2}}\, dx = \lim_{b \to \infty} \int_1^b (x+1)^{-5/2}\, dx = \lim_{b \to \infty} -\frac{2}{3}(x+1)^{-3/2}\Big|_0^b$$

$$= \lim_{b \to \infty} \left[-\frac{2}{3(b+1)^{3/2}} + \frac{2}{3} \right] = \frac{2}{3}.$$

9. $$A = \int_{-\infty}^2 e^{2x}\, dx = \lim_{a \to -\infty} \int_a^2 e^{2x}\, dx = \lim_{a \to -\infty} \tfrac{1}{2} e^{2x}\Big|_a^2 = \lim_{a \to -\infty} \left(\tfrac{1}{2} e^4 - \tfrac{1}{2} e^{2a} \right) = \tfrac{1}{2} e^4.$$

11. Using symmetry, the required area is given by

$$2\int_0^\infty \frac{x}{(1+x^2)^2}\, dx = 2 \lim_{b \to \infty} \int_0^\infty \frac{x}{(1+x^2)^2}\, dx.$$

To evaluate the indefinite integral $\displaystyle\int \frac{x}{(1+x^2)^2}\,dx$, put $u = 1 + x^2$ so that

$du = 2x\,dx$ or $x\,dx = \frac{1}{2}\,du$.

Then $\displaystyle\int \frac{x}{(1+x^2)^2}\,dx = \frac{1}{2}\int \frac{du}{u^2} = -\frac{1}{2u} + C = -\frac{1}{2(1+x^2)} + C.$

Therefore, $\displaystyle 2\lim_{b\to\infty}\int_0^b \frac{x}{(1+x^2)}\,dx = \lim_{b\to\infty} -\frac{1}{(1+x^2)^2}\bigg|_0^b = \lim_{b\to\infty}\left[-\frac{1}{(1+b^2)} + 1\right] = 1,$

or 1 sq unit.

13. a. $I(b) = \displaystyle\int_0^b \sqrt{x}\,dx = \frac{2}{3}x^{3/2}\bigg|_0^b = \frac{2}{3}b^{3/2}.$

   b. $\displaystyle\lim_{b\to\infty} I(b) = \lim_{b\to\infty}\frac{2}{3}b^{3/2} = \infty.$

15. $\displaystyle\int_1^\infty \frac{3}{x^4}\,dx = \lim_{b\to\infty}\int_1^b 3x^{-4}\,dx = \lim_{b\to\infty}\left(-\frac{1}{x^3}\right)\bigg|_1^b = \lim_{b\to\infty}\left(-\frac{1}{b^3} + 1\right) = 1.$

17. $A = \displaystyle\int_4^\infty \frac{2}{x^{3/2}}\,dx = \lim_{b\to\infty}\int_4^b 2x^{-3/2}\,dx = \lim_{b\to\infty} -4x^{-1/2}\bigg|_4^b = \lim_{b\to\infty}\left(-\frac{4}{\sqrt{b}} + 2\right) = 2.$

19. $\displaystyle\int_1^\infty \frac{4}{x}\,dx = \lim_{b\to\infty}\int_1^b \frac{4}{x}\,dx = \lim_{b\to\infty} 4\ln x\big|_1^b = \lim_{b\to\infty}(4\ln b) = \infty.$

21. $\displaystyle\int_{-\infty}^0 (x-2)^{-3}\,dx = \lim_{a\to-\infty}\int_a^0 (x-2)^{-3}\,dx = \lim_{a\to-\infty} -\frac{1}{2(x-2)^2}\bigg|_a^0 = -\frac{1}{8}.$

23. $\displaystyle\int_1^\infty \frac{1}{(2x-1)^{3/2}}\,dx = \lim_{b\to\infty}\int_1^b (2x-1)^{-3/2}\,dx = \lim_{b\to\infty} -\frac{1}{(2x-1)^{1/2}}\bigg|_1^b$

   $\displaystyle = \lim_{b\to\infty}\left(-\frac{1}{\sqrt{2b-1}} + 1\right) = 1.$

25. $\displaystyle\int_0^\infty e^{-x}\,dx = \lim_{b\to\infty}\int_0^b e^{-x}\,dx = \lim_{b\to\infty} -e^{-x}\big|_0^b = \lim_{b\to\infty}(-e^{-b} + 1) = 1.$

27. $\int_{-\infty}^{0} e^{2x}\,dx = \lim_{a\to-\infty} \tfrac{1}{2}e^{2x}\Big|_{a}^{0} = \lim_{a\to-\infty}\left(\tfrac{1}{2}-\tfrac{1}{2}e^{2a}\right) = \tfrac{1}{2}.$

29. $\int_{1}^{\infty} \dfrac{e^{\sqrt{x}}}{\sqrt{x}}\,dx = \lim_{b\to\infty}\int_{1}^{b}\dfrac{e^{\sqrt{x}}}{\sqrt{x}}\,dx = \lim_{b\to\infty} -2e^{\sqrt{x}}\Big|_{1}^{b}$ (Integrate by substitution: $u = \sqrt{x}$.)

$\qquad = \lim_{b\to\infty}(2e^{\sqrt{b}} - 2e) = \infty,$ and so it diverges.

31. $\int_{-\infty}^{0} xe^{x}\,dx = \lim_{a\to-\infty}\int_{a}^{0} xe^{x}\,dx = \lim_{a\to-\infty}(x-1)e^{x}\Big|_{a}^{0} = \lim_{a\to-\infty}[-1+(a-1)e^{a}] = -1.$
Note: We have used integration by parts to evaluate the integral.

33. $\int_{-\infty}^{\infty} x\,dx = \lim_{a\to-\infty}\tfrac{1}{2}x^{2}\Big|_{a}^{0} + \lim_{b\to\infty}\tfrac{1}{2}x^{2}\Big|_{0}^{b}$ both of which diverge and so the integral
diverges.

35. $\int_{-\infty}^{\infty} x^{3}(1+x^{4})^{-2}\,dx = \int_{-\infty}^{0} x^{3}(1+x^{4})^{-2}\,dx + \int_{0}^{\infty} x^{3}(1+x^{4})^{-2}\,dx$

$\qquad = \lim_{a\to-\infty}\int_{a}^{0} x^{3}(1+x^{4})^{-2}\,dx + \lim_{b\to\infty}\int_{0}^{b} x^{3}(1+x^{4})^{-2}\,dx$

$\qquad = \lim_{a\to-\infty}\left[-\dfrac{1}{4}(1+x^{4})^{-1}\Big|_{a}^{0}\right] + \lim_{b\to\infty}\left[-\dfrac{1}{4}(1+x^{4})^{-1}\Big|_{0}^{b}\right]$

$\qquad = \lim_{a\to-\infty}\left[-\dfrac{1}{4}+\dfrac{1}{4(1+a^{4})}\right] + \lim_{b\to\infty}\left[-\dfrac{1}{4(1+b^{4})}+\dfrac{1}{4}\right]$

$\qquad = -\tfrac{1}{4}+\tfrac{1}{4} = 0.$

37. $\int_{-\infty}^{\infty} xe^{1-x^{2}}\,dx = \lim_{a\to-\infty}\int_{a}^{0} xe^{1-x^{2}}\,dx + \lim_{b\to\infty}\int_{0}^{b} xe^{1-x^{2}}\,dx$

$\qquad = \lim_{a\to-\infty} -\tfrac{1}{2}e^{1-x^{2}}\Big|_{a}^{0} + \lim_{b\to\infty} -\tfrac{1}{2}e^{1-x^{2}}\Big|_{0}^{b}$

$\qquad = \lim_{a\to-\infty}\left(-\tfrac{1}{2}e+\tfrac{1}{2}e^{1-a^{2}}\right) + \lim_{b\to\infty}\left(-\tfrac{1}{2}e^{1-b^{2}}+\tfrac{1}{2}e\right) = 0.$

39. $\int_{-\infty}^{\infty} \dfrac{e^{-x}}{1+e^{-x}}\,dx = \lim_{a\to-\infty} -\ln(1+e^{-x})\Big|_{a}^{0} + \lim_{b\to\infty} -\ln(1+e^{-x})\Big|_{0}^{b} = \infty,$ and it is divergent.

41. First, we find the indefinite integral $I = \displaystyle\int \frac{1}{x \ln^3 x}\,dx$.

Let $u = \ln x$ so that $du = \dfrac{1}{x}\,dx$. Therefore,

$$I = \int \frac{du}{u^3} = -\frac{1}{2u^2} + C = -\frac{1}{2\ln^2 x} + C.$$

So $\displaystyle\int_e^\infty \frac{1}{x \ln^3 x}\,dx = \lim_{b\to\infty} \int_e^b \frac{1}{x \ln^3 x}\,dx$

$$= \lim_{b\to\infty}\left[-\frac{1}{2\ln^2 x}\Big|_e^b \right] = \lim_{b\to\infty}\left(-\frac{1}{2(\ln b)^2} + \frac{1}{2} \right) = \frac{1}{2}$$

and so the given integral is convergent.

43. We want the present value PV of a perpetuity with $m = 1$, $P = 1500$, and $r = 0.08$.

We find $\quad PV = \dfrac{(1)(1500)}{0.08} = 18{,}750$, or $\$18{,}750$.

45.
$$PV = \int_0^\infty (10{,}000 + 4000t)e^{-rt}\,dt = 10{,}000\int_0^\infty e^{-rt}\,dt + 4000\int_0^\infty te^{-rt}\,dt$$

$$= \lim_{b\to\infty}\left(-\frac{10{,}000}{r}e^{-rt}\Big|_0^b \right) + 4000\left(\frac{1}{r^2} \right)(-rt - 1)e^{-rt}\Big|_0^b$$

(Integrating by parts.)

$$= \frac{10{,}000}{r} + \frac{4000}{r^2} = \frac{10{,}000r + 4000}{r^2} \quad \text{dollars.}$$

47. True. $\displaystyle\int_a^\infty f(x)\,dx = \int_a^b f(x)\,dx + \int_b^\infty f(x)\,dx$. So if $\displaystyle\int_a^\infty f(x)\,dx$ exists then

$$\int_b^\infty f(x)\,dx = \int_a^\infty f(x)\,dx - \int_a^b f(x)\,dx.$$

49. False. Let
$$f(x) = \begin{cases} e^{2x} & \text{if } -\infty < x \le 0 \\ e^{-x} & \text{if } 0 < x < \infty \end{cases}.$$

Then $\displaystyle\int_{-\infty}^{\infty} f(x)\,dx = \int_{-\infty}^{0} e^{2x}\,dx + \int_{0}^{\infty} e^{-x}\,dx = \frac{1}{2}+1=\frac{3}{2}$.

But $\displaystyle 2\int_{0}^{\infty} f(x)\,dx = 2\int_{0}^{\infty} e^{-x}\,dx = 2$.

51. $\displaystyle\int_{0}^{\infty} e^{-px}\,dx = \lim_{b\to\infty}\int_{a}^{b} e^{-px}\,dx = \lim_{b\to\infty}\left[-\frac{1}{p}e^{-px}\Big|_{a}^{b}\right]=\lim_{b\to\infty}\left(-\frac{1}{p}e^{-pb}+\frac{1}{p}e^{-pa}\right)$

$\displaystyle = \frac{1}{pe^{pa}}$ if $p>0$ and is divergent if $p<0$.

CHAPTER 7 REVIEW EXERCISES, page 595

1. Let $u = 2x$ and $dv = e^{-x}\,dx$ so that $du = 2\,dx$ and $v = -e^{-x}$. Then

$$\int 2xe^{-x}\,dx = uv - \int v\,du = -2xe^{-x} + 2\int e^{-x}\,dx$$

$$= -2xe^{-x} - 2e^{-x} + C = -2(1+x)e^{-x} + C.$$

3. Let $u = \ln 5x$ and $dv = dx$, so that $du = \frac{1}{x}\,dx$ and $v = x$. Then

$$\int \ln 5x\,dx = x\ln 5x\,dx - \int dx = x\ln 5x - x + C = x(\ln 5x - 1) + C.$$

5. Let $u = x$ and $dv = e^{-2x}\,dx$ so that $du = dx$ and $v = -\frac{1}{2}e^{-2x}$. Then

$$\int_{0}^{1} xe^{-2x}\,dx = -\frac{1}{2}xe^{-2x}\Big|_{0}^{1} + \frac{1}{2}\int_{0}^{1} e^{-2x}\,dx = -\frac{1}{2}e^{-2} - \frac{1}{4}e^{-2x}\Big|_{0}^{1}$$

$$= -\frac{1}{2}e^{-2} - \frac{1}{4}e^{-2} + \frac{1}{4} = \frac{1}{4}(1 - 3e^{-2}).$$

7. $\displaystyle f(x) = \int f'(x)\,dx = \int \frac{\ln x}{\sqrt{x}}\,dx$. To evaluate the integral, we integrate by parts

with $u = \ln x$, $dv = x^{-1/2}\,dx$, $du = \frac{1}{x}\,dx$ and $v = 2x^{1/2}\,dx$. Then

$$\int \frac{\ln x}{x^{1/2}}\,dx = 2x^{1/2}\ln x - \int 2x^{-1/2}\,dx = 2x^{1/2}\ln x - 4x^{1/2} + C$$

$$= 2x^{1/2}(\ln x - 2) + C = 2\sqrt{x}(\ln x - 2) + C.$$

But $f(1) = -2$ and this gives $2\sqrt{1}(\ln 1 - 2) + C = -2$, or $C = 2$. Therefore,
$f(x) = 2\sqrt{x}(\ln x - 2) + 2$.

9. Using Formula 4 with $a = 3$ and $b = 2$, we obtain

$$\int \frac{x^2}{(3+2x)^2}\,dx = \frac{1}{8}\left[3+2x-\frac{9}{3+2x}-6\ln|3+2x|\right]+C.$$

11. Use Formula 24 with $a = 4$ and $n = 2$, obtaining

$$\int x^2 e^{4x}\,dx = \frac{1}{4}x^2 e^{4x} - \frac{1}{2}\int xe^{4x}\,dx.$$

Use Formula 23 to obtain

$$\int x^2 e^{4x}\,dx = \frac{1}{4}x^2 e^{4x} - \frac{1}{2}\left[\frac{1}{16}(4x-1)e^{4x}\right]+C$$

$$= \frac{1}{32}(8x^2 - 4x + 1)e^{4x} + C.$$

13. Use Formula 17 with $a = 2$ obtaining

$$\int \frac{dx}{x^2\sqrt{x^2-4}} = \frac{\sqrt{x^2-4}}{4x}+C.$$

15. $\displaystyle\int_0^\infty e^{-2x}\,dx = \lim_{b\to\infty}\int_0^b e^{-2x}\,dx = \lim_{b\to\infty}\left(-\frac{1}{2}e^{-2x}\right)\Big|_0^b = \lim_{b\to\infty}\left(-\frac{1}{2}e^{-2b}+\frac{1}{2}\right)=\frac{1}{2}.$

17. $\displaystyle\int_3^\infty \frac{2}{x}\,dx = \lim_{b\to\infty}\int_3^b \frac{2}{x}\,dx = \lim_{b\to\infty} 2\ln x\Big|_3^b = \lim_{b\to\infty}(2\ln b - 2\ln 3) = \infty.$

19. $\displaystyle\int_2^\infty \frac{dx}{(1+2x)^2} = \lim_{b\to\infty}\int_2^b (1+2x)^{-2}\,dx = \lim_{b\to\infty}(\tfrac{1}{2})(-1)(1+2x)^{-1}\Big|_2^b$

$$= \lim_{b\to\infty}\left(-\frac{1}{2(1+2b)}+\frac{1}{2(5)}\right)=\frac{1}{10}.$$

21. $\Delta x = \frac{b-a}{n} = \frac{3-1}{4} = \frac{1}{2}; x_0 = 1, x_1 = \frac{3}{2}, x_2 = 2, x_3 = \frac{5}{2}, x_4 = 3.$

Trapezoidal Rule:

$$\int_1^3 \frac{dx}{1+\sqrt{x}} \approx \frac{\frac{1}{2}}{2}\left[\frac{1}{2}+\frac{2}{1+\sqrt{1.5}}+\frac{2}{1+\sqrt{2}}+\frac{2}{1+\sqrt{2.5}}+\frac{1}{1+\sqrt{3}}\right] \approx 0.8421.$$

Simpson's Rule

$$\int_1^3 \frac{dx}{1+\sqrt{x}} \approx \frac{\frac{1}{2}}{3}\left[\frac{1}{2}+\frac{4}{1+\sqrt{1.5}}+\frac{2}{1+\sqrt{2}}+\frac{4}{1+\sqrt{2.5}}+\frac{1}{1+\sqrt{3}}\right] \approx 0.8404.$$

23. $\Delta x = \frac{1-(-1)}{4} = \frac{1}{2}$; $x_0 = -1$, $x_1 = -\frac{1}{2}$, $x_2 = 0$, $x_3 = \frac{1}{2}$, $x_4 = 1$.

Trapezoidal Rule:
$$\int_{-1}^{1} \sqrt{1+x^4}\, dx \approx \frac{0.5}{2}\left[\sqrt{2} + 2\sqrt{1+(-0.5)^4} + 2 + 2\sqrt{1+(0.5)^4} + \sqrt{2}\right]$$
$$\approx 2.2379.$$

Simpson's Rule:
$$\int_{-1}^{1} \sqrt{1+x^4}\, dx \approx \frac{0.5}{3}\left[\sqrt{2} + 4\sqrt{1+(-0.5)^4} + 2 + 4\sqrt{1+(0.5)^4} + \sqrt{2}\right]$$
$$\approx 2.1791.$$

25. The producer's surplus is given by $PS = \bar{p}\bar{x} - \int_0^{\bar{x}} s(x)\, dx$, where $\bar{x}$ is found by solving the equation $2\sqrt{25+x^2} = 13$. Then $\sqrt{25+x^2} = 13$, $25 + x^2 = 169$, and $x = \pm 12$. So $\bar{x} = 12$. Therefore, $PS = (26)(12) - 2\int_0^{12}(25+x^2)^{1/2}\, dx$.

Using Formula 7 with $a = 5$, we obtain
$$PS = (26)(12) - 2\int_0^{12}(25+x^2)^{1/2}\, dx$$
$$= 312 - 2\left(\frac{x}{2}(25+x^2)^{1/2} + \frac{25}{2}\ln\left|x + (25+x^2)^{1/2}\right|\right)\Big|_0^{12}$$
$$= 312 - 2[6(13) + \frac{25}{2}\ln(12+13) - \frac{25}{2}\ln 5] \approx 115.76405,$$
or $\$1,157,641$.

27. If $p = 30$, we have $2\sqrt{325-x^2} = 30$, $\sqrt{325-x^2} = 15$, or $325 - x^2 = 225$, $x^2 = 100$, or $x = \pm 10$. So the equilibrium point is $(10, 30)$.
$$CS = \int_0^{10} 2\sqrt{325-x^2}\, dx - (30)(10).$$
To evaluate the integral using Simpson's Rule with $n = 10$, we have
$$\Delta x = \frac{10-0}{10} = 1; \quad x_0 = 0, x_1 = 1, x_2 = 2, ..., x_{10} = 10.$$
$$2\int_0^{10}\sqrt{325-x^2}\, dx$$
$$\approx \frac{2}{3}\left[\sqrt{325} + 4\sqrt{325-1} + 2\sqrt{325-4} + \cdots + 4\sqrt{325-81} + \sqrt{325-100}\right]$$
Therefore, $CS \approx 341.0 - 300 \approx 41.1$, or $\$41,100$.

29. We want the present value of a perpetuity with $m = 1$, $P = 10{,}000$, and $r = 0.09$. We find

$$PV = \frac{(1)(10{,}000)}{0.09} \approx 111{,}111 \quad \text{or approximately } \$111{,}111.$$

CHAPTER 8

EXERCISES 8.1, page 605

1. $f(0, 0) = 2(0) + 3(0) - 4 = -4.$ $f(1, 0) = 2(1) + 3(0) - 4 = -2.$
    $f(0, 1) = 2(0) + 3(1) - 4 = -1.$ $f(1, 2) = 2(1) + 3(2) - 4 = 4.$
    $f(2,-1) = 2(2) + 3(-1) - 4 = -3.$

3. $f(1, 2) = 1^2 + 2(1)(2) - 1 + 3 = 7;\ f(2, 1) = 2^2 + 2(2)(1) - 2 + 3 = 9$
    $f(-1, 2) = (-1)^2 + 2(-1)(2) - (-1) + 3 = 1$ $f(2, -1) = 2^2 + 2(2)(-1) - 2 + 3 = 1.$

5. $g(s,t) = 3s\sqrt{t} + t\sqrt{s} + 2;\ \ g(1,2) = 3(1)\sqrt{2} + 2\sqrt{1} + 2 = 4 + 3\sqrt{2}$
    $g(2, 1) = 3(2)\sqrt{1} + \sqrt{2} + 2 = 8 + \sqrt{2};$
    $g(0, 4) = 0 + 0 + 2 = 2,\ g(4,9) = 3(4)\sqrt{9} + 9\sqrt{4} + 2 = 56.$

7. $h(1,e) = \ln e - e \ln 1 = \ln e = 1;\ h(e,1) = e \ln 1 - \ln e = -1;$
    $h(e,e) = e \ln e - e \ln e = 0.$

9. $g(r,s,t) = re^{s/t};\ g(1,1,1) = e,\ g(1,0,1) = 1,\ g(-1,-1,-1) = -e^{-1/(-1)} = -e.$

11. The domain of f is the set of all ordered pairs (x, y) where x and y are real numbers.

13. All real values of u and v except those satisfying the equation $u = v$.

15. The domain of g is the set of all ordered pairs (r,s) satisfying $rs \geq 0$, that is the set of all ordered pairs where both $r \geq 0$ and $s \geq 0$, or in which both $r \leq 0$ and $s \leq 0$.

17. The domain of h is the set of all ordered pairs (x, y) such that $x + y > 5$.

19. The level curves of $z = f(x, y) = 2x + 3y$ for $z = -2, -1, 0, 1, 2,$ follow.

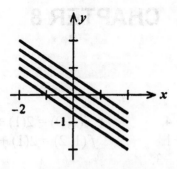

21. The level curves of $f(x,y) = 2x^2 + y$ for $z = -2, -1, 0, 1, 2,$ are shown below.

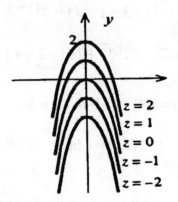

23. The level curves of $f(x,y) = \sqrt{16 - x^2 - y^2}$ for $z = 0, 1, 2, 3, 4$ follow.

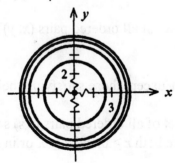

25. $V = f(1.5,4) = \pi(1.5)^2(4) = 9\pi$, or 9π cu ft

27. $R(4, 0.1) = \dfrac{4k}{(0.1)^4} = 40{,}000\,k$ dynes.

29. $R(x,y) = xp + yq = x(200 - \frac{1}{5}x - \frac{1}{10}y) + y(160 - \frac{1}{10}x - \frac{1}{4}y)$

$$= -\frac{1}{5}x^2 - \frac{1}{4}y^2 - \frac{1}{5}xy + 200x + 160y.$$

$R(100,60) = -\frac{1}{5}(10{,}000) - \frac{1}{4}(3600) - \frac{1}{5}(6000) + 200(100) + 160(60)$

$$= 25{,}500,$$

and this says that the revenue from the sales of 100 units of the finished and 60 units of the unfinished furniture per week is \$25,500,

$R(60,100) = -\frac{1}{5}(3600) - \frac{1}{4}(10{,}000) - \frac{1}{5}(6000) + 200(60) + 160(100)$

$$= 23{,}580,$$

and this says that the revenue from the sales of 60 units of the finished and 100 units of the unfinished furniture per week is \$23,580.

31. $R(300, 200) = -0.005(90000) - 0.003(40000) - 0.002(60000) + 20(300)$

$$+ 15(200)$$

$$= 8310, \text{ or } \$8310.$$

$R(200, 300) = -0.005(40000) - 0.003(90000) - 0.002(60000) + 20(200)$

$$+ 15(300)$$

$$= 7910, \text{ or } \$7910.$$

33. a. The domain of S is the set of all ordered pairs (W,H) such that W and H are nonnegative real numbers.

b. $S = 0.007184(70)^{0.425}(178)^{0.725} \approx 1.87$ sq meters.

35. $A = f(10{,}000, 0.1, 3) = 10{,}000e^{(0.1)(3)} = 10{,}000e^{0.3}$, or \$13,498.59.

37. $B = f(80{,}000, 0.09, 30, 60) = 80{,}000 \left[\dfrac{\left(1 + \dfrac{0.09}{12}\right)^{60} - 1}{\left(1 + \dfrac{0.09}{12}\right)^{360} - 1} \right] = 3295.89.$

Therefore, they owe $80{,}000 - 3295.89$, or \$76,704.11.

$B = f(80{,}000, 0{,}09, 30, 240) = 80{,}000 \left[\dfrac{\left(1 + \dfrac{0.09}{12}\right)^{240} - 1}{\left(1 + \dfrac{0.09}{12}\right)^{360} - 1} \right] = 29{,}185.38.$

Therefore, they owe $80{,}000 - 29{,}185.38$, or \$50,814.62.

39. $f(20, 40, 5) = \sqrt{\dfrac{2(20)(40)}{5}} = \sqrt{320} \approx 17.9$, or 18 bicycles.

41. False. Let $f(x,y) = xy$. Then
$$f(ax, ay) = (ax)(ay) = a^2 xy \neq axy = a\, f(x,y).$$

43. True. If $c > 0$, then $z = f(x, y) = c$ and the point (x, y, c) on the graph of f is c units above the xy-plane. Similarly, if $c < 0$, then $z = f(x, y) = c$ and the point (x, y, c) on the graph of f lies $|c|$ units below the xy-plane.

EXERCISES 8.2, page 620

1. $f_x = 2, \ f_y = 3$

3. $g_x = 4x, \ g_y = 4$

5. $f_x = -\dfrac{4y}{x^3}; \ f_y = \dfrac{2}{x^2}.$

7. $g(u,v) = \dfrac{u-v}{u+v}; \ \dfrac{\partial g}{\partial u} = \dfrac{(u+v)(1)-(u-v)(1)}{(u+v)^2} = \dfrac{2v}{(u+v)^2}.$

$\dfrac{\partial g}{\partial v} = \dfrac{(u+v)(-1)-(u-v)(1)}{(u+v)^2} = -\dfrac{2u}{(u+v)^2}.$

9. $f(s,t) = (s^2 - st + t^2)^3; \ f_s = 3(s^2 - st + t^2)^2(2s - t)$ and $f_t = 3(s^2 - st + t^2)^2(-s + 2t)$

11. $f(x,y) = (x^2 + y^2)^{2/3}; \ f_x = \tfrac{2}{3}(x^2 + y^2)^{-1/3}(2x) = \tfrac{4}{3}x(x^2 + y^2)^{-1/3}.$ Similarly,
$f_y = \tfrac{4}{3}y(x^2 + y^2)^{-1/3}.$

13. $f(x,y) = e^{xy+1}; \ f_x = ye^{xy+1}, \ f_y = xe^{xy+1}.$

15. $f(x,y) = x\ln y + y\ln x; \ f_x = \ln y + \dfrac{y}{x}, \ f_y = \dfrac{x}{y} + \ln x.$

17. $g(u,v) = e^u \ln v. \ g_u = e^u \ln v, \ g_v = \dfrac{e^u}{v}.$

19. $f(x,y,z) = xyz + xy^2 + yz^2 + zx^2; f_x = yz + y^2 + 2xz,$
$f_y = xz + 2xy + z^2, f_z = xy + 2yz + x^2.$

21. $h(r,s,t) = e^{rst}; h_r = ste^{rst}, h_s = rte^{rst}, h_t = rse^{rst}.$

23. $f(x,y) = x^2y + xy^2; f_x(1,2) = 2xy + y^2\big|_{(1,2)} = 8; f_y(1,2) = x^2 + 2xy\big|_{(1,2)} = 5.$

25. $f(x,y) = x\sqrt{y} + y^2 = xy^{1/2} + y^2; f_x(2,1) = \sqrt{y}\,\big|_{(2,1)} = 1,$

$f_y(2,1) = \dfrac{x}{2\sqrt{y}} + 2y\,\big|_{(2,1)} = 3.$

27. $f(x,y) = \dfrac{x}{y}; f_x(1,2) = \dfrac{1}{y}\bigg|_{(1,2)} = \dfrac{1}{2}, f_y(1,2) = -\dfrac{x}{y^2}\bigg|_{(1,2)} = -\dfrac{1}{4}.$

29. $f(x,y) = e^{xy}. f_x(1,1) = ye^{xy}\big|_{(1,1)} = e, f_y(1,1) = xe^{xy}\big|_{(1,1)} = e.$

31. $f(x,y,z) = x^2yz^3; f_x(1,0,2) = 2xyz^3\big|_{(1,0,2)} = 0; f_y(1,0,2) = x^2z^3\big|_{(1,0,2)} = 8.$
$f_z(1,0,2) = 3x^2yz^2\big|_{(1,0,2)} = 0.$

33. $f(x,y) = x^2y + xy^3; f_x = 2xy + y^3, f_y = x^2 + 3xy^2.$
Therefore, $f_{xx} = 2y, f_{xy} = 2x + 3y^2 = f_{yx}, f_{yy} = 6xy.$

35. $f(x,y) = x^2 - 2xy + 2y^2 + x - 2y; f_x = 2x - 2y + 1, f_y = -2x + 4y - 2; f_{xx} = 2,$
$f_{xy} = -2, f_{yx} = -2, f_{yy} = 4.$

37. $f(x,y) = (x^2 + y^2)^{1/2}; f_x = \tfrac{1}{2}(x^2 + y^2)^{-1/2}(2x) = x(x^2 + y^2)^{-1/2};$
$f_y = y(x^2 + y^2)^{-1/2}.$
$f_{xx} = (x^2 + y^2)^{-1/2} + x(-\tfrac{1}{2})(x^2 + y^2)^{-3/2}(2x) = (x^2 + y^2)^{-1/2} - x^2(x^2 + y^2)^{-3/2}$
$= (x^2 + y^2)^{-3/2}(x^2 + y^2 - x^2) = \dfrac{y^2}{(x^2 + y^2)^{3/2}}.$

$$f_{xy} = x(-\tfrac{1}{2})(x^2+y^2)^{-3/2}(2y) = -\frac{xy}{(x^2+y^2)^{3/2}} = f_{yx}.$$

$$f_{yy} = (x^2+y^2)^{-1/2} + y(-\tfrac{1}{2})(x^2+y^2)^{-3/2}(2y) = (x^2+y^2)^{-1/2} - y^2(x^2+y^2)^{-3/2}$$

$$= (x^2+y^2)^{-3/2}(x^2+y^2-y^2) = \frac{x^2}{(x^2+y^2)^{3/2}}.$$

39.　$f(x,y) = e^{-x/y}; \; f_x = -\dfrac{1}{y}e^{-x/y}; \; f_y = \dfrac{x}{y^2}e^{-x/y}; \; f_{xx} = \dfrac{1}{y^2}e^{-x/y};$

$$f_{xy} = -\frac{x}{y^3}e^{-x/y} + \frac{1}{y^2}e^{-x/y} = \left(\frac{-x+y}{y^3}\right)e^{-x/y} = f_{yx}.$$

$$f_{yy} = -\frac{2x}{y^3}e^{-x/y} + \frac{x^2}{y^4}e^{-x/y} = \frac{x}{y^3}\left(\frac{x}{y}-2\right)e^{-x/y}.$$

41.　a.　$f(x,y) = 20x^{3/4}y^{1/4}. \; f_x(256,16) = 15\left(\dfrac{y}{x}\right)^{1/4}\Big|_{(256,16)}$

$$= 15\left(\frac{16}{256}\right)^{1/4} = 15\left(\frac{2}{4}\right) = 7.5.$$

$$f_y(256,16) = 5\left(\frac{x}{y}\right)^{3/4}\Big|_{(256,16)} = 5\left(\frac{256}{16}\right)^{3/4} = 5(80) = 40.$$

　　b.　Yes.

43.　$p(x,y) = 200 - 10(x-\tfrac{1}{2})^2 - 15(y-1)^2. \; \dfrac{\partial p}{\partial x}(0,1) = -20(x-\tfrac{1}{2})\big|_{(0,1)} = 10;$

At the location (0,1) in the figure, the price of land is changing at the rate of \$10 per sq ft per mile change to the right.

$$\frac{\partial p}{\partial y}(0,1) = -30(y-1)\big|_{(0,1)} = 0;$$

At the location (0,1) in the figure, the price of land is constant per mile change upwards.

45.　$f(p,q) = 10{,}000 - 10p - e^{0.5q}; \; g(p,q) = 50{,}000 - 4000q - 10p.$

$$\frac{\partial f}{\partial q} = -0.5e^{0.5q} < 0 \text{ and } \frac{\partial g}{\partial p} = -10 < 0$$

and so the two commodities are complementary commodities.

47.　$R(x, y) = -0.2x^2 - 0.25y^2 - 0.2xy + 200x + 160y.$

$$\frac{\partial R}{\partial x}(300, 250) = -0.4x - 0.2y + 200\big|_{(300,250)}$$

$$= -0.4(300) - 0.2(250) + 200 = 30$$

and this says that at a sales level of 300 finished and 250 unfinished units the revenue is increasing at the rate of $30 per week per unit increase in the finished units.

$$\frac{\partial R}{\partial y}(300, 250) = -0.5y - 0.2x + 160\big|_{(300,250)}$$

$$= -0.5(250) - 0.2(300) + 160 = -25$$

and this says that at a level of 300 finished and 250 unfinished units the revenue is decreasing at the rate of $25 per week per increase in the unfinished units.

49.　$V = \dfrac{30.9T}{P}$. $\dfrac{\partial V}{\partial T} = \dfrac{30.9}{P}$ and $\dfrac{\partial V}{\partial P} = -\dfrac{30.9T}{P^2}$.

Therefore, $\dfrac{\partial V}{\partial T}\bigg|_{T=300, P=800} = \dfrac{30.9}{800} = 0.039$, or 0.039 liters/degree.

$$\dfrac{\partial V}{\partial P}\bigg|_{T=300, P=800} = -\dfrac{(30.9)(300)}{800^2} = -0.015$$

or -0.015 liters/mm of mercury.

51.　$V = \dfrac{kT}{P}$ and $\dfrac{\partial V}{\partial T} = \dfrac{k}{P}$; $T = \dfrac{VP}{k}$ and $\dfrac{\partial T}{\partial P} = \dfrac{V}{k} = \dfrac{T}{P}$; and

$$P = \dfrac{kT}{V} \text{ and } \dfrac{\partial P}{\partial V} = -\dfrac{kT}{V^2} = -kT \cdot \dfrac{P^2}{(kT)^2} = -\dfrac{P^2}{kT}$$

Therefore $\dfrac{\partial V}{\partial T} \cdot \dfrac{\partial T}{\partial P} \cdot \dfrac{\partial P}{\partial V} = \dfrac{k}{P} \cdot \dfrac{T}{P} \cdot -\dfrac{P^2}{kT} = -1.$

53.	True. This is a consequence of the definition of $f_x(a,b)$ as the rate of change of f in the x-direction at (a,b) with y held fixed.

55.	False. Let $f(x,y) = xy^{5/3}$. Then $f_{xy} = \dfrac{5}{3}y^{2/3} = f_{yx}$. So both f_{xy} and f_{yx} exist at $(0,0)$. But $f_{yy} = \dfrac{10x}{9y^{1/3}}$ is not defined at $(0,0)$.

USING TECHNOLOGY EXERCISES 8.2, page 623

1.	1.3124; 0.4038	3. −1.8889; 0.7778	5. −0.3863; −0.8497

EXERCISES 8.3, page 634

1.	$f(x,y) = 1 - 2x^2 - 3y^2$. To find the critical point(s) of f, we solve the system
$$\begin{cases} f_x = -4x = 0 \\ f_y = -6y = 0 \end{cases}$$
obtaining $(0,0)$ as the only critical point of f. Next,
$$f_{xx} = -4, f_{xy} = 0, \text{ and } f_{yy} = -6.$$
In particular, $f_{xx}(0,0) = -4, f_{xy}(0,0) = 0$, and $f_{yy}(0,0) = -6$, giving
$$D(0,0) = (-4)(-6) - 0^2 = 24 > 0.$$
Since $f_{xx}(0,0) < 0$, the Second Derivative Test implies that $(0,0)$ gives rise to a relative maximum of f. Finally, the relative maximum of f is $f(0,0) = 1$.

3.	To find the critical points of f, we solve the system
$$\begin{cases} f_x = 2x - 2 = 0 \\ f_y = -2y + 4 = 0 \end{cases}$$
obtaining $x = 1$ and $y = 2$ so that $(1,2)$ is the only critical point.
$$f_{xx} = 2, f_{xy} = 0, \text{ and } f_{yy} = -2.$$
So $D(x,y) = f_{xx}f_{yy} - f_{xy}^2 = -4$. In particular, $D(1,2) = -4 < 0$ and so $(1,2)$ affords a saddle point of f and $f(1,2) = 4$.

5.	$f(x,y) = x^2 + 2xy + 2y^2 - 4x + 8y - 1$. To find the critical point(s) of f, we solve the system $\quad\begin{cases} f_x = 2x + 2y - 4 = 0 \\ f_y = 2x + 4y + 8 = 0 \end{cases}$

obtaining $(8,-6)$ as the critical point of f. Next, $f_{xx} = 2, f_{xy} = 2, f_{yy} = 4$. In particular, $f_{xx}(8,-6) = 2, f_{xy}(8,-6) = 2, f_{yy}(8,-6) = 4$, giving $D = 2(4) - 4 = 4 > 0$. Since $f_{xx}(8,-6) > 0$, $(8,-6)$ gives rise to a relative minimum of f. Finally, the relative minimum value of f is $f(8,-6) = -41$.

7. $f(x,y) = 2x^3 + y^2 - 9x^2 - 4y + 12x - 2.$. To find the critical points of f, we solve the system
$$\begin{cases} f_x = 6x^2 - 18x + 12 = 0 \\ f_y = 2y - 4 = 0 \end{cases}$$
The first equation is equivalent to $x^2 - 3x + 2 = 0$, or $(x - 2)(x - 1) = 0$ which gives $x = 1$ or 2. The second equation of the system gives $y = 2$. Therefore, there are two critical points, $(1,2)$ and $(2,2)$. Next, we compute
$$f_{xx} = 12x - 18 = 6(2x - 3), f_{xy} = 0, f_{yy} = 2.$$

At the point $(1,2)$:
$$f_{xx}(1,2) = 6(2 - 3) = -6, f_{xy}(1,2) = 0, \text{ and } f_{yy}(1,2) = 2.$$

Therefore, $D = (-6)(2) - 0 = -12 < 0$ and we conclude that $(1,2)$ gives rise to a saddle point of f. At the point $(2,2)$:
$$f_{xx}(2,2) = 6(4 - 3) = 6, f_{xy}(2,2) = 0, \text{ and } f_{yy}(2,2) = 2.$$

Therefore, $D = (6)(2) - 0 = 12 > 0$. Since $f_{xx}(2,2) > 0$, we see that $(2,2)$ gives rise to a relative minimum with value $f(2,2) = -2$.

9. To find the critical points of f, we solve the system
$$\begin{cases} f_x = 3x^2 - 2y + 7 = 0 \\ f_y = 2y - 2x - 8 = 0 \end{cases}$$
Adding the two equations gives $3x^2 - 2x - 1 = 0$, or $(3x + 1)(x - 1) = 0$. Therefore, $x = -1/3$ or 1. Substituting each of these values of x into the second equation gives $y = 8/3$ and $y = 5$, respectively. Therefore, $(-\frac{1}{3}, \frac{11}{3})$ and $(1,5)$ are critical points of f.
Next, $f_{xx} = 6x, f_{xy} = -2$, and $f_{yy} = 2$. So $D(x,y) = 12x - 4 = 4(3x - 1)$. Then
$$D(-\tfrac{1}{3}, \tfrac{11}{3}) = 4(-1-1) = -8 < 0$$

and so $(-\frac{1}{3}, \frac{11}{3})$ gives a saddle point. Next, $D(1,5) = 4(3 - 1) = 8 > 0$ and since $f_{xx}(1,5) = 6 > 0$, we see that $(1,5)$ gives rise to a relative minimum.

11. To find the critical points of f, we solve the system
$$\begin{cases} f_x = 3x^2 - 3y = 0 \\ f_y = -3x + 3y^2 = 0 \end{cases}$$
The first equation gives $y = x^2$ which when substituted into the second equation gives $-3x + 3x^4 = 3x(x^3 - 1) = 0$. Therefore, $x = 0$ or 1. Substituting these values of x into the first equation gives $y = 0$ and $y = 1$, respectively. Therefore, $(0,0)$ and $(1,1)$ are critical points of f. Next, we find $f_{xx} = 6x$, $f_{xy} = -3$, and $f_{yy} = 6y$. So $D = f_{xx}f_{yy} - f_{xy}^2 = 36xy - 9$. Since $D(0,0) = -9 < 0$, we see that $(0,0)$ gives a saddle point of f. Next, $D(1,1) = 36 - 9 = 27 > 0$ and since $f_{xx}(1,1) = 6 > 0$, we see that $f(1,1) = -3$ is a relative minimum value of f.

13. Solving the system of equations
$$\begin{cases} f_x = y - \frac{4}{x^2} = 0 \\ f_y = x - \frac{2}{y^2} = 0 \end{cases}$$
we obtain $y = \frac{4}{x^2}$. Therefore, $x - 2\left(\frac{x^4}{16}\right) = 0$ and $8x - x^4 = x(8 - x^3) = 0$, and $x = 0$, or $x = 2$. Since $x = 0$ is not in the domain of f, $(2,1)$ is the only critical point of f. Next, $f_{xx} = \frac{8}{y^3}$, $f_{xy} = 1$, and $f_{yy} = \frac{4}{y^3}$. Therefore,

$$D(2,1) = \frac{32}{x^3y^3} - 1 \bigg|_{(2,1)} = 4 - 1 = 3 > 0 \text{ and } f_{xx}(2,1) = 1 > 0. \text{ Therefore, the relative}$$

minimum value of f is $f(2,1) = 2 + 4/2 + 2/1 = 6$.

15. Solving the system of equations $f_x = 2x = 0$ and $f_y = -2ye^{y^2} = 0$, we obtain $x = 0$ and $y = 0$. Therefore, $(0,0)$ is the only critical point of f. Next,
$$f_{xx} = 2, f_{xy} = 0, f_{yy} = -2e^{y^2} - 4y^2e^{y^2}.$$
Therefore, $D(0,0) = -4e^{y^2}(1 + 2y^2)\bigg|_{(0,0)} = -4(1) < 0$, and we conclude that $(0,0)$ is a saddle point.

17. $f(x,y) = e^{x^2+y^2}$
Solving the system

$$\begin{cases} f_x = 2xe^{x^2+y^2} = 0 \\ f_y = 2ye^{x^2+y^2} = 0 \end{cases}$$

we see that $x = 0$ and $y = 0$ (recall that $e^{x^2+y^2} \neq 0$). Therefore, $(0,0)$ is the only critical point of f. Next, we compute

$$f_{xx} = 2e^{x^2+y^2} + 2x(2x)e^{x^2+y^2} = 2(1+2x^2)e^{x^2+y^2}$$
$$f_{xy} = 2x(2y)e^{x^2+y^2} = 4xye^{x^2+y^2}$$
$$f_{yy} = 2(1+2y^2)e^{x^2+y^2}.$$

In particular, at the point $(0,0)$, $f_{xx}(0,0) = 2$, $f_{xy}(0,0) = 0$, and $f_{yy}(0,0) = 2$. Therefore, $D = (2)(2) - 0 = 4 > 0$. Furthermore, since $f_{xx}(0,0) > 0$, we conclude that $(0,0)$ gives rise to a relative minimum of f. The relative minimum value of f is $f(0,0) = 1$.

19. $f(x,y) = \ln(1+x^2+y^2)$. We solve the system of equations

$$f_x = \frac{2x}{1+x^2+y^2} = 0 \quad \text{and} \quad f_y = \frac{2y}{1+x^2+y^2} = 0,$$

obtaining $x = 0$ and $y = 0$. Therefore, $(0,0)$ is the only critical point of f. Next,

$$f_{xx} = \frac{(1+x^2+y^2)2-(2x)(2x)}{(1+x^2+y^2)^2} = \frac{2+2y^2-2x^2}{(1+x^2+y^2)^2}$$

$$f_{yy} = \frac{(1+x^2+y^2)2-(2y)(2y)}{(1+x^2+y^2)^2} = \frac{2+2x^2-2y^2}{(1+x^2+y^2)^2}$$

$$f_{xy} = -2x(1+x^2+y^2)^{-2}(2y) = -\frac{4xy}{(1+x^2+y^2)^2}.$$

Therefore, $D(x,y) = \dfrac{(2+2y^2-2x^2)(2+2x^2-2y^2)}{(1+x^2+y^2)^4} - \dfrac{16x^2y^2}{(1+x^2+y^2)^4}.$

Since $D(0,0) = \frac{4}{1} > 0$ and $f_{xx}(0,0) = 2 > 0$, $f(0,0) = 0$ is a relative minimum value.

21. $P(x) = -0.2x^2 - 0.25y^2 - 0.2xy + 200x + 160y - 100x - 70y - 4000$
$= -0.2x^2 - 0.25y^2 - 0.2xy + 100x + 90y - 4000.$

Then $\begin{cases} P_x = -0.4x - 0.2y + 100 = 0 \\ P_y = -0.5y - 0.2x + 90 = 0 \end{cases}$

implies that $\begin{cases} 4x+2y=1000 \\ 2x+5y=\ 900 \end{cases}$. Solving, we find $x = 200$ and $y = 100$.

Next, $P_{xx} = -0.4, P_{yy} = -0.5, P_{xy} = -0.2$, and

$D(200,100) = (-0.4)(-0.5) - (-0.2)^2 > 0$. Since $P_{xx}(200, 100) < 0$, we conclude that $(200,100)$ is a relative maximum of P. Thus, the company should manufacture 200 finished and 100 unfinished units per week. The maximum profit is

$$P(200,100) = -0.2(200)^2 - 0.25(100)^2 - 0.2(100)(200) + 100(200) + 90(100) - 4000$$
$$= 10,500, \text{ or } \$10,500 \text{ dollars.}$$

23. $p(x,y) = 200 - 10(x - \frac{1}{2})^2 - 15(y-1)^2$. Solving the system of equations

$$\begin{cases} p_x = -20(x - \frac{1}{2}) = 0 \\ p_y = -30(y-1) = 0 \end{cases}$$

we obtain $x = 1/2$, $y = 1$. We conclude that the only critical point of f is $(\frac{1}{2},1)$.

Next, $\quad p_{xx} = -20, p_{xy} = 0, p_{yy} = -30$

so $\quad D(\frac{1}{2},1) = (-20)(-30) = 600 > 0$.

Since $p_{xx} = -20 < 0$, we conclude that $f(\frac{1}{2},1)$ gives a relative maximum. So we conclude that the price of land is highest at $(\frac{1}{2},1)$.

25. We want to minimize

$$f(x,y) = D^2 = (x-5)^2 + (y-2)^2 + (x+4)^2 + (y-4)^2 + (x+1)^2 + (y+3)^2.$$

Next, $\begin{cases} f_x = 2(x-5)+2(x+4)+2(x+1) = 6x = 0, \\ f_y = 2(y-2)+2(y-4)+2(y+3) = 6y-6 = 0 \end{cases}$

and we conclude that $x = 0$ and $y = 1$. Also,

$\quad f_{xx} = 6, f_{xy} = 0, f_{yy} = 6$ and $D(x,y) = (6)(6) = 36 > 0$.

Since $f_{xx} > 0$, we conclude that the function is minimized at $(0,1)$ and so $(0,1)$ gives the desired location.

27. The volume is given by $V = xyz = xz(108 - 2x - 2z) = 108xz - 2x^2z - 2xz^2$. Solving the system of equations,

$$V_x = 108z - 4xz - 2z^2 = 0 \text{ and}$$
$$V_z = 108x - 2x^2 - 4xz = 0,$$

we obtain

$$(108 - 4x - 2z)z = 0, \text{ or } 108 - 4x - 2z = 0$$

and $(108 - 4z - 2x)x = 0, \text{ or } 108 - 2x - 4z = 0.$ Thus

$$\begin{cases} 108 - 4x - 2z = 0 \\ 216 - 4x - 8z = 0 \end{cases}$$

gives $-108 + 6z = 0,$ or $z = 18.$ So $x = \frac{1}{4}(108 - 36) = 18$ and

$y = 108 - 2x - 2z = 108 - 72 = 36,$ and $(18,18)$ is the critical point of V. Next,
$$V_{xx} = -4z, \ V_{zz} = -4x, \ V_{xz} = 108 - 4x - 4z, \text{ and}$$
$$D(18,18) = -4(18)(-4)(18) - [108 - 4(18) - 4(18)]^2 > 0$$
and $V_{xx}(18,18) < 0.$ We conclude that the dimensions yielding the maximum volume are $18" \times 36" \times 18".$

29. Since $V = xyz,$ $z = \dfrac{48}{xy}.$ Then the amount of material used in the box is given by

$$S = xy + 2xz + 3yz = xy + \frac{48}{xy}(2x + 3y) = xy + \frac{96}{y} + \frac{144}{x}. \text{ Solving the system of}$$

equations $\begin{cases} S_x = y - \frac{144}{x^2} = 0 \\ S_y = x - \frac{96}{y^2} = 0 \end{cases},$ we have $y = \dfrac{144}{x^2}.$ Therefore

$x - \dfrac{96x^4}{144^2} = 0,\ 144^2 x - 96x^4 = 0,\ 96x(216 - x^3) = 0,$ and $x = 0$ or $x = 6.$ Then

$y = \dfrac{144}{36} = 4.$ Next, $S_{xx} = \dfrac{288}{x^3},\ S_{yy} = \dfrac{192}{y^3},$ and $S_{xy} = 1.$

At the point $(6,4)$:

$$D(x,y) = \frac{(288)(192)}{x^3 y^3} - 1 \Bigg|_{(6,4)} = \frac{288(192)}{216(64)} - 1 = 3 > 0,$$

and $S_{xx} > 0.$ Therefore, we conclude that the function is minimized when its dimensions are $6" \times 4" \times 2".$

31. False. Let $f(x,y) = -x^2 - y^2 + 4xy.$ Then setting
$$f_x(x,y) = -2x + 4y = 0$$
$$f_y(x,y) = -2y + 4x = 0$$

we find that $(0,0)$ is the only critical point of f. Next,

$f_{xx} = -2$, $f_{xy} = 4$, and $f_{yy} = -2$.

Since $D(x,y) = f_{xx}f_{yy} - f_{xy}^2 = (-2)(-2) - 4^2 = -12 < 0$,

we see that $(0, 0, 0)$ is a saddle point.

EXERCISES 8.4, page 644

1. a. We first summarize the data:

x	y	x^2	Xy
1	4	1	4
2	6	4	12
3	8	9	24
4	11	16	44
10	29	30	84

The normal equations are $4b + 10m = 29$
$10b + 30m = 84.$

Solving this system of equations, we obtain $m = 2.3$ and $b = 1.5$. So an equation is
$y = 2.3x + 1.5$.

b. The scatter diagram and the least squares line for this data follow:

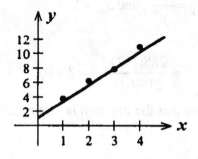

3. a. We first summarize the data:

x	y	x^2	xy
1	4.5	1	4.5
2	5	4	10
3	3	9	9
4	2	16	8
4	3.5	16	14
6	1	36	6
20	19	82	51.5

The normal equations are $6b + 20m = 19$

$$20b + 82m = 51.5.$$

The solutions are $m \approx -0.7717$ and $b \approx 5.7391$ and so a required equation is
$y = -0.772x + 5.739$.

b. The scatter diagram and the least-squares line for these data follow.

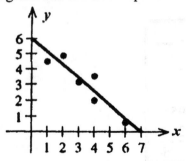

5. a. We first summarize the data:

x	y	x^2	xy
1	3	1	3
2	5	4	10
3	5	9	15
4	7	16	28
5	8	25	40
15	28	55	96

The normal equations are $55m + 15b = 96$
$$15m + 5b = 28.$$

Solving, we find $m = 1.2$ and $b = 2$, so that the required equation is $y = 1.2x + 2$.
 b.   The scatter diagram and the least-squares line for the given data follow.

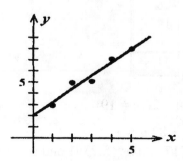

7.   a. We first summarize the data:

x	y	x^2	xy
4	0.5	16	2
4.5	0.6	20.25	2.7
5	0.8	25	4
5.5	0.9	30.25	4.95
6	1.2	36	7.2
25	4	127.5	20.85

The normal equations are $5b + \quad 25m = \ 4$
$$25b + 127.5m = 20.85.$$
The solutions are $m = 0.34$ and $b = -0.9$, and so a required equation is
$y = 0.34x - 0.9$.

b. The scatter diagram and the least-squares line for these data follow.

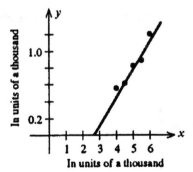

In units of a thousand / In units of a thousand

c. If $x = 6.4$, then $y = 0.34(6.4) - 0.9 = 1.276$ and so 1276 completed applications might be expected.

9. a. We first summarize the data:

x	y	x^2	xy
1	436	1	436
2	438	4	876
3	428	9	1284
4	430	16	1720
5	426	25	2130
15	2158	55	6446

The normal equations are $5b + 15m = 2158$
$15b + 55m = 6446.$

Solving this system, we find $m = -2.8$ and $b = 440.$
Thus, the equation of the least-squares line is $y = -2.8x + 440.$

b. The scatter diagram and the least-squares line for this data are shown in the figure that follows.

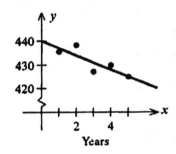

Years

c. Two years from now, the average SAT verbal score in that area will be
$y = -2.8(7) + 440 = 420.4$.

11. a. We first summarize the data:

x	y	x^2	xy
0	168	0	0
10	213	100	2130
20	297	400	5940
30	374	900	11220
40	427	1600	17080
51	467	2601	23817
151	1946	5601	60187

The normal equations are $6b + 151m = 1946$

$$151b + 5601m = 60187.$$

The solutions are $m = 6.2264$ and $b = 167.636$ and so a required equation is
$y = 6.226x + 167.6$.
b. In 2000, $x = 60$, $y = 6.226(63) + 167.6 \approx 60$. Hence, the expected size of the average farm will be 560 acres.

13. a. We first summarize the data:

x	y	x^2	xy
1	20	1	20
2	24	4	48
3	26	9	78
4	28	16	112
5	32	25	160
15	130	55	418

The normal equations are $5b + 15m = 130$
$$15b + 55m = 418.$$
The solutions are $m = 2.8$ and $b = 17.6$, and so an equation of the line is
$$y = 2.8x + 17.6.$$

b. When $x = 8$, $y = 2.8(8) + 17.6 = 40$. Hence, the state subsidy is expected to be $40 million for the eighth year.

15. a. We first summarize the data:

x	y	x^2	xy
1	16.7	1	16.7
3	26	9	78
5	33.3	25	166.5
7	48.3	49	338.1
9	57	81	513
11	65.8	121	723.8
13	74.2	169	964.6
15	83.3	225	1249.5
64	404.6	680	4050.2

The normal equations are $\quad 8b + 64m = \quad 404.6$

$$64b + 680m = 4050.2.$$

The solutions are $m = 4.8417$ and $b = 11.8417$ and so a required equation is $y = 4.842x + 11.842$.

b. In 1993, $x = 19$, and so $y = 4.842(19) + 11.842 = 103.84$. Hence the estimated number of cans produced in 1993 is 103.8 billion.

17. a. We first summarize the data:

x	y	x^2	xy
0	21.7	0	0
1	32.1	1	32.1
2	45.0	4	90
3	58.3	9	174.9
4	69.6	16	278.4
10	226.7	30	575.4

The normal equations are
$$5b + 10m = 226.7$$
$$10b + 30m = 575.4.$$
The solutions are $m = 12.2$ and $b = 20.9$ and so a required equation is $y = 12.2x + 20.9$.

b. In 2003, $x = 5$ so $y = 12.2(5) + 20.9 = 81.9$, or 81.9 million computers are expected to be connected to the internet in Europe in that year.

19. a. We first summarize the data:

x	y	x^2	xy
4.25	178	18.0625	756.5
10	667	100	6670
14	1194	196	16716
15.5	1500	240.25	23250
17.8	1388	316.84	24706.4
19.5	1640	380.25	31980
81.05	6567	1251.4025	104,078.9

The normal equations are
$$6b + 81.05m = 6567$$
$$81.05b + 1251.4025m = 104,078.9.$$

The solutions are $m = 98.1761$ and $b = -231.696$ and so a required equation is
$$y = 98.176x - 231.7.$$
b. If $x = 20$, then $y = 98.176(20) - 231.7 = 1731.82$. Hence, if the health-spending in the U.S. were in line with OECD countries, it should only have been $1732 per capita.

21. True. The error involves the sum of the squares of the form $[f(x_i) - y_i]^2$ where f is the least-squares function and y_i is a data point. Thus, the error is zero if and only if $f(x_i) = y_i$ for each $1 \le i \le n$.

USING TECHNOLOGY EXERCISES 8.4, page 647

1. $y = 2.3596x + 3.8639$ 3. $y = -1.1948x + 3.5525$

5. a. $y = 13.321x + 72.57$ b. 192 million tons

1. We form the Lagrangian function $F(x,y,\lambda) = x^2 + 3y^2 + \lambda(x + y - 1)$. We solve the
    system
    $$\begin{cases} F_x = 2x + \lambda = 0 \\ F_y = 6y + \lambda = 0 \\ F_\lambda = x + y - 1 = 0. \end{cases}$$
    Solving the first and the second equations for x and y in terms of λ we obtain
    $x = -\frac{\lambda}{2}$ and $y = -\frac{\lambda}{6}$ which, upon substitution into the third equation, yields
    $-\frac{\lambda}{2} - \frac{\lambda}{6} - 1 = 0$ or $\lambda = -\frac{3}{2}$. Therefore, $x = \frac{3}{4}$ and $y = \frac{1}{4}$ which gives the point
    $(\frac{3}{4}, \frac{1}{4})$ as the sole critical point of F. Therefore, $(\frac{3}{4}, \frac{1}{4}) = \frac{3}{4}$ is a minimum of F.

3. We form the Lagrangian function $F(x,y,\lambda) = 2x + 3y - x^2 - y^2 + \lambda(x + 2y - 9)$. We
    then solve the system

    $$\begin{cases} F_x = 2 - 2x + \lambda = 0 \\ F_y = 3 - 2y + 2\lambda = 0. \\ F_\lambda = x + 2y - 9 = 0 \end{cases}$$

    Solving the first equation λ, we obtain $\lambda = 2x - 2$. Substituting into the second
    equation, we have $3 - 2y + 4x - 4 = 0$, or $4x - 2y - 1 = 0$.   Adding this
    equation
    to the third equation in the system, we have $5x - 10 = 0$, or $x = 2$. Therefore,
    $y = 7/2$ and $f(2, \frac{7}{2}) = -\frac{7}{4}$ is the maximum value of f.

5. Form the Lagrangian function $F(x,y,\lambda) = x^2 + 4y^2 + \lambda(xy - 1)$. We then solve the
    system $\begin{cases} F_x = 2x + \lambda y = 0 \\ F_y = 8y + \lambda x = 0. \\ F_\lambda = xy - 1 = 0 \end{cases}$
    Multiplying the first and second equations by x and y, respectively, and
    subtracting the resulting equations, we obtain $2x^2 - 8y^2 = 0$, or $x = \pm 2y$.
    Substituting this into the third equation gives $2y^2 - 1 = 0$ or $y = \pm\frac{\sqrt{2}}{2}$. We
    conclude that $f(-\sqrt{2}, -\frac{\sqrt{2}}{2}) = f(\sqrt{2}, \frac{\sqrt{2}}{2}) = 4$ is the minimum value of f.

7.   We form the Lagrangian function
$$F(x,y,\lambda) = x + 5y - 2xy - x^2 - 2y^2 + \lambda(2x + y - 4).$$
Next, we solve the system
$$\begin{cases} F_x = 1 - 2y - 2x + 2\lambda = 0 \\ F_y = 5 - 2x - 4y + \lambda = 0 \\ F_\lambda = 2x + y - 4 = 0 \end{cases}$$
Solving the last two equations for x and y in terms of λ, we obtain
$$y = \tfrac{1}{3}(1 + \lambda) \quad \text{and} \quad x = \tfrac{1}{6}(11 - \lambda)$$
which, upon substitution into the first equation, yields
$$1 - \tfrac{2}{3}(1 + \lambda) - \tfrac{1}{3}(11 - \lambda) + 2\lambda = 0$$
or $1 - \tfrac{2}{3} - \tfrac{2}{3}\lambda - \tfrac{11}{3} + \tfrac{\lambda}{3} + 2\lambda = 0$
or $\lambda = 2$. Therefore, $x = 3/2$ and $y = 1$. The maximum of f is
$$f(\tfrac{3}{2}, 1) = \tfrac{3}{2} + 5 - 2(\tfrac{3}{2}) - (\tfrac{3}{2})^2 - 2 = -\tfrac{3}{4}.$$

9.   Form the Lagrangian $F(x,y,\lambda) = xy^2 + \lambda(9x^2 + y^2 - 9)$. We then solve
$$\begin{cases} F_x = y^2 + 18\lambda x = 0 \\ F_y = 2xy + 2\lambda y = 0 \\ F_\lambda = 9x^2 + y^2 - 9 = 0. \end{cases}$$
The first equation gives $\lambda = -\dfrac{y^2}{18x}$. Substituting into the second gives
$$2xy + 2y\left(-\frac{y^2}{18x}\right) = 0, \text{ or } 18x^2 y - y^3 = y(18x^2 - y^2) = 0,$$
giving $y = 0$ or $y = \pm 3\sqrt{2}x$. If $y = 0$, then the third equation gives $9x^2 - 9 = 0$ or $x = \pm 1$. If $y = \pm 3\sqrt{3}/3$. Therefore, the points $(-1,0), (-\sqrt{3}/3, -\sqrt{6})$, $(-\sqrt{3}/3, \sqrt{6}), (\sqrt{3}/3, -\sqrt{6})$ and $(\sqrt{3}/3, \sqrt{6})$ give rise to extreme values of f subject to the given constraint. Evaluating $f(x,y)$ at each of these points, we see that
$$f(\sqrt{3}/3, -\sqrt{6}) = (\sqrt{3}/3, \sqrt{6}) = 2\sqrt{3}$$
is the maximum value of f.

11.   We form the Lagrangian function $F(x,y,\lambda) = xy + \lambda(x^2 + y^2 - 16)$. To find the critical points of F, we solve the system

$$\begin{cases} F_x = y + 2\lambda x = 0 \\ F_y = x + 2\lambda y = 0 \\ F_\lambda = x^2 + y^2 - 16 = 0 \end{cases}$$

Solving the first equation for λ and substituting this value into the second

Equation yields $x - 2\left(\dfrac{y}{2x}\right)y = 0$, or $x^2 = y^2$. Substituting the last equation into

the third equation in the system, yields $x^2 + x^2 - 16 = 0$, or $x^2 = 8$, that is,

$x = \pm 2\sqrt{2}$. The corresponding values of y are $y = \pm 2\sqrt{2}$. Therefore the critical

points of F are

$$(-2\sqrt{2}, -2\sqrt{2}), (-2\sqrt{2}, 2\sqrt{2}), (2\sqrt{2}, -2\sqrt{2})(2\sqrt{2}, 2\sqrt{2}).$$

Evaluating f at each of these values, we find that $f(-2\sqrt{2}, 2\sqrt{2}) = -8$ and f

$(2\sqrt{2}, -2\sqrt{2}) = -8$ are relative minimum values and $f(-2\sqrt{2}, -2\sqrt{2}) = 8$ and f

$(2\sqrt{2}, 2\sqrt{2}) = 8$, are relative maximum values.

13. We form the Lagrangian function $F(x,y,\lambda) = xy^2 + \lambda(x^2 + y^2 - 1)$. Next, we solve
the system

$$\begin{cases} F_x = \quad y^2 + 2x\lambda = 0 \\ F_y = 2xy + 2y\lambda = 0 \quad . \\ F_\lambda = x^2 + y^2 - 1 = 0 \end{cases}$$

We find that $x = \pm\sqrt{3}/3$ and $y = \pm\sqrt{6}/3$ and $x = \pm 1$, $y = 0$. Evaluating f at each

of the critical points $(-\frac{\sqrt{3}}{3}, -\frac{\sqrt{6}}{3}), (-\frac{\sqrt{3}}{3}, \frac{\sqrt{6}}{3})(\frac{\sqrt{3}}{3}, -\frac{\sqrt{6}}{3})(\frac{\sqrt{3}}{3}, \frac{\sqrt{6}}{3}), (-1,0)$, and $(1,0)$,

we find that $f(-\frac{\sqrt{3}}{3}, -\frac{\sqrt{6}}{3}) = \quad -\frac{2\sqrt{3}}{9}$ and $f(-\frac{\sqrt{3}}{3}, \frac{\sqrt{6}}{3}) = -\frac{2\sqrt{3}}{9}$ are relative minimum

values and $f(\frac{\sqrt{3}}{3}, -\frac{\sqrt{6}}{3}) = \frac{2\sqrt{3}}{9}$ and $f(\frac{\sqrt{3}}{3}, \frac{\sqrt{6}}{3}) = \frac{2\sqrt{3}}{9}$ are relative maximum values.

15. Form the Lagrangian function $F(x,y,z,\lambda) = x^2 + y^2 + z^2 + \lambda(3x + 2y + z - 6)$. We
solve the system

$$\begin{cases} F_x = 2x + 3\lambda = 0 \\ F_y = 2y + 2\lambda = 0 \\ F_z = 2x + \lambda = 0 \\ F_\lambda = 3x + 2y + z - 6 = 0. \end{cases}$$

The third equation give $\lambda = -2z$. Substituting into the first two equations gives
$$\begin{cases} 2x - 6z = 0 \\ 2y - 4z = 0. \end{cases}$$
So $x = 3z$ and $y = 2z$. Substituting into the third equation yields
$9z + 4z + z - 6 = 0$, or $z = 3/7$. Therefore, $x = 9/7$ and $y = 6/7$. Therefore,
$f\left(\frac{9}{7}, \frac{6}{7}, \frac{3}{7}\right) = \frac{18}{7}$ is the minimum value of F.

17.   We want to maximize P subject to the constraint $x + y = 200$. The Lagrangian function is
$$F(x, y, \lambda) = -0.2x^3 - 0.25y^2 - 0.2xy + 100x + 90y - 4000 + \lambda(x + y - 200).$$
Next, we solve
$$\begin{cases} F_x = -0.4x - 0.2y + 100 + \lambda = 0 \\ F_y = -0.5y - 0.2x + 90 + \lambda = 0 \\ F_\lambda = x + y - 200 = 0. \end{cases}$$
Subtracting the first equation from the second yields
$0.2x - 0.3y - 10 = 0$, or $2x - 3y - 100 = 0$.
Multiplying the third equation in the system by 2 and subtracting the resulting equation from the last equation, we find
$-5y + 300 = 0$ or $y = 60$.
So $x = 140$ and the company should make 140 finished and 60 unfinished units.

19.   Suppose each of the sides made of pine board is x feet long and those of steel are y feet long. Then $xy = 800$. The cost is $C = 12x + 3y$ and is to be minimized subject to the condition $xy = 800$. We form the Lagrangian function
$$F(x, y, \lambda) = 12x + 3y + \lambda(xy - 800).$$
We solve the system
$$\begin{cases} F_x = 12 + \lambda y = 0 \\ F_y = 3 + \lambda x = 0 \\ F_\lambda = xy - 800 = 0. \end{cases}$$
Multiplying the first equation by x and the second equation by y and subtracting the resulting equations, we obtain $12x - 3y = 0$, or $y = 4x$. Substituting this into the third equation of the system, we obtain
$$4x^2 - 800 = 0, \text{ or } x = \pm\sqrt{200} = \pm 10\sqrt{2}.$$

Since x must be positive, we take $x = 10\sqrt{2}$. So $y = 40\sqrt{2}$. So the dimensions are approximately 14.14ft by 56.56 ft.

21. Refer to the following figure.

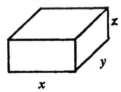

We form the Lagrangian function $F(x,y,\lambda) = xyz + \lambda(3xy + 2xz + 2yz - 36)$. Then, we solve the system

$$\begin{cases} F_x = yz + 3\lambda y + 2\lambda z = 0 \\ F_y = xz + 3\lambda x + 2\lambda z = 0 \\ F_z = xy + 2\lambda x + 2\lambda y = 0 \\ F_\lambda = 3xy + 2xz + 2yz - 36 = 0 \end{cases}$$

Multiplying the first, second, and third equation by x, y, and z, respectively, we obtain

$$\begin{cases} xyz + 3\lambda xy + 2\lambda xz = 0 \\ xyz + 3\lambda xy + 2\lambda yz = 0 \\ xyz + 2\lambda xz + 2\lambda yz = 0. \end{cases}$$

Subtracting the second equation from the first and the third equation from the second, yields

$$\begin{cases} 2\lambda(x - y)z = 0 \\ \lambda x(3y - 2z) = 0. \end{cases}$$

Solving this system, we find that $x = y$ and $x = 3/2y$. Substituting these values into the third equation, we find that

$$3y^2 + 2y(\tfrac{3}{2})y + 2y(\tfrac{3}{2}) - 36 = 0$$

and $y = \pm 2$. We reject the negative root, and find that $x = 2$, $y = 2$, and $z = 3$ provides the desired relative maximum and the dimensions are $2' \times 2' \times 3'$.

23. We want to maximize $f(x,y) = 90x^{1/4}y^{3/4}$ subject to $x + y = 60{,}000$. We form the Lagrangian function $F(x,y,\lambda) = 90x^{1/4}y^{3/4} + \lambda(x + y - 60{,}000)$. Now set

$$\begin{cases} F_x = \frac{45}{2}x^{-3/4}y^{3/4} + \lambda = 0 \\ F_y = \frac{135}{2}x^{1/4}y^{-1/4} + \lambda = 0 \\ F_\lambda = x + y - 60{,}000 = 0. \end{cases}$$

Eliminating λ in the first two equations gives

$$\frac{45}{2}\left(\frac{y}{x}\right)^{3/4} - \frac{135}{2}\left(\frac{x}{y}\right)^{1/4} = 0$$

$$\frac{y}{x} - 3 = 0, \text{ or } y = 3x.$$

Substituting this value into the third equation in the system, we find $x + 3x = 60{,}000$ and $x = 15{,}000$ and $y = 45{,}000$. So the company should spend $15,000 on newspaper advertisements and $45,000 on television advertisements.

25. We want to minimize $C = 2xy + 8xz + 6yz$ subject to $xyz = 12000$. Form the Lagrangian function $F(x,y,z,\lambda) = 2xy + 8xz + 6yz + \lambda(xyz - 12000)$. Next, we solve the system

$$\begin{cases} F_x = 2y + 8z + \lambda yz = 0 \\ F_y = 2x + 6z + \lambda xz = 0 \\ F_z = 8x + 6y + 3xy = 0 \\ F_\lambda = xyz - 12{,}000 = 0. \end{cases}$$

Multiplying the first, second, and third equations by x, y, and z, we obtain

$$\begin{cases} 2xy + 8yz + 3xyz = 0 \\ 2xy + 6yz + \lambda xyz = 0 \\ 8xz + 6yz + \lambda xyz = 0. \end{cases}$$

The first two equations imply that $z(8x - 6y) = 0$ or since $z \neq 0$, we have $x = (3/4)y$. The second and third equations imply that $x(8z - 2y) = 0$ or $x = (1/4)y$. Substituting these values into the third equation of the system, we have $(\frac{3}{4}y)(y)(\frac{1}{4}y) = 12{,}000$ or $y^3 = 64{,}000$, or $y = 40$.
Therefore, $x = 30$ and $z = 10$. So the heating cost is
$$C = 2(30)(40) + 8(30)(10) + 6(40)(10) = 7200,$$
or $7200 as obtained earlier.

27. False. See Example 1, Section 8.5

1. $f(x,y) = x^2 + 2y; \quad df = 2x\,dx + 2\,dy$

3. $f(x,y) = 2x^2 - 3xy + 4x; \quad df = (4x - 3y + 4)dx - 3x\,dy$

5. $f(x,y) = \sqrt{x^2 + y^2}; \quad df = \frac{1}{2}(x^2 + y^2)^{-1/2}(2x)\,dx + \frac{1}{2}(x^2 + y^2)^{-1/2}(2y)\,dy$

$$= \frac{x}{\sqrt{x^2 + y^2}}\,dx + \frac{y}{\sqrt{x^2 + y^2}}\,dy.$$

7. $f(x,y) = \dfrac{5y}{x - y};$

$$df = \frac{\partial}{\partial x}[5y(x - y)^{-1}]dx + \frac{\partial}{\partial x}\left(\frac{5y}{x - y}\right)dy$$

$$= 5y(-1)(x - y)^{-2}\,dx + \frac{(x - y)(5) - 5y(-1)}{(x - y)^2}\,dy$$

$$= -\frac{5y}{(x - y)^2}\,dx + \frac{5x}{(x - y)^2}\,dy.$$

9. $f(x,y) = 2x^5 - ye^{-3x}; \quad df = (10x^4 + 3ye^{-3x})dx - e^{-3x}dy.$

11. $f(x,y) = x^2e^y + y\ln x; \quad df = (2xe^y + \dfrac{y}{x})dx + (x^2e^y + \ln x)dy.$

13. $f(x,y,z) = xy^2z^3; \quad df = y^2z^3\,dx + 2xyz^3\,dy + 3xy^2z^2\,dz.$

15. $f(x,y,z) = \dfrac{x}{y + z}; \quad df = \dfrac{1}{y + z}\,dx + x(-1)(y + z)^{-2}\,dy + x(-1)(y + z)^{-2}\,dz$

$$= \frac{1}{y + z}\,dx - \frac{x}{(y + z)^2}\,dy - \frac{x}{(y + z)^2}\,dz.$$

17. $f(x,y,z) = xyz + xe^{yz}; \quad df = (yz + e^{yz})dx + (xz + xze^{yz})dy + (xy + xye^{yz})dz$

$$= (yz + e^{yz})dx + xz(1 + e^{yz})dy + xy(1 + e^{yz})dz.$$

19. $\Delta z \approx dz = (8x - y)dx - x\,dy\big|_{\substack{x=1,\ y=2 \\ dx=0.01,\ dy=0.02}} = (8-2)(0.01) - 0.02 = 0.04.$

21. $\Delta z \approx dz = \frac{2}{3}x^{-1/3}y^{1/2}dx + \frac{1}{2}x^{2/3}y^{-1/2}\,dy\big|_{\substack{x=8,\ y=9 \\ dx=-0.03,\ dy=0.03}}$

$\qquad = \frac{2}{3}(\frac{3}{2})(-0.03) + \frac{1}{2}(\frac{4}{3})(0.03) = -0.03 + 0.02 = -0.01.$

23. $f_x = \dfrac{(x-y)-x(1)}{(x-y)^2} = -\dfrac{y}{(x-y)^2}, \quad f_y = \dfrac{(x-y)(0)-x(-1)}{(x-y)^2} = \dfrac{x}{(x-y)^2}.$

Therefore, $\Delta z \approx dz = \dfrac{-y\,dx + x\,dy}{(x-y)^2}.$

Here $x = -3$, $y = -2$, $dx = -0.02$, and $dy = 0.02$. So the approximate change in $z = f(x,y)$ is

$$\Delta z \approx \frac{-(-2)(-0.02) + (-3)(0.02)}{(-1)^2} = -0.1.$$

25. $\Delta z \approx dz = 2e^{-y}\,dx - 2xe^{-y}\,dy.$ With $x = 4$, $y = 0$, $dx = 0.03$ and $dy = 0.03$, we have
$\Delta z \approx 2e^0(0.03) - 2(4)(e^0)(0.03) = -0.18.$

27. $f(x,y) = xe^{xy} - y^2$; $\ f_x = xye^{xy} + e^{xy} = e^{xy}(1+xy); \ f_y = x^2 e^{xy} - 2y;$

and $\qquad \Delta z \approx dz = e^{xy}(1+xy)dx + (x^2 e^{xy} - 2y)dy.$

With $x = -1$, $y = 0$, $dx = 0.03$, and $y = 0.03$, we have
$\Delta z \approx e^0(1)(0.03) + (1)(e^0)(0.03) = 0.06.$

29. $f(x,y) = x\ln x + y\ln x$

$f_x = \ln x + x\left(\dfrac{1}{x}\right) + \dfrac{y}{x} = \ln x + 1 + \dfrac{y}{x}, \ f_y = \ln x.$

So $\Delta z \approx dz = \left(\ln x + 1 + \dfrac{y}{x}\right)dx + \ln x\,dy.$

Using $x = 2$, $y = 3$, $dx = -0.02$, and $dy = -0.11$, we have
$\Delta z \approx (\ln 2 + 1 + 1.5)(-0.02) + (\ln 2)(-0.11) \approx -0.1401.$

31. $\Delta P \approx dP = (-0.04x + y + 39)dx + (-30y + x + 25)dy.$
With $x = 4000$, $y = 150$, $dx = 500$, and $dy = -10$, we have

$$\Delta P \approx [-0.04(4000) + 150 + 39](500) + [-30(150) + 4000 + 25](-10)$$
$$= 19,250 \text{ or } \$19,250 \text{ per month.}$$

33. $R(x,y) = -x^2 - 0.5y^2 + xy + 8x + 3y + 20$
$$dR = -2x\,dx - y\,dy + y\,dx + x\,dy + 8dx + 3dy$$
$$= (-2x + y + 8)\,dx = (-y + x + 3)\,dy.$$
If $x = 10$, $y = 15$, $dx = 1$, and $dy = -1$,

$$dR = [-2(10) + 15 + 8](1) + (-15 + 10 + 3)(-1) = 3(1) + (-2)(-1) = 5.$$
So the revenue is expected to increase by \$5000 per month

35. $R = \dfrac{x}{y}$; $\Delta R \approx dR = \dfrac{1}{y}dx - \dfrac{x}{y^2}dy$

Therefore,
$$\Delta R = \frac{1}{4}(2) - \frac{60}{4^2}(-0.2) = \frac{1}{2} + \frac{15}{4}\left(\frac{1}{5}\right)$$
$$= \frac{1}{2} + \frac{3}{4} = \frac{5}{4} = 1.25$$
or \$1.25.

37. $V = \pi r^2 h$; $\Delta V \approx dV = 2\pi r h\,dr + \pi r^2\,dh$.
If $r = 8$, $h = 20$, $dr = (\pm 0.1)$, then
$$dV = 2\pi(8)(20)(\pm 0.1) + \pi(8)^2(\pm 0.1) = 320\pi(\pm 0.1) + 64\pi(\pm 0.1) = \pm 38.4\,\pi \text{ cc}.$$

39. Differentiating implicitly, we have
$$-\frac{1}{R^2}dR = -\frac{1}{R_1^2}dR_1 - \frac{1}{R_2^2}dR_2 - \frac{1}{R_3^2}dR_3$$
or $|dR| \le \left(\dfrac{R}{R_1}\right)^2 |dR_1| + \left(\dfrac{R}{R_2}\right)^2 |dR_2| + \left(\dfrac{R}{R_3}\right)^2 |dR_3|$.
With $R_1 = 100$, $R_2 = 200$, $R_3 = 300$, $|dR_1| \le 1$, $|dR_2| \le 2$, and $|dR_3| \le 3$,
we have
$$\frac{1}{R} = \frac{1}{100} + \frac{1}{200} + \frac{1}{300} = \frac{6 + 3 + 2}{600} = \frac{11}{600}.$$

$$|dR| \le \left(\frac{6}{11}\right)^2 (1) + \left(\frac{3}{11}\right)^2 (2) + \left(\frac{2}{11}\right)^2 (3) = 0.5455, \text{ or } 0.55 \text{ ohms.}$$

EXERCISES 8.7, page 676

1. $\displaystyle\int_1^2\int_0^1 (y+2x)\,dy\,dx = \int_1^2 \tfrac{1}{2}y^2 + 2xy\Big|_{y=0}^{y=1}\,dx = \int_1^2 (\tfrac{1}{2}+2x)\,dx = \tfrac{1}{2}x + x^2\Big|_1^2 = 5 - \tfrac{3}{2} = \tfrac{7}{2}.$

3. $\displaystyle\int_{-1}^1\int_0^1 xy^2\,dy\,dx = \int_{-1}^1 \tfrac{1}{3}xy^3\Big|_0^1\,dx = \int_{-1}^1 \tfrac{1}{3}x\,dx = \tfrac{x^2}{6}\Big|_{-1}^1 = \tfrac{1}{6} - (\tfrac{1}{6}) = 0.$

5. $\displaystyle\int_{-1}^2\int_1^{e^3} \frac{x}{y}\,dy\,dx = \int_{-1}^2 x\ln y\Big|_1^{e^3}\,dx = \int_{-1}^2 x\ln e^3\,dx = \int_{-1}^2 3x\,dx = \tfrac{3}{2}x^2\Big|_{-1}^2$

$\qquad = \tfrac{3}{2}(4) - \tfrac{3}{2}(1) = \tfrac{9}{2}.$

7. $\displaystyle\int_{-2}^0\int_0^1 4xe^{2x^2+y}\,dx\,dy = \int_{-2}^0 e^{2x^2+y}\Big|_{x=0}^{x=1}\,dy = \int_{-2}^0 (e^{2+y} - e^y)\,dy = (e^{2+y} - e^y)\Big|_{-2}^0$

$\qquad = [(e^2 - 1) - (e^0 - e^{-2}) = e^2 - 2 + e^{-2} = (e^2 - 1)(1 - e^{-2}).$

9. $\displaystyle\int_0^1\int_1^e \ln y\,dy\,dx = \int_0^1 y\ln y - y\Big|_{y=1}^{y=e}\,dx = \int_0^1 dx = 1.$

11. $\displaystyle\int_0^1\int_0^x (x+2y)\,dy\,dx = \int_0^1 (xy + y^2)\Big|_{y=0}^{y=x}\,dx = \int_0^1 2x^2\,dx = \tfrac{2}{3}x^3\Big|_0^1 = \tfrac{2}{3}.$

13. $\displaystyle\int_1^3\int_0^{x+1} (2x+4y)\,dy\,dx = \int_1^3 2xy + 2y^2\Big|_{y=0}^{y=x+1}\,dx = \int_1^3 [2x(x+1) + 2(x+1)^2]\,dx$

$\qquad = \int_1^3 (4x^2 + 6x + 2)\,dx = (\tfrac{4}{3}x^3 + 3x^2 + 2x)\Big|_1^3$

$\qquad = (36 + 27 + 6) - (\tfrac{4}{3} + 3 + 2) = \tfrac{188}{3}.$

15. $\displaystyle\int_0^4\int_0^{\sqrt{y}} (x+y)\,dx\,dy = \int_0^4 \tfrac{1}{2}x^2 + xy\Big|_{x=0}^{x=\sqrt{y}}\,dy = \int_0^4 (\tfrac{1}{2}y + y^{3/2})\,dy$

$\qquad = (\tfrac{1}{4}y^2 + \tfrac{2}{5}y^{5/2})\Big|_0^4 = 4 + \tfrac{64}{5} = \tfrac{84}{5}.$

$$\int_0^1 \int_y^{2-y} xy\,dx\,dy = \int_0^1 \tfrac{1}{2}x^2 y\Big|_{x=y}^{x=2-y} dy = \int_0^1 \tfrac{1}{2}(2-y)^2 y - \tfrac{1}{2}y^3]\,dy$$

$$= \int_0^1 (2y - 2y^2)\,dy = (y^2 - \tfrac{2}{3}y^3)\Big|_0^1 = \tfrac{1}{3}.$$

21. The area of R is $1/2$. Therefore, the average value of f is

$$2\int_0^1 \int_0^x xe^y\,dy\,dx = 2\int_0^1 xe^y\Big|_0^x dx = 2\int_0^1 (xe^x - x)\,dx = 2(xe^x - e^x - \tfrac{1}{2}x^2)\Big|_0^1$$
$$= 2(e - e - \tfrac{1}{2} + 1) = 1.$$

23. The population is

$$2\int_0^5 \int_{-2}^0 \frac{10,000e^y}{1+0.5x}\,dy\,dx = 20,000\int_0^5 \frac{e^y}{1+0.5x}\Big|_{y=-2}^{y=0} dx$$

$$= 20,000(1-e^{-2})\int_0^5 \frac{1}{1+0.5x}\,dx = 20,000(1-e^{-2})2\,\ln(1+0.5x)\Big|_0^5$$
$$= 40,000(1 - e^{-2})\ln 3.5 \approx 43,329.$$

25. The average weekly profit is

$$\frac{1}{(20)(20)}\int_{100}^{120}\int_{180}^{200}(-0.2x^2 - 0.25y^2 - 0.2xy + 100x + 90y - 4000)\,dx\,dy$$

$$= \frac{1}{400}\int_{100}^{120} -\tfrac{1}{15}x^3 - 0.25y^2x - 0.1x^2y + 50x^2 + 90xy - 4000x\Big|_{x=180}^{x=200} dy$$

$$= \frac{1}{400}\int_{100}^{120}(-144,533.33 - 5y^2 - 760y + 380,000 + 1800y - 80,000)\,dy$$

$$= \frac{1}{400}\int_{100}^{120}(155,466.67 - 5y^2 + 1040y)\,dy$$

$$= \frac{1}{400}(155,466.67y - \tfrac{5}{3}y^3 + 520y^2)\Big|_{100}^{120}$$

$$= \frac{1}{400}(3,109,333.40 - 1,213,333.30 + 2,288,000) \approx 10,460\text{ , or }\$10,460/\text{wk.}$$

27. True. $\iint\limits_{R} g(x,y)\,dA$ gives the volume of the solid bounded above by the surface

$z = g(x,y)$. $\iint\limits_{R} f(x,y)\,dA$ gives the volume of the solid bounded above by the surface $z = f(x, y)$. Therefore,

$$\iint_R g(x, y)\, dA - \iint_R f(x, y)\, dA = \iint_R [g(x, y) - f(x, y)]\, dA$$

gives the volume of the solid bounded above by $z = g(x, y)$ and below by
$z = f(x, y)$.

CHAPTER 8 REVIEW EXERCISES, page 687

1. $f(0,1) = 0; f(1,0) = 0, f(1,1) = \dfrac{1}{1+1} = \dfrac{1}{2}.$

   $f(0,0)$ does not exist because the point $(0,0)$ does not lie in the domain of f.

3. $h(1,1,0) = 1 + 1 = 2$, $h(-1,1,1) = -e - 1 = -(e + 1)$,
   $h(1,-1,1) = -e - 1 = -(e + 1)$.

5. $D = \{(x, y)\,|\,y \neq -x\}$

7. The domain of f is the set of all ordered triplets (x, y, z) of real numbers such that
   $z \geq 0$ and $x \neq 1$, $y \neq 1$, and $z \neq 1$.

9. $z = y - x^2$

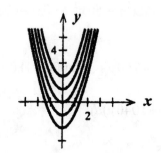

11. $z = e^{xy}$

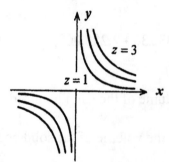

13. $f(x,y) = x\sqrt{y} + y\sqrt{x}$; $f_x = \sqrt{y} + \dfrac{y}{2\sqrt{x}}$; $f_y = \dfrac{x}{2\sqrt{y}} + \sqrt{x}$

15. $f(x,y) = \dfrac{x-y}{y+2x}$. $f_x = \dfrac{(y+2x)-(x-y)(2)}{(y+2x)^2} = \dfrac{3y}{(y+2x)^2}$.

$f_y = \dfrac{(y+2x)(-1)-(x-y)}{(y+2x)^2} = \dfrac{-3x}{(y+2x)^2}$.

17. $h(x,y) = (2xy+3y^2)^5$; $h_x = 10y(2xy+3y^2)^4$; $h_y = 10(x+3y)(2xy+3y^2)^4$.

19. $f(x,y) = (x^2+y^2)e^{x^2+y^2}$;

$f_x = 2xe^{x^2+y^2} + (x^2+y^2)(2x)e^{x^2+y^2} = 2x(x^2+y^2+1)e^{x^2+y^2}$.

$f_y = 2ye^{x^2+y^2} + (x^2+y^2)(2y)e^{x^2+y^2} = 2y(x^2+y^2+1)e^{x^2+y^2}$.

21. $f(x,y) = \ln\left(1+\dfrac{x^2}{y^2}\right)$. $f_x = \dfrac{\frac{2x}{y^2}}{1+\frac{x^2}{y^2}} = \dfrac{2x}{x^2+y^2}$; $f_y = \dfrac{-\frac{2x^2}{y^3}}{1+\frac{x^2}{y^2}} = -\dfrac{2x^2}{y(x^2+y^2)}$.

23. $f(x,y) = x^4 + 2x^2y^2 - y^4$; $f_x = 4x^3 + 4xy^2$, $f_y = 4x^2y - 4y^3$;

$f_{xx} = 12x^2 + 4y^2$, $f_{xy} = 8xy = f_{yx}$, $f_{yy} = 4x^2 - 12y^2$.

25. $g(x,y) = \dfrac{x}{x+y^2}$; $g_x = \dfrac{(x+y^2)-x}{(x+y^2)^2} = \dfrac{y^2}{(x+y^2)^2}$, $g_y = \dfrac{-2xy}{(x+y^2)^2}$.

Therefore, $g_{xx} = -2y^2(x+y^2)^{-3} = -\dfrac{2y^2}{(x+y^2)^3}$,

$$g_{yy} = \dfrac{(x+y^2)^2(-2x)+2xy(2)(x+y^2)2y}{(x+y^2)^4} = \dfrac{2x(x^2+y^2)[-x-y^2+4y^2]}{(x+y^2)^4}$$

$$= \dfrac{2x(3y^2-x)}{(x+y^2)^3}.$$

and $g_{xy} = \dfrac{(x+y^2)2y-y^2(2)(x+y^2)2y}{(x+y^2)^4} = \dfrac{2(x+y^2)[xy+y^3-2y^3]}{(x+y^2)^4}$

$$= \dfrac{2y(x-y^2)}{(x+y^2)^3} = g_{yx}.$$

27. $h(s,t) = \ln\left(\dfrac{s}{t}\right)$. Write $h(s,t) = \ln s - \ln t$. Then

$$h_s = \frac{1}{s}, \ h_t = -\frac{1}{t}.$$

Therefore, $h_{ss} = -\dfrac{1}{s^2}, \ h_{st} = h_{ts} = 0, \ h_{tt} = \dfrac{1}{t^2}.$

29. $f(x,y) = 2x^2 + y^2 - 8x - 6y + 4$; To find the critical points of f, we solve the

system $\begin{cases} f_x = 4x - 8 = 0 \\ f_y = 2y - 6 = 0 \end{cases}$

obtaining $x = 2$ and $y = 3$. Therefore, the sole critical point of f is $(2,3)$. Next,
$f_{xx} = 4, f_{xy} = 0, f_{yy} = 2$.
Therefore, $D = f_{xx}(2,3)f_{yy}(2,3) - f_{xy}(2,3)^2 = 8 > 0.$
Since $f_{xx}(2,3) > 0$, we see that $f(2,3) = -13$ is a relative minimum.

31. $f(x,y) = x^3 - 3xy + y^2$. We solve the system of equations $\begin{cases} f_x = 3x^2 - 3y = 0 \\ f_y = -3x + 2y = 0 \end{cases}$

obtaining $x^2 - y = 0$, or $y = x^2$. Then $-3x + 2x^2 = 0$, and $x(2x - 3) = 0$, and $x = 0$,
or $x = 3/2$ and $y = 0$, or $y = 9/4$. Therefore, the critical points are $(0,0)$ and $(\frac{3}{2}, \frac{9}{4})$.
Next, $f_{xx} = 6x, f_{xy} = -3$, and $f_{yy} = 2$ and $D(x,y) = 12x - 9 = 3(4x - 3)$. Therefore,
$D(0,0) = -9$ so $(0,0)$ is a saddle point.
$$D(\tfrac{3}{2}, \tfrac{9}{4}) = 3(6-3) = 9 > 0, \quad \text{and} \quad f_{xx}(\tfrac{3}{2}, \tfrac{9}{4}) > 0$$
and therefore, $f(\tfrac{3}{2}, \tfrac{9}{4}) = \tfrac{27}{8} - \tfrac{81}{8} + \tfrac{81}{16} = -\tfrac{27}{16}$
is the relative minimum value.

33. $f(x,y) = f(x,y) = e^{2x^2 + y^2}$. To find the critical points of f, we solve the system

$$\begin{cases} f_x = 4xe^{2x^2 + y^2} = 0 \\ f_y = 2ye^{2x^2 + y^2} = 0 \end{cases}$$

giving $(0,0)$ as the only critical point of f. Next,
$$f_{xx} = 4(e^{2x^2 + y^2} + 4x^2 e^{2x^2 + y^2}) = 4(1 + 4x^2)e^{2x^2 + y^2}$$
$$f_{xy} = 8xye^{2x^2 + y^2}$$
$$f_{yy} = 2(1 + 2y^2)e^{2x^2 + y^2}.$$

Therefore,

$$D = f_{xx}(0,0)f_{yy}(0,0) - f_{xy}^2(0,0) = (4)(2) - 0 = 8 > 0$$

and so $(0,0)$ gives a relative minimum of f since $f_{xx}(0,0) > 0$. The minimum value of f is $f(0,0) = e^0 = 1$.

35. We form the Lagrangian function $F(x,y,\lambda) = -3x^2 - y^2 + 2xy + \lambda(2x + y - 4)$.
Next, we solve the system

$$\begin{cases} F_x = 6x + 2y + 2\lambda = 0 \\ F_y = -2y + 2x + \lambda = 0 . \\ F_\lambda = 2x + y - 4 = 0 \end{cases}$$

Multiplying the second equation by 2 and subtracting the resultant equation from the first equation yields $6y - 10x = 0$ so $y = 5x/3$. Substituting this value of y into the third equation of the system gives $2x + \frac{5}{3}x - 4 = 0$.

So $x = \frac{12}{11}$ and consequently $y = \frac{20}{11}$. So $(\frac{12}{11}, \frac{20}{11})$ gives the maximum value for f subject to the given constraint.

37. The Lagrangian function is $F(x,y,\lambda) = 2x - 3y + 1 + \lambda(2x^2 + 3y^2 - 125)$. Next, we solve the system of equations

$$\begin{cases} F_x = 2 + 4\lambda x = 0 \\ F_y = -3 + 6\lambda y = 0 \\ F_\lambda = 2x^2 + 3y^2 - 125 = 0. \end{cases}$$

Solving the first equation for x gives $x = -1/2\lambda$. The second equation gives $y = 1/2\lambda$. substituting these values of x and y into the third equation gives

$$2\left(-\frac{1}{2\lambda}\right)^2 + 3\left(\frac{1}{2\lambda}\right)^2 - 125 = 0$$

$$\frac{1}{2\lambda^2} + \frac{3}{4\lambda^2} - 125 = 0$$

$$2 + 3 - 500\lambda^2 = 0, \text{ or } \lambda = \pm\frac{1}{10}.$$

Therefore, $x = \pm 5$ and $y = \pm 5$ and so the critical points of f are $(-5,5)$ and $(5,-5)$.
Next, we compute

$$f(-5,5) = 2(-5) - 3(5) + 1 = -24.$$
$$f(5,-5) = 2(5) - 3(-5) + 1 = 26.$$

So f has a maximum value of 26 at $(5,-5)$ and a minimum value of -24 at $(-5,5)$.

      8 Calculus of Several Variables

39. $df = \frac{3}{2}(x^2 + y^4)^{1/2}(2x)dx + \frac{3}{2}(x^2 + y^4)^{1/2}(4y^3)dy$.

At $(3,2)$, $df = 9(9+16)^{1/2}dx + 48(9+16)^{1/2}dy = 45dx + 240dy$.

41. $\Delta f \approx df = (4xy^3 + 6y^2x - 2y)dx + (6x^2y^2 + 6yx^2 - 2x)dy$.

With $x = 1$, $y = -1$, $dx = 0.02$, and $dy = 0.02$, we find
$\Delta f \approx (-4+6+2)(0.02) + (6-6-2)(0.02) = 0.04$.

43. $\int_{-1}^{2}\int_{2}^{4}(3x - 2y)dx\, dy = \int_{-1}^{2}\frac{3}{2}x^2 - 2xy\Big|_{x=2}^{x=4}dy = \int_{-1}^{2}[(24-8y)-(6-4y)]dy$

$= \int_{-1}^{2}(18-4y)dy = (18y - 2y^2)\Big|_{-1}^{2} = (36-8)-(-18-2) = 48$.

45. $\int_{0}^{1}\int_{x^3}^{x^2}2x^2y\, dy\, dx = \int_{0}^{1}x^2y^2\Big|_{y=x^3}^{y=x^2}dx = \int_{0}^{1}x^2(x^4 - x^6)dx$

$= \int_{0}^{1}(x^6 - x^8)dx = \frac{1}{7} - \frac{1}{9} = \frac{2}{63}$.

47. $\int_{0}^{2}\int_{1}^{x}(4x^2 + y^2)dy\, dx = \int_{0}^{2}4x^2y + \frac{1}{3}y^3\Big|_{y=0}^{y=1}dx = \int_{0}^{2}(4x^2 + \frac{1}{3})dx$

$= (\frac{4}{3}x^3 + \frac{1}{3}x)\Big|_{0}^{2} = \frac{32}{3} + \frac{2}{3} = \frac{34}{3}$.

49. The area of R is

$\int_{0}^{2}\int_{x^2}^{2x}dy\, dx = \int_{0}^{2}y\Big|_{y=x^2}^{y=2x}dx = \int_{0}^{2}(2x - x^2)dx = (x^2 - \frac{1}{3}x^3)\Big|_{0}^{2} = \frac{4}{3}$.

Then

$AV = \frac{1}{4/3}\int_{0}^{2}\int_{x^2}^{2x}(xy+1)dy\, dx = \frac{3}{4}\int_{0}^{2}\frac{xy^2}{2}+y\Big|_{x^2}^{2x}dx$

$= \frac{3}{4}\int_{0}^{2}(-\frac{1}{2}x^5 + 2x^3 - x^2 + 2x)dx = \frac{3}{4}(-\frac{1}{12}x^6 + \frac{1}{2}x^4 - \frac{1}{3}x^3 + x^2)\Big|_{0}^{2}$

$= \frac{3}{4}(-\frac{16}{3} + 8 - \frac{8}{3} + 4) = 3$.

51. $f(p,q) = 900 - 9p - e^{0.4q}$; $g(p,q) = 20{,}000 - 3000q - 4p$.

We compute $\dfrac{\partial f}{\partial q} = -0.4e^{0.4q}$ and $\dfrac{\partial g}{\partial p} = -4$. Since $\dfrac{\partial f}{\partial q} < 0$ and $\dfrac{\partial g}{\partial p} < 0$

for all $p > 0$ and $q > 0$, we conclude that compact disc players and audio discs are complementary commodities.

53. We want to maximize the function $R(x,y) = -x^2 - 0.5y^2 + xy + 8x + 3y + 20$.
To find the critical point of R, we solve the system
$$R_x = -2x + y + 8 = 0$$
$$R_y = -y + x + 3 = 0.$$
Adding the two equations, we obtain $-x + 11 = 0$, or $x = 11$. So $y = 14$.
Therefore, $(11,14)$ is a critical point of R. Next, we compute
$$R_{xx} = -2, R_{xy} = 1, R_{yy} = -1.$$
So $D(x,y) = R_{xx}R_{yy} - R_{xy}^2 = 2 - 1 = 1$.

In particular $D(11,14) = 1 > 0$. Since $R_{xx}(11,14) = -2 < 0$, we see that $(11,14)$ gives a relative maximum of R. The nature of the problem suggests that this in fact an absolute maximum. So the company should spend \$11,000 on advertising and employ 14 agents in order to maximize its revenue.

55. Refer to the following diagram.

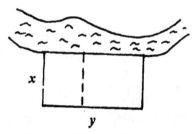

We want to minimize $C(x,y) = 3(2x) + 2(x) + 3y = 8x + 3y$ subject to
$xy = 303,750$.   The Lagrangian function is
$$F(x,y,\lambda) = 8x + 3y + \lambda(xy - 303,750).$$
Next, we solve the system
$$\begin{cases} F_x = 8 + \lambda y = 0 \\ F_y = 3 + \lambda x = 0 \\ F_\lambda = xy - 303,750 = 0 \end{cases}$$
Solving the first equation for y gives $y = -8/\lambda$. The second equation gives $x = -3/\lambda$. Substituting this value into the third equation gives

$$\left(-\frac{3}{\lambda}\right)\left(-\frac{8}{\lambda}\right) = 303,750 \text{ or } \lambda^2 = \frac{24}{303,750} = \frac{4}{50,625},$$

or $\lambda = \pm\frac{2}{225}$. Therefore, $x = 337.5$ and $y = 900$ and so the required dimensions of the pasture are 337.5 yd by 900 yd.

8 Calculus of Several Variables

CHAPTER 9

EXERCISES 9.1, page 679

1. $y' = 2x$. Substituting this value into the given differential equation yields
$$x(2x) + x^2 = 3x^2 \qquad [xy' + y = 3x^2]$$
and the given differential equation is satisfied. Therefore $y = x^2$ is a solution of the differential equation.

3. $y = \frac{1}{2} + ce^{-x^2}$, $y' = -2cxe^{-x^2}$. Substituting this value into the differential equation gives
$$-2cxe^{-x^2} + 2x(\frac{1}{2} + ce^{-x^2}) = x \qquad [y' + 2xy = x]$$
and the differential equation is satisfied. So $y = \frac{1}{2} + ce^{-x^2}$ is a solution of the differential equation.

5. $y = e^{-2x}$, $y' = -2e^{-2x}$. $y'' = 4e^{-2x}$. Substituting these values into the differential equation yields
$$4e^{-2x} - 2e^{-2x} - 2e^{-2x} = 0, \quad [y'' + y' - 2y = 0]$$
and so the differential equation is satisfied. Therefore, $y = e^{-2x}$ is a solution of the differential equation.

7. $y = C_1 e^{-2x} + C_2 x e^{-2x}$.
$y' = -2C_1 e^{-2x} + C_2 e^{-2x} - 2C_2 x e^{-2x}$
$y'' = 4C_1 e^{-2x} - 2C_2 e^{-2x} - 2C_2 e^{-2x} + 4C_2 x e^{-2x} = 4C_1 e^{-2x} - 4C_2 e^{-2x} + 4C_2 x e^{-2x}$.
Substituting these values into the differential equation, we find
$$4C_1 e^{-2x} - 4C_2 e^{-2x} + 4C_2 x e^{-2x} - 8C_1 e^{-2x} + 4C_2 e^{-2x} - 8C_2 x e^{-2x} + 4C_1 e^{-2x} + 4C_2 x e^{-2x}$$
$$= 0$$
and so the equation is satisfied and $y = C_1 e^{-2x} + C_2 x e^{-2x}$ is a solution of the given equation.

9. $y = \dfrac{C_1}{x} + C_2 \dfrac{\ln x}{x} = C_1 x^{-1} + C_2 x^{-1} \ln x$

$$y' = -C_1 x^{-2} + C_2(-x^{-2}\ln x + x^{-2}) = -C_1 x^{-2} + C_2 x^{-2}\,(1 - \ln x)$$
$$y'' = 2C_1 x^{-3} + C_2[-2x^{-3}(1 - \ln x) - x^{-3}] = 2C_1 x^{-3} + C_2 x^{-3}(2\ln x - 3).$$

Substituting into the differential equation yields

$$x^2[2C_1 x^{-3} + C_2 x^{-3}(2\ln x - 3)] + 3x[-C_1 x^{-2} + C_2 x^{-2}(1 - \ln x)] + C_1 x^{-1} + C_2 x^{-1}\ln x$$
$$= 2C_1 x^{-1} + C_2 x^{-1}(2\ln x - 3) - 3C_1 x^{-1} + 3C_2 x^{-1}(1 - \ln x) + C_1 x^{-1} + C_2 x^{-1}\ln x$$
$$= 0,$$

$$[x^2 y'' + 3xy' + y = 0]$$

and so the equation is satisfied. Therefore, $y = \dfrac{C_1}{x} + \dfrac{\ln x}{x}$ is a solution of the differential equation.

11. $y = C - Ae^{-kt}; y' = kAe^{-kt}$.

Substituting these values into the differential equation, we find
$$kAe^{-kt} = k[C - (C - Ae^{-kt})] = kAe^{-kt}, \qquad [y' = k(C - y)]$$
and so the equation is satisfied.

13. $y = Cx^2 - 2x,\ y' = 2Cx - 2$.

Substituting into the differential equation gives
$$2Cx - 2 - 2 \cdot \frac{1}{x}(cx^2 - 2x) = 2Cx - 2 - 2Cx + 4 = 2$$
and so the equation is satisfied. Next,
$$y(1) = 10 \text{ implies } C - 2 = 10, \text{ or } C = 12.$$
Therefore, a particular solution is $y = 12x^2 - 2x$.

15. $y = \dfrac{C}{x} = Cx^{-1},\ y' = -Cx^{-2} = -\dfrac{C}{x^2}$.

Substituting into the differential equation gives
$$-\frac{C}{x^2} + \left(\frac{1}{x}\right)\left(\frac{C}{x}\right) = 0,$$
and so the equation is satisfied. Next,
$$y(1) = 1 \text{ implies } C = 1 \text{ and so a particular solution is } y = \frac{1}{x}.$$

17. $y = \dfrac{Ce^x}{x} + \dfrac{1}{2}xe^x = Cx^{-1}e^x + \dfrac{1}{2}xe^x.$

$y' = C(x^{-1}e^x - x^{-2}e^x) + \dfrac{1}{2}(e^x + xe^x).$

Substituting these values into the differential equation gives

$$C\left(\dfrac{e^x}{x} - \dfrac{e^x}{x^2}\right) + \dfrac{1}{2}e^x + \dfrac{1}{2}xe^x + \left(\dfrac{1}{x} - 1\right)\left(\dfrac{Ce^x}{x} + \dfrac{1}{2}xe^x\right) = e^x,$$

and so the equation is satisfied. Next,

$$y(1) = -\dfrac{1}{2}e \text{ implies } \dfrac{Ce}{1} + \dfrac{1}{2}e = -\dfrac{1}{2}e \text{ or } C = -1.$$

Therefore, a particular solution is $y = -\dfrac{e^x}{x} + \dfrac{1}{2}xe^x.$

19. Let $Q(t)$ denote the amount of the substance present at time t. Since the substance decays at a rate directly proportional to the amount present, we have

$$\dfrac{dQ}{dt} = -kQ,$$

where k (positive) is the constant of proportion. The side condition is $Q(0) = Q_0$.

21. Let $A(t)$ denote the total investment at time t. Then

$$\dfrac{dA}{dt} = k(C - A),$$

where k is the constant of proportion.

23. Since the rate of decrease of the concentration of the drug at time t is proportional to the concentration $C(t)$ at any time t, we have

$$\dfrac{dC}{dt} = -kC,$$

where k is the (positive) constant of proportion. The initial condition is $C(0) = C_0$.

25. The rate of decrease of the temperature is dy/dt. Since this is proportional to the difference between the temperature y and C, we have

$$\dfrac{dy}{dt} = -k(y - C),$$

where k is a constant of proportionality. Furthermore, the initial temperature of y_0

degrees translates into the condition $y(0) = y_0$.

27. Since the relative growth rate of one organ, $(dx/dt)/x$ is proportional to the relative growth rate of the other, $(dy/dt)/y$, we have

$$\frac{1}{x}\cdot\frac{dx}{dt} = k\frac{1}{y}\cdot\frac{dy}{dt},$$

where k is a constant of proportionality.

29. True. $y = x^2 + 2x + x^{-1}$, $y' = 2x + 2 - \dfrac{1}{x^2}$;

$$x\left(2x + 2 - \frac{1}{x^2}\right) + x^2 + 2x + \frac{1}{x} = 2x^2 + 2x - \frac{1}{x} + x^2 + 2x + \frac{1}{x} = 3x^2 + 4x$$

31. False. $y = 2 + ce^{-x^3}$, $y' = -3cx^2 e^{-x^3}$

$$-3cx^2 e^{-x^3} + 3x^2(2 + ce^{-x^3}) = -3cx^2 e^{-x^3} + 6x^2 + 3cx^2 e^{-x^3} = 6x^2 \ne x.$$

33. False. Consider the solution in Problem 29. Next,

$$y = 2f(x) = 2x^2 + 4x + 2x^{-1}.$$

Then $y' = 4x + 4 - \dfrac{2}{x^2}$. Substituting these values, we have

$$x\left(4x + 4 - \frac{2}{x^2}\right) + 2x^2 + 4x + \frac{2}{x} = 6x^2 + 8x \ne 3x^2 + 4x.$$

EXERCISES 9.2, page 704

1. $\dfrac{dy}{dx} = \dfrac{x+1}{y^2}$, $y^2\,dy = (x+1)\,dx$. So

$$\int y^2\,dy = \int (x+1)\,dx, \quad \tfrac{1}{3}y^3 = \tfrac{1}{2}x^2 + x + C.$$

3. $\dfrac{dy}{dx} = \dfrac{e^x}{y^2}$, $y^2\,dy = e^x\,dx$. So

$$\int y^2\,dy = \int e^x\,dx, \quad \tfrac{1}{3}y^3 = e^x + C.$$

5. $\dfrac{dy}{dx} = 2y, \dfrac{dy}{y} = 2\,dx.$

$\displaystyle\int \dfrac{dy}{y} = \int 2\,dx, \ \ln|y| = 2x + C_1$

so $y = e^{2x+C_1} = e^{C_1}e^{2x} = ce^{2x}$, where $c = e^{C_1}$.

7. $\dfrac{dy}{dx} = xy^2; \dfrac{dy}{y^2} = x\,dx. \ \displaystyle\int \dfrac{dy}{y^2} = \int x\,dx, \ -\dfrac{1}{y} = \dfrac{1}{2}x^2 + C.$

9. $\dfrac{dy}{dx} = -2(3y+4); \dfrac{dy}{3y+4} = -2\,dx;$

$\displaystyle\int \dfrac{dy}{3y+4} = \int -2\,dx, \ \tfrac{1}{3}\ln|3y+4| = -2x + c_1$

$3y+4 = c_2 e^{-6x}$, where $c_2 = e^{c_1}$

or $y = -\tfrac{4}{3} + ce^{-6x}$, where $c = \tfrac{1}{3}c_2$.

11. $\dfrac{dy}{dx} = \dfrac{x^2+1}{3y^2}, \ 3y^2\,dy = (x^2+1)\,dx.$

$\displaystyle\int 3y^2\,dy = \int (x^2+1)\,dx.$ Therefore, $y^3 = \tfrac{1}{3}x^3 + x + C.$

13. $\dfrac{dy}{dx} = \sqrt{\dfrac{y}{x}} = \dfrac{y^{1/2}}{x^{1/2}}, \ \dfrac{dy}{y^{1/2}} = \dfrac{dx}{x^{1/2}}.$

$\displaystyle\int y^{-1/2}\,dy = \int x^{-1/2}\,dx; \ 2y^{1/2} = 2x^{1/2} + C_1,$

or $y^{1/2} - x^{1/2} = C$, where $C = \tfrac{1}{2}C_1.$

15. $\dfrac{dy}{dx} = \dfrac{y\ln x}{x}, \dfrac{dy}{y} = \dfrac{\ln x}{x}\,dx$

$\displaystyle\int \dfrac{dy}{y} = \int \dfrac{\ln x}{x}\,dx, \ \ln|y| = \tfrac{1}{2}(\ln x)^2 + C.$

17. $\dfrac{dy}{dx} = \dfrac{2x}{y}, \displaystyle\int y\,dy = \int 2x\,dx, \ \dfrac{1}{2}y^2 = x^2 + C,$

$y(1) = -2$ implies $\frac{1}{2}(-2)^2 = 1^2 + C$, or $C = 1$. Therefore, the solution is
$y^2 = 2x^2 + 2$.

19. $\frac{dy}{dx} = 2 - y$, $\int \frac{dy}{2-y} = dx$

$-\ln|2 - y| = x + C$. The condition $y(0) = 3$ implies
$-\ln|2 - 3| = 0 + C$, or $C = -\ln 1 = 0$. Therefore, the solution is
$-\ln|2 - y| = x$, $\ln|2 - y| = -x$, $2 - y = e^{-x}$, or $y = 2 - e^{-x}$.

21. $\frac{dy}{dx} = 3xy - 2x = x(3y - 2)$

$\int \frac{dy}{3y-2} = \int x \, dx$, $\frac{1}{3}\ln|3y - 2| = \frac{1}{2}x^2 + C$

The condition $y(0) = 1$ gives $\frac{1}{3}\ln 1 = 0 + c$, or $c = 0$.
Therefore, the solution is $\frac{1}{3}\ln|3y - 2| = \frac{1}{2}x^2$,

$\qquad \ln|3y - 2| = \frac{3}{2}x^2$, $3y - 2 = e^{(3/2)x^2}$, $\qquad$ or $\qquad y = \frac{2}{3} + \frac{1}{3}e^{(3/2)x^2}$.

23. $\frac{dy}{dx} = \frac{xy}{x^2+1}$. Separating variables and integrating, we have

$\qquad \int \frac{dy}{y} = \int \frac{x}{x^2+1} dx$

$\qquad \ln|y| - \ln(x^2 + 1)^{1/2} = \ln C$

$\qquad \ln \frac{|y|}{\sqrt{x^2+1}} = C$ and $y = C\sqrt{x^2 + 1}$.

Next, using the condition $y(0) = 1$ gives $y(0) = C = 1$. Therefore, the solution is
$y = \sqrt{x^2 + 1}$.

25. $\frac{dy}{dx} = xye^x$, $\int \frac{dy}{y} = \int xe^x \, dx$

$\ln|y| = (x - 1)e^x + C$.

The condition $y(1) = 1$ gives $\ln 1 = 0 + C$, or $C = 0$. Therefore, the solution is $\ln|y| = (x-1)e^x$.

27. $\dfrac{dy}{dx} = 3x^2 e^{-y}$, $e^y\,dy = 3x^2\,dx$, $e^y = x^3 + C$, $y = \ln(x^3 + C)$

$y(0) = 1$ implies that $1 = \ln C$ or $C = e$. Therefore, $y = \ln(x^3 + e)$.

29. Let $y = f(x)$. Then

$$\frac{dy}{dx} = \frac{3x^2}{2y}, \quad \int 2y\,dy = \int 3x^2\,dx$$

$$y^2 = x^3 + C.$$

The given condition implies that $9 = 1 + C$, or $C = 8$. Therefore, a required equation defining the function is

$$y^2 = x^3 + 8.$$

31. $\dfrac{dQ}{dt} = -kQ$, $\displaystyle\int \frac{dQ}{Q} = \int -k\,dt$.

$\ln|Q| = -kt + C_1$, $Q = e^{-kt + C_1} = Ce^{-kt}$, where $C = e^{C_1}$.

The condition $Q(0) = Q_0$ implies $Ce^0 = C = Q_0$. Therefore, $Q(t) = Q_0 e^{-kt}$.

33. $\dfrac{dA}{dt} = k\left(\frac{C}{k} - A\right)$, $\displaystyle\int \frac{dA}{\frac{C}{k} - A} = \int k\,dt$; $-\ln\left|\frac{C}{k} - A\right| = kt + d$ (Assume $\left(\frac{C}{k} - A\right) > 0$

$\ln\left|\frac{C}{k} - A\right| = -kt - d_1$.

Therefore, $\frac{C}{k} - A = d_2 e^{-kt}$, or $A = \frac{C}{k} - d_2 e^{-kt}$ where $d_2 = e^{-d_1}$.

35. $\dfrac{dS}{dt} = k(D - S)$, $\displaystyle\int \frac{dS}{D - S} = \int k\,dt$

$-\ln(D - S) = kt + d_1$ (where $D - S > 0$ and d_1 is a constant).

$\ln(D - S) = -kt - d_1$

$\quad D - S = d_2 e^{-kt}$, where $d_2 = e^{-d_1}$

$\quad\quad S = D - d_2 e^{-kt}$.

The condition $S(0) = S_0$ gives

$\quad D - d_2 = S_0$, or $d_2 = D - S_0$.

Therefore, $S(t) = D - (D - S_0)e^{-kt}$.

37. **True.** Rewrite it in the form $\dfrac{dy}{dx} = y(x-1) + 2(x-1) = (y+2)(x-1)$ and it is evident that the equation is separable.

39. **True.** The equation can be rewritten as $f(x)g(y)dx + F(x)G(y)dy = 0$ or
$$\frac{dy}{dx} = -\frac{f(x)g(y)}{F(x)G(y)} = \phi(x)\psi(y) \quad \text{where } \phi(x) = -\frac{f(x)}{F(x)} \text{ and } \psi(y) = \frac{g(y)}{G(y)}$$
and is evidently separable.

41. **False.** It cannot be written in the form $\dfrac{dy}{dx} = f(x)g(y)$.

EXERCISES 9.3, page 713

1. $\dfrac{dy}{dx} = -ky, \; y = y_0 e^{-kt}$.

3. Solving $\dfrac{dQ}{dt} = kQ$, we find $Q = Q_0 e^{kt}$. Here $k = 0.02$ and $Q_0 = 4.5$. Therefore, $Q(t) = 4.5e^{0.02t}$. At the beginning of the year 2005, the population will be
$$Q(25) = 4.5e^{0.02(25)} \approx 7.4, \text{ or } 7.4 \text{ billion.}$$

5. $\dfrac{dL}{L} = k\,dx$. Integrating, we have

   $\ln L = kx + C_1$, or $L = L_0 e^{kx}$, where $L_0 = e^{C_1}$.
   Using the given condition $L(\tfrac{1}{2}) = \tfrac{1}{2}L_0$, we have
   $\tfrac{1}{2}L_0 = L_0 e^{k/2}$, $e^{k/2} = \tfrac{1}{2}$, $\tfrac{1}{2}k = \ln\tfrac{1}{2}$, $k = -2\ln 2 \approx -1.3863$.
   Therefore, $L = L_0 e^{-1.3863x}$. We want to find x so that $L = \tfrac{1}{4}L_0$; that is, so that
   $\tfrac{1}{4}L_0 = L_0 e^{-1.3863x}$, $e^{-1.3863x} = \tfrac{1}{4}$,
   or $-1.3863x = -\ln 4 = -1.3863$. So $x = 1$. Therefore, 1/2 inch additional material is needed.

7. $\dfrac{dQ}{dt}=kQ^2$; $\displaystyle\int\dfrac{dQ}{Q^2}=\int k\,dt$ and $-\dfrac{1}{Q}=kt+C$. Therefore, $Q=-\dfrac{1}{kt+C}$.

Now, $Q=50$ when $t=0$, and, therefore, $50=-\dfrac{1}{C}$, and $C=-\dfrac{1}{50}$. Next,

$Q=-\dfrac{1}{kt-\frac{1}{50}}$.   Since $Q=10$ when $t=1$,

$\qquad 10=-\dfrac{1}{k-\frac{1}{50}}$, $10\left(k-\tfrac{1}{50}\right)=-1$, $10k-\tfrac{1}{5}=-1$.

$\qquad 10k=-1+\tfrac{1}{5}=-\tfrac{4}{5}$. $k=-\tfrac{4}{50}=-\tfrac{2}{25}$.

Therefore, $Q(t)=\dfrac{1}{\frac{2}{25}t+\frac{1}{50}}=\dfrac{1}{\frac{4t+1}{50}}=\dfrac{50}{4t+1}$,

and $\qquad Q(2)=\dfrac{50}{8+1}\approx 5.56$ grams.

9. Let C be the temperature of the surrounding medium and let T be the temperature
at any time t. Then $\dfrac{dT}{dt}=k(C-T)$. The solution of the differential equation is

$\qquad T=C-Ae^{-kt}$ [see Equation (10)]

With $C=72$, we have $T=72-Ae^{-kt}$. Next, $T(0)=212$ gives
$\qquad 212=72-A$, or $A=-140$.
Therefore, $T=72+140e^{-kt}$. Using the condition $T(2)=140$, we have

$\qquad 72+140e^{-2k}=140$, $e^{-2k}=\dfrac{68}{140}$, or $k=-\dfrac{1}{2}\ln\left(\dfrac{68}{140}\right)\approx 0.3611$.

So $T(t)=72+140e^{-0.3611t}$. When $T=110$, we have
$\qquad 72+140e^{-0.3611t}=110$, $140e^{-0.3611t}=38$,

or $e^{-0.3611t}=\dfrac{38}{140}$, $-0.3611t=\ln\left(\dfrac{38}{140}\right)$, and $t\approx 3.6$, or 3.6 minutes.

11. The differential equation is $\dfrac{dQ}{dt}=k(40-Q)$, whose solution is

$\qquad Q(t)=40-Ae^{-kt}$ (see Example 2). $Q(0)=0$ gives $A=40$ and so
$\qquad Q(t)=40(1-e^{-kt})$. Next, $Q(2)=10$ gives $40=(1-e^{-2k})=10$,

$1 - e^{-2k} = 0.25$, $e^{-2k} = 0.75$ and $k = -\frac{1}{2}\ln 0.75 \approx 0.1438$.

So $Q(t) = 40(1 - e^{-0.1438t})$.

The number of claims the average trainee can process after six weeks is

$$Q(6) = 40(1 - e^{-0.1438(6)}) \approx 23, \text{ or } 23 \text{ claims.}$$

13. $P(25) = -\dfrac{0.5}{0.008} + \left(289 + \dfrac{0.5}{0.008}\right)e^{0.008(25)} \approx 290.5$ million.

15. Let $Q(t)$ denote the number of people who have heard the rumor. Then

$$\frac{dQ}{dt} = kQ(400 - Q).$$

The solution is $Q(t) = \dfrac{400}{1 + Ae^{-400kt}}$. The condition $Q(0) = 10$ gives

$10 = \dfrac{400}{1 + A}$, $10 + 10A = 400$, or $A = 39$. Therefore, $Q(t) = \dfrac{400}{1 + 39e^{-400kt}}$.

Next, the condition, $Q(2) = 80$ gives

$$\frac{400}{1 + 39e^{-400kt}} = 80, \quad 1 + 39e^{-900k} = 5,$$

$$39e^{-800k} = 4, \quad \text{and} \quad k = -\frac{1}{800}\ln\left(\frac{4}{39}\right) \approx 0.0028466.$$

Therefore, $Q(t) = \dfrac{400}{1 + 39e^{-1.1386t}}$. In particular, the number of people who wil have

heard the rumor after a week is

$$Q(7) = \frac{400}{1 + 39e^{-1.1386(7)}} \approx 395, \quad \text{or } 395 \text{ people.}$$

17. Separating variables and integrating, we obtain

$$\int \frac{dQ}{Q(C - \ln Q)} = \int k\, dt$$

$-\ln|C - \ln Q| = kt + C_1$, $C - \ln Q = C_2 e^{-kt}$ $(C_2 = e^{-C_1})$

$\ln Q = C - C_2 e^{-kt}$, $Q = e^C e^{-C_2 e^{-kt}}$.

Using the condition $Q(0) = Q_0$, we have

$$Q_0 = e^C e^{-C_2}, \quad \ln Q_0 = C - C_2, \text{ or } C_2 = C - \ln Q_0.$$

So $Q(t) = e^C e^{-(C-\ln Q_0)e^{-kt}}$.

19. The differential equation governing this process is
$$\frac{dx}{dt} = 6 - \frac{3x}{20}.$$
Separating varibles and integrating, we obtain
$$\frac{dx}{dt} = \frac{120-3x}{20} = \frac{3(40-x)}{20},$$

$$\frac{dx}{40-x} = \frac{3}{20} dt.$$

$$-\ln|40-x| = \frac{3}{20}t + C_1.$$

$$40-x = Ce^{-3t/20}.$$

Therefore, $x = 40 - Ce^{-3t/20}$. The initial condition $x(0) = 0$ implies
$$0 = 40 - C, \text{ and } C = 40.$$

Therefore, $x(t) = 40(1 - e^{-3t/20})$. The amount of salt present at the end of 20 minutes is
$$x(20) = 40(1 - e^{-3}) \approx 38 \text{ lbs}.$$
The amount of salt present in the long run is
$$\lim_{t \to \infty} 40(1 - e^{-3t/20}) = 40 \text{ lbs}.$$

EXERCISES 9.4, page 722

1. a. $x_0 = 0$, $b = 1$, and $n = 4$.

Therefore, $h = \frac{1}{4}$ and $x_0 = 0$, $x_1 = \frac{1}{4}$, $x_2 = \frac{1}{2}$, $x_3 = \frac{3}{4}$, and $x_4 = b = 1$. Also $F(x,y) = x + y$ and $y_0 = y(0) = 1$.

$\tilde{y}_0 = y_0 = 1$

$\tilde{y}_1 = \tilde{y}_0 + hF(x_0, \tilde{y}_0) = 1 + \frac{1}{4}(0 + 1) = \frac{5}{4}$.

$\tilde{y}_2 = \tilde{y}_1 + hF(x_1, \tilde{y}_1) = \frac{5}{4} + \frac{1}{4}\left(\frac{1}{4} + \frac{5}{4}\right) = \frac{13}{8}$

$\tilde{y}_3 = \tilde{y}_2 + hF(x_2, \tilde{y}_2) = \frac{13}{8} + \frac{1}{4}\left(\frac{1}{2} + \frac{13}{8}\right) = \frac{69}{32}$.

$\tilde{y}_4 = \tilde{y}_3 + hF(x_3, \tilde{y}_3) = \frac{69}{32} + \frac{1}{4}\left(\frac{3}{4} + \frac{69}{32}\right) = \frac{369}{128}$.

Therefore, $y(1) = \frac{369}{128} = 2.8828$.

b. $n = 6$. Therefore, $h = \frac{1}{6}$.

So $x_0 = 0$, $x_1 = \frac{1}{6}$, $x_2 = \frac{2}{6}$, $x_3 = \frac{3}{6}$, $x_4 = \frac{4}{6}$, $x_5 = \frac{5}{6}$, $x_6 = 1$.
Therefore,

$$\tilde{y}_0 = y_0 = 1$$
$$\tilde{y}_1 = \tilde{y}_0 + hF(x_0, \tilde{y}_0) = 1 + \frac{1}{6}(0+1) = \frac{7}{6}.$$
$$\tilde{y}_2 = \tilde{y}_1 + hF(x_1, \tilde{y}_1) = \frac{7}{6} + \frac{1}{6}\left(\frac{1}{6} + \frac{7}{6}\right) = \frac{50}{36} = \frac{25}{18}.$$
$$\tilde{y}_3 = \tilde{y}_2 + hF(x_2, \tilde{y}_2) = \frac{25}{18} + \frac{1}{6}\left(\frac{2}{6} + \frac{25}{18}\right) = \frac{181}{108}.$$
$$\tilde{y}_4 = \tilde{y}_3 + hF(x_3, \tilde{y}_3) = \frac{181}{108} + \frac{1}{6}\left(\frac{3}{6} + \frac{181}{108}\right) = \frac{1321}{648}.$$
$$\tilde{y}_5 = \tilde{y}_4 + hF(x_4, \tilde{y}_4) = \frac{1321}{648} + \frac{1}{6}\left(\frac{4}{6} + \frac{1321}{648}\right) = \frac{9679}{3888}.$$
$$\tilde{y}_6 = \tilde{y}_5 + hF(x_5, \tilde{y}_5) = \frac{9679}{3888} + \frac{1}{6}\left(\frac{5}{6} + \frac{9679}{3888}\right) = \frac{70993}{23328}.$$
Therefore, $y(1) = \frac{70993}{23328} \approx 3.043$.

3. a. Here $x_0 = 0$ and $b = 2$. Taking $n = 4$, we have $h = \frac{2}{4} = \frac{1}{2}$ and
    $x_0 = 0$, $x_1 = \frac{1}{2}$, $x_2 = 1$, $x_3 = \frac{3}{2}$, and $x_4 = 2$.
    Also, $F(x, y) = 2x - y + 1$ and $y(0) = y_0 = 2$. Therefore,
    $$\tilde{y}_0 = y_0 = 2$$
    $$\tilde{y}_1 = \tilde{y}_0 + hF(x_0, \tilde{y}_0) = 2 + \frac{1}{2}(0 - 2 + 1) = \frac{3}{2}$$
    $$\tilde{y}_2 = \tilde{y}_1 + hF(x_1, \tilde{y}_1) = \frac{3}{2} + \frac{1}{2}\left(1 - \frac{3}{2} + 1\right) = \frac{7}{4}.$$
    $$\tilde{y}_3 = \tilde{y}_2 + hF(x_2, \tilde{y}_2) = \frac{7}{4} + \frac{1}{2}\left(2 - \frac{7}{4} + 1\right) = \frac{19}{8}.$$
    $$\tilde{y}_4 = \tilde{y}_3 + hF(x_3, \tilde{y}_3) = \frac{19}{8} + \frac{1}{2}\left(3 - \frac{19}{8} + 1\right) = \frac{51}{16}.$$
    Therefore, $y(2) = \frac{51}{16} \approx 3.1875$.

    b. $n = 6$. Therefore, $h = \frac{2}{6} = \frac{1}{3}$, and so
    $x_0 = 0$, $x_1 = \frac{1}{3}$, $x_2 = \frac{2}{3}$, $x_3 = 1$, $x_4 = \frac{4}{3}$, $x_5 = \frac{5}{3}$, $x_6 = 2$.
    Therefore,
    $$\tilde{y}_0 = y_0 = 2$$
    $$\tilde{y}_1 = \tilde{y}_0 + hF(x_0, \tilde{y}_0) = 2 + \frac{1}{3}(0 - 2 + 1) = \frac{5}{3}$$
    $$\tilde{y}_2 = \tilde{y}_1 + hF(x_1, \tilde{y}_1) = \frac{5}{3} + \frac{1}{3}\left(\frac{2}{3} - \frac{5}{3} + 1\right) = \frac{5}{3}$$
    $$\tilde{y}_3 = \tilde{y}_2 + hF(x_2, \tilde{y}_2) = \frac{5}{3} + \frac{1}{3}\left(\frac{4}{3} - \frac{5}{3} + 1\right) = \frac{17}{9}$$
    $$\tilde{y}_4 = \tilde{y}_3 + hF(x_3, \tilde{y}_3) = \frac{17}{9} + \frac{1}{3}\left(2 - \frac{17}{9} + 1\right) = \frac{61}{27}.$$
    $$\tilde{y}_5 = \tilde{y}_4 + hF(x_4, \tilde{y}_4) = \frac{61}{27} + \frac{1}{3}\left(\frac{8}{3} - \frac{61}{27} + 1\right) = \frac{221}{81}.$$
    $$\tilde{y}_6 = \tilde{y}_5 + hF(x_5, \tilde{y}_5) = \frac{221}{81} + \frac{1}{3}\left(\frac{10}{3} - \frac{221}{81} + 1\right) = \frac{793}{243}.$$
    Therefore, $y(2) = \frac{793}{243} \approx 3.2634$.

5. a. Here $x_0 = 0$ and $b = 0.5$. Taking $n = 4$, we have $h = \dfrac{0.5}{4} \approx 0.125$. Therefore,

$x_0 = 0$, $x_1 = 0.125$, $x_2 = 0.25$, $x_3 = 0.375$, and $x_4 = 0.5$. Also, $F(x, y) = -2xy^2$ and $y(0) = y_0 = 1$. So

$\tilde{y}_0 = y_0 = 1$

$\tilde{y}_1 = \tilde{y}_0 + hF(x_0, \tilde{y}_0) = 1 + 0.125(0) = 1$

$\tilde{y}_2 = \tilde{y}_1 + hF(x_1, \tilde{y}_1) = 1 + 0.125[-2(0.125)(1)] = 0.96875.$

$\tilde{y}_3 = \tilde{y}_2 + hF(x_2, \tilde{y}_2) = 0.96875 + 0.125[-2(0.25)(0.96875)^2] \approx 0.910095.$

$\tilde{y}_4 = \tilde{y}_3 + hF(x_3, \tilde{y}_3) = 0.910095 + 0.125[-2(0.375)(0.910095)^2]$
$\quad = 0.832445.$

Therefore, $y(0.5) = 0.8324$.

b. $n = 6$. Therefore, $h = \dfrac{0.5}{6} \approx 0.83333$. So

$x_0 = 0$, $x_1 = 0.083333$, $x_2 = 0.166667$, $x_3 = 0.245000$, and $x_4 = 0.333333$, $x_5 = 0.416666$, and $x_5 = 0.5$.

Therefore,

$\tilde{y}_0 = y_0 = 1$

$\tilde{y}_1 = \tilde{y}_0 + hF(x_0, \tilde{y}_0) = 1 + 0.083333(0) = 1$

$\tilde{y}_2 = \tilde{y}_1 + hF(x_1, \tilde{y}_1) = 1 + 0.083333[-2(0.083333)(1)] \approx 0.986111.$

$\tilde{y}_3 = \tilde{y}_2 + hF(x_2, \tilde{y}_2) = 0.986111 + 0.083333[-2(0.1666670(0.986111)^2] \approx 0.920772.$

$\tilde{y}_4 = \tilde{y}_3 + hF(x_3, \tilde{y}_3) = 0.95100 - 0.083333[-2(0.245000)(0.959100)^2] \approx 0.959100.$

$\tilde{y}_5 = \tilde{y}_4 + hF(x_4, \tilde{y}_4) = 0.921539 + 0.08333[-2(0.333333)(0.921539)^2] \approx 0.873671.$

$\tilde{y}_6 = \tilde{y}_5 + hF(x_5, \tilde{y}_5) = 0.874360 + 0.083333[-2(0.416666)(0.874360)^2] \approx 0.820665.$

Therefore, $y(0.5) \approx 0.8207$.

7. a. $x_0 = 1$, $b = 1.5$, $n = 4$, and $h = 0.125$. Therefore,

$x_0 = 1$, $x_1 = 1.125$, $x_2 = 1.25$, $x_3 = 1.375$, $x_4 = 1.5$.

Also, $F(x, y) = \sqrt{x + y}$, $y(1) = 1$. Therefore,

$\tilde{y}_0 = y_0 = 1$

$\tilde{y}_1 = \tilde{y}_0 + hF(x_0, \tilde{y}_0) = 1 + 0.125\sqrt{1 + 1} \approx 1.7667767.$

$$\tilde{y}_2 = \tilde{y}_1 + hF(x_1, \tilde{y}_1) = 1.1767767 + 0.125\sqrt{1.125 + 1.767767}$$
$$\approx 1.3664218.$$

$$\tilde{y}_3 = \tilde{y}_2 + hF(x_2, \tilde{y}_2) = 1.3664218 + 0.125\sqrt{1.25 + 1.3664218}$$
$$\approx 1.5686138.$$

$$\tilde{y}_4 = \tilde{y}_3 + hF(x_3, \tilde{y}_3) = 1.5686138 + 0.125\sqrt{1.375 + 1.5686138}$$
$$\approx 1.7830758.$$

Therefore, $y(1.5) \approx 1.7831$.

b. $n = 6$. Therefore, $h = \dfrac{0.5}{6} \approx 0.0833333$. So

$x_0 = 1$, $x_1 = 1.0833333$, $x_2 = 1.1666666$, $x_3 = 1.25$, $x_4 = 1.3333332$, $x_5 = 1.4166665$, and $x_5 = 1.5$. So

$$\tilde{y}_0 = y_0 = 1$$

$$\tilde{y}_1 = \tilde{y}_0 + hF(x_0, \tilde{y}_0) = 1 + 0.0833333\sqrt{1 + 1} \approx 1.1178511.$$

$$\tilde{y}_2 = \tilde{y}_1 + hF(x_1, \tilde{y}_1) = 1.1178511 + 0.0833333\sqrt{1.0833333 + 1.1178511} \approx 1.2414876.$$

$$\tilde{y}_3 = \tilde{y}_2 + hF(x_2, \tilde{y}_2) = 1.2414876 + 0.0833333\sqrt{1.166666 + 1.2414876} \approx 1.3708061.$$

$$\tilde{y}_4 = \tilde{y}_3 + hF(x_3, \tilde{y}_3) = 1.3708061 + 0.0833333\sqrt{1.25 + 1.3708061} \approx 1.5057136.$$

$$\tilde{y}_5 = \tilde{y}_4 + hF(x_4, \tilde{y}_4) = 1.5057136 + 0.0833333\sqrt{1.333332 + 1.5057136} \approx 1.6461258.$$

$$\tilde{y}_6 = \tilde{y}_5 + hF(x_5, \tilde{y}_5) = 1.6461258 + 0.0833333\sqrt{1.4166665 + 1.6461258} \approx 1.791966.$$

Therefore $y(1.5) \approx 1.7920$.

9. a. Here $x_0 = 0$ and $b = 1$. Taking $n = 4$, we have $h = 0.25$. Therefore,
    $x_0 = 0$, $x_1 = 0.25$, $x_2 = 0.5$, $x_3 = 0.75$, and $x_4 = 1$.

   Also, $F(x, y) = \dfrac{x}{y}$ and $y_0 = y(0) = 1$. Therefore,

   $$\tilde{y}_0 = y_0 = 1$$
   $$\tilde{y}_1 = \tilde{y}_0 + hF(x_0, \tilde{y}_0) = 1 + 0.25(0) = 1$$
   $$\tilde{y}_2 = \tilde{y}_1 + hF(x_1, \tilde{y}_1) = 1 + \left(\frac{0.25}{1}\right) \approx 1.0625$$

   $$\tilde{y}_3 = \tilde{y}_2 + hF(x_2, \tilde{y}_2) = 1.0625 + 0.25\left(\frac{0.5}{1.0625}\right) \approx 1.180147.$$

$$\tilde{y}_4 = \tilde{y}_3 + hF(x_3, \tilde{y}_3) = 1.180147 + 0.25\left(\frac{0.75}{1.180147}\right) \approx 1.339026.$$

Therefore, $y(1) \approx 1.3390$.

b. $n = 6$. Therefore, $h = \frac{1}{6} \approx 0.166667$ and

$x_0 = 0,\ x_1 \approx 0.166667,\ x_2 \approx 0.333333,\ x_3 \approx 0.500000, x_4 \approx 0.666667,$

$x_5 \approx 0.833333$, and $x_6 = 1$.

Therefore,

$\tilde{y}_0 = y_0 = 1$

$\tilde{y}_1 = \tilde{y}_0 + hF(x_0, \tilde{y}_0) = 1 + 0.1666671(0) = 1.$

$$\tilde{y}_2 = \tilde{y}_1 + hF(x_1, \tilde{y}_1) = 1 + 0.166667\left(\frac{0.166667}{1}\right) \approx 1.027778.$$

$$\tilde{y}_3 = \tilde{y}_2 + hF(x_2, \tilde{y}_2) = 1.0277778 + 0.166667\left(\frac{0.333333}{1.027778}\right) \approx 1.081832.$$

$$\tilde{y}_4 = \tilde{y}_3 + hF(x_3, \tilde{y}_3) = 1.081832 + 0.166667\left(\frac{0.500000}{1.081832}\right) \approx 1.158862.$$

$$\tilde{y}_5 = \tilde{y}_4 + hF(x_4, \tilde{y}_4) = 1.158862 + 0.166667\left(\frac{0.666667}{1.158862}\right) \approx 1.254742.$$

$$\tilde{y}_6 = \tilde{y}_5 + hF(x_5, \tilde{y}_5) = 1.254742 + 0.166667\left(\frac{0.833333}{1.254742}\right) \approx 1.365433.$$

Therefore, $y(1) \approx 1.3654$.

11. Here $x_0 = 0,\ b = 1$. With $n = 5$, we have $h = 1.5 = 0.2$ and

$x_0 = 0,\ x_1 = 0.2,\ x_2 = 0.4,\ x_3 = 0.6,\ x_4 = 0.8$, and $x_5 = 1$.

$F(x, y) = \frac{1}{2}xy$ and $y(0) = 1$. Therefore,

$\tilde{y}_0 = y_0 = 1$

$\tilde{y}_1 = \tilde{y}_0 + hF(x_0, \tilde{y}_0) = 1 + 0.2[0.5(0)(1)] = 1.$

$\tilde{y}_2 = \tilde{y}_1 + hF(x_1, \tilde{y}_1) = 1 + 0.2[0.5(0.2)(1)] = 1.02$

$\tilde{y}_3 = \tilde{y}_2 + hF(x_2, \tilde{y}_2) = 1.02 + 0.2[0.5(0.4)(1.02)] = 1.0608.$

$\tilde{y}_4 = \tilde{y}_3 + hF(x_3, \tilde{y}_3) = 1.0608 + 0.2[0.5(0.6)(1.0608)] = 1.124448.$

$\tilde{y}_5 = \tilde{y}_4 + hF(x_4, \tilde{y}_4) = 1.124448 + 0.2[0.5(0.8)(1.124448)] = 1.21440384.$

The solutions are summarized below.

x	0	0.2	0.4	0.6	0.8	1
$\tilde{y}_n$	1	1	1.02	1.0608	1.1245	1.2144

13. $x_0 = 0$, $b = 1$, $n = 5$, and, therefore, $h = 0.2$. So

$x_0 = 0$, $x_1 = 0.2$, $x_2 = 0.4$, $x_3 = 0.6$, $x_4 = 0.8$, $x_5 = 1$ and
$F(x,y) = 2x - y + 1$, $y(0) = 2$

$\tilde{y}_0 = y_0 = 2$

$\tilde{y}_1 = \tilde{y}_0 + hF(x_0, \tilde{y}_0) = 2 + 0.2[2(0) - 2 + 1] = 1.8.$

$\tilde{y}_2 = \tilde{y}_1 + hF(x_1, \tilde{y}_1) = 1.8 + 0.2[2(0.2) - 1.8 + 1] = 1.72$

$\tilde{y}_3 = \tilde{y}_2 + hF(x_2, \tilde{y}_2) = 1.72 + 0.2[2(0.4) - 1.72 + 1] = 1.736.$

$\tilde{y}_4 = \tilde{y}_3 + hF(x_3, \tilde{y}_3) = 1.736 + 0.2[2(0.6) - 1.736 + 1] = 1.8288.$

$\tilde{y}_5 = \tilde{y}_4 + hF(x_4, \tilde{y}_4) = 1.8288 + 0.2[2(0.8) - 1.8288 + 1] = 1.98304$

$\tilde{y}_6 = \tilde{y}_5 + hF(x_5, \tilde{y}_5) =$

The solutions are summarized below.

x	0	0.2	0. 4	0.6	0.8	1
$\tilde{y}_n$	2	1.8	1.72	1.736	1.8288	1.9830

15. Here $x_0 = 0$, $b = 0.5$, and with $n = 5$, we have $h = 0.1$. So

$x_0 = 0$, $x_1 = 0.1$, $x_2 = 0.2$, $x_3 = 0.3$, $x_4 = 0.4$, and $x_5 = 0.5$.

$F(x,y) = x^2 + y$ and $y(0) = 1$. Therefore,

$\tilde{y}_0 = y_0 = 1$

$\tilde{y}_1 = \tilde{y}_0 + hF(x_0, \tilde{y}_0) = 1 + 0.1(0 + 1) = 1.1$

$y_2 = \tilde{y}_1 + hF(x_1, \tilde{y}_1) = 1.1 + 0.1(0.1)^2 + 1.1] \approx 1.211.$

$\tilde{y}_3 = \tilde{y}_2 + hF(x_2, \tilde{y}_2) = 1.211 + 0.1[(0.2)^2 + 1.211] \approx 1.3361.$

$\tilde{y}_4 = \tilde{y}_3 + hF(x_3, \tilde{y}_3) = 1.3361 + 0.1[(0.3)^2 + 1.3361] \approx 1.47871.$

$\tilde{y}_5 = \tilde{y}_4 + hF(x_4, \tilde{y}_4) = 1.47871 + 0.1[(0.4)^2 + 1.573481] \approx 1.64258.$

The solutions are summarized below.

x	0	0.1	0. 2	0.3	0.4	0.5
$\tilde{y}_n$	1	1.1	1.211	1.3361	1.4787	1.64258

1. $y = C_1 e^{2x} + C_2 e^{-3x}; \; y' = 2C_1 e^{2x} - 3C_2 e^{-3x}; \; y'' = 4C_1 e^{2x} + 9C_2 e^{-3x}$.

   Substituting these values into the differential equation, we have
   $$4C_1 e^{2x} + 9C_2 e^{-3x} + 2C_2 e^{2x} - 3C_2 e^{-3x} - 6(C_1 e^{2x} + C_2 e^{-3x}) = 0$$
   $$[y'' + y' - 6y = 0]$$
   and the differential equation is satisfied.

3. $y = Cx^{-4/3}$. So $y' = -\frac{4}{3} Cx^{-7/3}$. Substituting these values into the given differential equation which can be written in the form
   $$\frac{dy}{dx} = -\frac{4xy^3}{3x^2 y^2} = -\frac{4y}{3x}, \text{ we find}$$
   $$-\frac{4}{3} Cx^{-7/3} = -\frac{4(Cx^{-4/3})}{3x} = -\frac{4}{3} Cx^{-7/3},$$
   we see that y is a solution of the differential equation.

5. $y = (9x + C)^{-1/3}$. So $y' = -\frac{1}{3}(9x + C)^{-4/3}(9) = -3(9x + C)^{-4/3}$. Substituting into the differential equation, we have
   $$-3(9x + C)^{-4/3} = -3[(9x + C)^{-1/3}]^{-4} = -3(9x + C)^{-4/3}$$
   $$[y' = -3y^4]$$
   and it is satisfied. Therefore, y is indeed a solution. Next, using the side condition, we find $y(0) = C^{-1/3} = \frac{1}{2},$ or $C = 8$.

   Therefore, the required solution is $y = (9x + 8)^{-1/3}$.

7. $\dfrac{dy}{4 - y} = 2\, dt$ implies that $-\ln|4 - y| = 2t + C_1$.

   $4 - y = Ce^{-2t},$ or $y = 4 - Ce^{-2t},$ where $C = e^{-C_1}$.

9. We have $\dfrac{dy}{dx} = 3x^2 y + y^2 = y^2(3x^2 + 1)$. Separating variables and integrating, we have
   $$\int y^{-2}\, dy = \int (3x^2 + 1)\, dx$$

$$-\frac{1}{y} = x^3 + x + C.$$

Using the side condition, we find $\frac{1}{2} = C$. So the solution is

$$y = -\frac{1}{x^3 + x + \frac{1}{2}} = -\frac{2}{2x^3 + 2x + 1}.$$

11. We have

$$\frac{dy}{dx} = -\frac{3}{2}x^2 y.$$

Separating variables and integrating, we have

$$\int \frac{dy}{y} = \int -\frac{3}{2}x^2\, dx$$

$$\ln|y| = -\frac{1}{2}x^3 + C$$

Using the condition $y(0) = 3$, we have $\ln 3 = C$. Therefore,

$$\ln|y| = -\frac{1}{2}x^3 + \ln 3 \quad \text{or} \quad y = e^{-(1/2)x^3 + \ln 3} = 3e^{-x^3/2}.$$

13. a. $x_0 = 0$, $b = 1$, and $n = 4$. Therefore, $h = 0.25$. So
   $x_0 = 0$, $x_1 = 0.25$, $x_2 = 0.5$, $x_3 = 0.75$, $x_4 = 1$.
   Also $F(x,y) = x + y^2$, $y(0) = 0$. Therefore,
   $\tilde{y}_0 = y_0 = 0$
   $\tilde{y}_1 = \tilde{y}_0 + hF(x_0, \tilde{y}_0) = 0 + 0.25(0) = 0$
   $\tilde{y}_2 = \tilde{y}_1 + hF(x_1, \tilde{y}_1) = 0 + 0.25(0.25 + 0) = 0.0625.$
   $\tilde{y}_3 = \tilde{y}_2 + hF(x_2, \tilde{y}_2) = 0.0625 + 0.25(0.5 + 0.0625^2) = 0.1884766.$
   $\tilde{y}_4 = \tilde{y}_3 + hF(x_3, \tilde{y}_3) = 0.1884766 + 0.25(0.75 + 0.884766^2) = 0.3848575.$
   Therefore, $y(1) = 0.3849$.
   b. $n = 6$. Therefore, $h = 0.1666667$.
   So $x_0 = 0$, $x_1 = 1.666667$, $x_2 = 0.3333334$, $x_3 = 0.5$, $x_4 = 0.6666667$,
      $x_5 = 0.8333334$, and $x_6 = 1$.
   So

   $\tilde{y}_0 = y_0 = 0$
   $\tilde{y}_1 = \tilde{y}_0 + hF(x_0, \tilde{y}_0) = 0 + 0.1666667(0 + 0) = 0$

$\tilde{y}_2 = \tilde{y}_1 + hF(x_1, \tilde{y}_1) = 1 + 0.1666667[0.1666667 + 0^2] = 0.0277778.$

$\tilde{y}_3 = \tilde{y}_2 + hF(x_2, \tilde{y}_2) = 0.0277778 + 0.1666667(0.3333334 + 0.0277778^2) = 0.0834619.$

$\tilde{y}_4 = \tilde{y}_3 + hF(x_3, \tilde{y}_3) = 0.0834619 + 0.1666667(0.5 + 0.0834619^2)$
$\approx 0.1679562.$

$\tilde{y}_5 = \tilde{y}_4 + hF(x_4, \tilde{y}_4) = 0.1679562 + 0.1666667(0.666667 + 0.1679562^2)$
$\approx 0.2837689.$

$\tilde{y}_6 = \tilde{y}_5 + hF(x_5, \tilde{y}_5) = 0.2837689 + 0.1666667(0.8333334 + 0.2837689^2)$
$\approx 0.4360785.$

Therefore, $y(1) \approx 0.4361$.

15. a. Here $x_0 = 0$, $b = 1$, and $n = 4$. So $h = 1/4 = 0.25$. So
$x_1 = 0$, $x_1 = 0.25$, $x_2 = 0.5$, $x_3 = 0.75$, and $x_4 = 1$.
$F(x, y) = 1 + 2xy^2$ and $y(0) = 0$. So
$\tilde{y}_0 = y_0 = 0$

$\tilde{y}_1 = \tilde{y}_0 + hF(x_0, \tilde{y}_0) = 0 + 0.25[1 + 0] = 0.25.$

$\tilde{y}_2 = \tilde{y}_1 + hF(x_1, \tilde{y}_1) = 0.25 + 0.25[1 + 2(0.05)(0.25)^2] = 0.507812.$

$\tilde{y}_3 = \tilde{y}_2 + hF(x_2, \tilde{y}_2) = 0.507813 + 0.25[1 + 2(0.5)(0.507813)^2] \approx 0.822281.$

$\tilde{y}_4 = \tilde{y}_3 + hF(x_3, \tilde{y}_3) = 0.822281 + 0.25[1 + 2(0.75)(0.822281)^2] \approx 1.32584.$

Therefore, $y(1) = 1.3258$.

b. $n = 6$. Therefore, $h = 1/6 = 0.1666667$.
So
$x_0 = 0$, $x_1 = 1.666667$, $x_2 = 0.3333334$, $x_3 = 0.5$, $x_4 = 0.6666667$, $x_5 = 0.8333334$,
and $x_6 = 1$. So
$\tilde{y}_0 = y_0 = 0$

$\tilde{y}_1 = \tilde{y}_0 + hF(x_0, \tilde{y}_0) = 0 + 0.1666667(1 + 0) = 0.166667.$

$\tilde{y}_2 = \tilde{y}_1 + hF(x_1, \tilde{y}_1) = 1 + 0.1666667[1 + 2(0.166667)(0.166667)^2 = 0.334877.$

$\tilde{y}_3 = \tilde{y}_2 + hF(x_2, \tilde{y}_2) = 0.334877 + 0.1666667[1 + 2(0.333333)(0.334877)^2]$
$\approx 0.514003.$

$\tilde{y}_4 = \tilde{y}_3 + hF(x_3, \tilde{y}_3) = 0.514004 + 0.1666667[1 + 2(0.50000)(0.515005)^2]$
$\approx 0.724703.$

$$\tilde{y}_5 = \tilde{y}_4 + hF(x_4, \tilde{y}_4) = 0.724704 + 0.1666667(1 + 2(0.666667)(0.724704)^2]$$
$$\approx 1.008081.$$

$$\tilde{y}_6 = \tilde{y}_5 + hF(x_5, \tilde{y}_5) = 1.008081 + 0.1666667[1 + 2(0.833333)(1.008081)^2]$$
$$\approx 1.457034.$$

Therefore, $y(1) = 1.4570$.

17. Here $x_0 = 0$, $b = 1$, so with $n = 5$, and $h = 1/5 = 0.2$ Then
$x_0 = 0$, $x_1 = 0.2$, $x_2 = 0.4$, $x_3 = 0.6$, $x_4 = 0.8$, and $x_5 = 1$.
Also, $F(x, y) = 2xy$ and $y_0 = y(0) = 1$.
Therefore,
$$\tilde{y}_0 = y_0 = 1$$
$$\tilde{y}_1 = \tilde{y}_0 + hF(x_0, \tilde{y}_0) = 1 + (0.2)(2)(0)(1) = 1$$
$$\tilde{y} = \tilde{y}_1 + hF(x_1, \tilde{y}_1) = 1 + (0.2)(2)(0.2)(1) = 1.08.$$
$$\tilde{y}_3 = \tilde{y}_2 + hF(x_2, \tilde{y}_2) = 1.08 + (0.2)(2)(0.4)(1.08) = 1.2528.$$
$$\tilde{y}_4 = \tilde{y}_3 + hF(x_3, \tilde{y}_3) = 1.2528 + (0.2)(2)(0.6)(1.2528) = 1.553472.$$
$$\tilde{y}_5 = \tilde{y}_4 + hF(x_4, \tilde{y}_4) = 1.553472 + (0.2)(2)(0.8)(1.553472) = 2.05058.$$
Therefore, $y(1) = 2.0506$.

19. $\dfrac{dS}{dT} = -kS$;   Thus, $S(t) = S_0 e^{-kt}$, $S(t) = 50,000 e^{-kt}$.

$$S(2) = 32,000 = 50,000 e^{-2k}, \ e^{-2k} = \frac{32}{50} = 0.64.$$

$-2k \ln e \approx \ln 0.64$,   or $k = 0.223144$.

a. Therefore, $S = 50,000 e^{-0.223144t} = 50,000(0.8)^t$.

b. $S(5) = 50,000(0.8)^5 = 16,384$, or \$16,384.

21. Separating variables and integrating, we have
$$\int \frac{dA}{rA + P} = \int dt$$

$$\frac{1}{r} \ln(rA + P) = t + C_1$$

$$\ln(rA + P) = rt + C_2 \quad (C_2 = C_1 r)$$

$$rA + P = Ce^{rt}.$$

So $\quad A = \dfrac{1}{r}(Ce^{rt} - P)$.

Using the condition $A(0) = 0$, we have

$$0 = \frac{1}{r}(C - P), \text{ or } C = P.$$

So $\quad A = \dfrac{P}{R}(e^{rt} - 1)$.

The size of the fund after five years is

$$A = \frac{50{,}000}{0.12}(e^{(0.12)(5)} - 1) \approx 342{,}549.50,$$

or approximately \$342,549.50.

23. According to Newton's Law of cooling,

$$\frac{dT}{dt} = k(350 - T)$$

where T is the temperature of the roast. We also have the conditions $T(0) = 68$ and $T(2) = 118$. Separating the variables in the differential equation and integrating, we have

$$\frac{dT}{350 - T} = k\,dt$$

$$\ln|350 - T| = kt + C_1$$

$$350 - T = Ce^{kt},$$

or $\quad\quad\quad\quad T = 350 - Ce^{kt}.$

Using the condition $T(0) = 68$, we have

$$350 - C = 68, \text{ or } C = 282.$$

So $\quad\quad\quad\quad T = 350 - 282e^{kt}.$

Next, we use the condition $T(2) = 118$ to find

$$118 = 350 - 282e^{2k}$$

$$e^{2k} = 0.822695035.$$

So $\quad\quad k = \frac{1}{2}\ln 0.822695035 \approx -0.097584.$

Therefore, $T = 350 - 282e^{-0.097584t}$.

We want to find t when $T = 150$. So we solve

$$150 = 350 - 282e^{-0.097584t}$$

$$e^{-0.097584t} = 0.709219858$$

$$-0.097584t = \ln 0.709219858 = -0.343589704,$$

or $\qquad t = 3.52096,$ or approximately 3.5 hours.

So the roast would have been $150°\,F$ at approximately 7:30 P.M.

25. $N = \dfrac{200}{1+49e^{-200kt}}$. When $t = 2, N = 40$, and

$$40 = \frac{200}{1+49e^{-400k}}$$

$$40 + 1960e^{-400k} = 200$$

$$e^{-400k} = \frac{160}{1960}(0.0816327)$$

$$-400\ln k = \ln 0.0816327$$

$$k = 0.00626.$$

Therefore, $N(5) = \dfrac{200}{1+49e^{-200(5)(0.00626)}} \approx 183,$ or 183 families.

CHAPTER 10

EXERCISES 10.1 , page 735

1. $f(x) = \dfrac{1}{3} > 0$ on [3,6]. Next, $\displaystyle\int_3^6 \dfrac{1}{3}\,dx = \dfrac{1}{3}x\Big|_3^6 = \dfrac{1}{3}(6-3) = 1.$

3. $f(x) \geq 0$ on [2,6]. Next $\displaystyle\int_2^6 \dfrac{2}{32}x\,dx = \dfrac{1}{32}x^2\Big|_2^6 = \dfrac{1}{32}(36-4) = 1,$
   and so f is a probability density function on [2,6].

5. $f(x) = \dfrac{2}{9}(3x - x^2)$ is nonnegative on [0,3] since both the factors x
   and 3 - x are nonnegative there. Next, we compute
   $$\int_0^3 \dfrac{2}{9}(3x - x^2)\,dx = \dfrac{2}{9}\left(\dfrac{3}{2}x^2 - \dfrac{1}{3}x^3\right)\Big|_0^3 = \dfrac{2}{9}\left(\dfrac{27}{2} - 9\right) = 1,$$
   and so f is a probability density function.

7. $f(x) = \dfrac{12 - x}{72}$ is nonnegative on [0,12]. Next, we see that
   $$\int_0^{12} \dfrac{12-x}{72}\,dx = \int_0^{12}\left(\dfrac{1}{6} - \dfrac{x}{72}\right)dx = \dfrac{1}{6}x - \dfrac{1}{144}x^2\Big|_0^{12} = 2 - 1 = 1,$$
   and conclude that f is a probability function.

9. $f(x) = \dfrac{8}{7x^2}$ is nonnegative on [1,8]. Next, we compute
   $$\int_1^8 \dfrac{8}{7x^2}\,dx = -\dfrac{8}{7x}\Big|_1^8 = -\dfrac{8}{7}\left(\dfrac{1}{8} - 1\right) = 1,$$
   and so f is a probability density function.

11. First $f(x) \geq 0$ on $[0,\infty)$ is self-evident. Next,
    $$\int_0^{\theta} \dfrac{x}{(x^2 + 1)^{3/2}}\,dx = \lim_{b\to\infty}\int_0^b x(x^2 + 1)^{-3/2}\,dx$$
    Let $I = \displaystyle\int x(x^2 + 1)^{-3/2}\,dx$. Integrate I by the method of substitution letting
    $u = x^2 + 1$ so that $du = 2x\,dx$, and so

10 Probability and Calculus

354

$$I = \frac{1}{2}\int u^{-3/2} du = \frac{1}{2}(-2u^{-1/2}) + C = -\frac{1}{\sqrt{u}} + C = -\frac{1}{\sqrt{x^2+1}} + C.$$

Therefore,

$$\int_0^\infty \frac{x}{(x^2+1)^{3/2}} dx = \lim_{b\to\infty}\left[-\frac{1}{\sqrt{x^2+1}}\Big|_0^b\right] = \lim_{b\to\infty}\left(-\frac{1}{\sqrt{b^2+1}} + 1\right) = 1,$$

and this completes the proof.

13. $\int_1^4 k\,dx = kx\Big|_1^4 = 3k = 1$ implies $k = \dfrac{1}{3}$.

15. $\int_0^4 k(4-x)\,dx = k\int_0^4 (4-x)\,dx = k(4x - \frac{1}{2}x^2)\Big|_0^4 = k(16-8) = 8k = 1$
implies that $k = 1/8$.

17. $\int_0^4 kx^{1/2}\,dx = \frac{2}{3}kx^{3/2}\Big|_0^4 = \frac{16}{3}k = 1$ implies $k = \frac{3}{16}$.

19. $\int_1^\infty \frac{k}{x^3}\,dx = \lim_{b\to\infty}\int_1^b kx^{-3}\,dx = \lim_{b\to\infty}-\frac{k}{2x^2}\Big|_1^b = \lim_{b\to\infty}\left(-\frac{k}{2b^2} + \frac{k}{2}\right) = \frac{k}{2} = 1$
implies $k = 2$.

21.   a. $P(2 \le x \le 4) = \int_2^4 \frac{1}{12}x\,dx = \frac{1}{24}x^2\Big|_2^4 = \frac{1}{24}(16-4) = \frac{1}{2}$.

   b. $P(1 \le x \le 4) = \int_1^4 \frac{1}{12}x\,dx = \frac{1}{24}x^2\Big|_1^4 = \frac{1}{24}(16-1) = \frac{5}{8}$.

   c. $P(x \ge 2) = \int_2^5 \frac{1}{12}x\,dx = \frac{1}{24}x^2\Big|_2^5 = \frac{1}{24}(25-4) = \frac{7}{8}$.

   d. $P(x = 2) = \int_2^2 \frac{1}{12}x\,dx = 0$.

23.   a. $P(-1 \le x \le 1) = \int_{-1}^1 \frac{3}{32}(4-x^2)\,dx = \frac{3}{32}(4x - \frac{1}{3}x^3)\Big|_{-1}^1$
$$= \frac{3}{32}[(4-\frac{1}{3}) - (-4+\frac{1}{3})] = \frac{11}{16}.$$

b. $P(x \le 0) = \int_{-2}^{0} \frac{3}{32}(4-x^2)dx = \frac{3}{32}(4x - \frac{1}{3}x^3)\Big|_{-2}^{0} = \frac{3}{32}[0-(-8+\frac{8}{3})] = \frac{1}{2}.$

c. $P(x > -1) = \int_{-1}^{2} \frac{3}{32}(4-x^2)dx = \frac{3}{32}(4x - \frac{1}{3}x^3)\Big|_{-1}^{2}$

$\qquad = \frac{3}{32}[(8-\frac{8}{3})-(-4+\frac{1}{3})] = \frac{27}{32}.$

d. $P(x = 0) = \int_{0}^{0} \frac{3}{32}(4-x^2)dx = 0.$

25. a. $P(x \ge 4) = \int_{4}^{9} \frac{1}{4}x^{-1/2} dx = \frac{1}{2}x^{1/2}\Big|_{4}^{9} = \frac{1}{2}(3-2) = \frac{1}{2}.$

b. $P(1 \le x \le 8) = \int_{1}^{8} \frac{1}{4}x^{-1/2} dx = \frac{1}{2}x^{1/2}\Big|_{1}^{8} = \frac{1}{2}(2\sqrt{2}-1) \approx 0.9142.$

c. $P(x = 3) = \int_{3}^{3} \frac{1}{4}x^{-1/2} dx = 0.$

d. $P(x \le 4) = \int_{1}^{4} \frac{1}{4}x^{-1/2} dx = \frac{1}{2}x^{1/2}\Big|_{1}^{4} = \frac{1}{2}(2-1) = \frac{1}{2}.$

27. a. $P(0 \le x \le 4) = \int_{0}^{4} 4xe^{-x^2} dx = -e^{-2x^2}\Big|_{0}^{4}$

$\qquad = -e^{-32} + 1 \approx 1.$

b. $P(x \ge 1) = \int_{1}^{\infty} 4xe^{-x^2} dx = \lim_{b \to \infty} \int_{1}^{b} 4xe^{-x^2} dx$

$\qquad = \lim_{b \to \infty} \left[-e^{-2x^2}\Big|_{1}^{b} \right] = \lim_{b \to \infty}(-e^{-2b^2} + e^{-2}) = e^{-2} \approx 0.1353.$

29. a. $P(\frac{1}{2} \le x \le 1) = \int_{1/2}^{1} x\,dx = \frac{1}{2}x^2\Big|_{1/2}^{1} = \frac{1}{2}(1) - \frac{1}{2}\left(\frac{1}{4}\right) = \frac{1}{2} - \frac{1}{8} = \frac{3}{8}.$

b. $P(\frac{1}{2} \le x \le \frac{3}{2}) = \int_{1/2}^{1} x\,dx + \int_{1}^{3/2} (2-x)\,dx = \frac{3}{8} + \left(2x - \frac{1}{2}x^2\right)\Big|_{1}^{3/2}$

(Using the results from (9))

$\qquad = \frac{3}{8} + \frac{1}{2}x(4-x)\Big|_{1}^{3/2} = \frac{3}{8} + \left[\frac{1}{2}\left(\frac{3}{2}\right)\left(4-\frac{3}{2}\right) - \frac{1}{2}(3)\right]$

$\qquad = \frac{3}{8} + \frac{3}{4}\left(\frac{5}{2}\right) - \frac{3}{2} = \frac{3}{4}.$

c. $P(x \geq 1) = \int_1^2 (2-x)\,dx = 2x - \frac{1}{2}x^2 \Big|_1^2 = (4-2) - \left(2 - \frac{1}{2}\right) = \frac{1}{2}.$

d. $P(x \leq \frac{3}{2}) = \int_0^1 x\,dx + \int_1^{3/2} (2-x)\,dx = \frac{1}{2}x^2 \Big|_0^1 + \left(2x - \frac{1}{2}x^2\right)\Big|_1^{3/2}$

$$= \frac{1}{2} + \left[2\left(\frac{3}{2}\right) - \frac{1}{2}\left(\frac{9}{4}\right)\right] - \left(2 - \frac{1}{2}\right)$$

$$= \frac{1}{2} + \left(3 - \frac{9}{8}\right) - \frac{3}{2} = \frac{1}{2} + \frac{15}{8} - \frac{3}{2} = \frac{7}{8}.$$

31.   a. $P(x \leq 100) = \int_0^{100} \frac{1}{100} e^{-x/100}\,dx = -e^{-x/100}\Big|_0^{100} = -e^{-1} + 1 \approx 0.6321.$

   b. $P(x \geq 120) = \int_{120}^\infty \frac{1}{100} e^{-x/100}\,dx = \lim_{b \to \infty} \int_{120}^b \frac{1}{100} e^{-x/100}\,dx$

$$= \lim_{b \to \infty} -e^{-x/100}\Big|_{120}^b = \lim_{b \to \infty}(-e^{-b/100} + e^{-(120/100)}) = e^{-1.2}$$

$$\approx 0.3012.$$

   c. $P(60 < x < 140) = \int_{60}^{140} \frac{1}{100} e^{-x/100}\,dx = -e^{-x/100}\Big|_{60}^{140} = -e^{-1.4} + e^{-0.6} \approx 0.3022.$

33.   a. $P(600 \leq t \leq 800) = \int_{600}^{800} 0.001 e^{-0.001t}\,dt = -e^{-0.001t}\Big|_{600}^{800} = -e^{-0.8} + e^{-0.6}$

$$\approx 0.0995.$$

   b. $P(t \geq 1200) = \int_{1200}^\infty 0.001 e^{-0.001t}\,dt = \lim_{b \to \infty} \int_{1200}^b 0.001 e^{-0.001t}\,dt$

$$= \lim_{b \to \infty} -e^{-0.001t}\Big|_{1200}^b = \lim_{b \to \infty}(-e^{-0.001b} + e^{-1.2}) = e^{-1.2} \approx 0.3012.$$

35. $P(t \geq 2) = \int_2^\infty \frac{1}{30} e^{-t/30}\,dt = \lim_{b \to \infty} \int_2^b \frac{1}{30} e^{-t/30}\,dt = \lim_{b \to \infty} -e^{-t/30}\Big|_2^b$

$$= \lim_{b \to \infty}(-e^{-b/30} + e^{-1/15}) = e^{-1/15} \approx 0.9355.$$

37. $P(1 \leq x \leq 2) = \int_1^2 \frac{2}{9} x(3-x)\,dx = \frac{2}{9}\left(\frac{3}{2}x^2 - \frac{1}{3}x^3\right)\Big|_1^2$

$$= \frac{2}{9}\left[(6 - \frac{8}{3}) - (\frac{3}{2} - \frac{1}{3})\right] = \frac{13}{27} \approx 0.4815.$$

$$P(x \geq 1) = \int_1^3 \tfrac{2}{9}x(3-x)dx = \tfrac{2}{9}(\tfrac{3}{2}x^2 - \tfrac{1}{3}x^3)\Big|_1^3$$
$$= \tfrac{2}{9}[(\tfrac{27}{2}-9)-(\tfrac{3}{2}-\tfrac{1}{3})] = \tfrac{20}{27} \approx 0.740741.$$

39. $P(t > 4) = \int_4^\infty 9(9+t^2)^{-3/2}\,dt = \lim_{b\to\infty} \int_4^b 9(9+t^2)^{-3/2}\,dt$

$$= \lim_{b\to\infty} \frac{t}{\sqrt{9+t^2}}\Big|_4^b = \lim_{b\to\infty}\left(\frac{b}{\sqrt{9+b^2}} - \frac{4}{5}\right) = 1 - \frac{4}{5} = \frac{1}{5}.$$

41. False. f must be nonnegative on $[a, b]$ as well.

USING TECHNOLOGY EXERCISES 10.1, page 737

1.

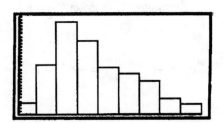

3. a.

x	0	1	2	3	4	5
$P(X{=}x)$	0.017	0.067	0.033	0.117	0.233	0.133

x	6	7	8	9	10
$P(X{=}x)$	0.167	0.1	0.05	0.067	0.017

b.

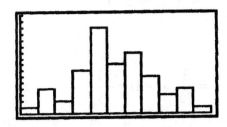

EXERCISES 10.2, page 752

1. $\mu = \int_3^6 \frac{1}{3} x \, dx = \frac{1}{6} x^2 \Big|_3^6 = \frac{1}{6}(36-9) = \frac{9}{2}.$

   $\text{Var}(x) = \int_3^6 \frac{1}{3} x^2 \, dx - \frac{81}{4} = \frac{1}{9} x^3 \Big|_3^6 - \frac{81}{4} = \frac{1}{9}(216-27) - \frac{81}{4} = \frac{3}{4}.$

   $\sigma = \sqrt{\dfrac{3}{4}} = \dfrac{\sqrt{3}}{2} \approx 0.8660.$

3. $\mu = \int_0^5 \frac{3}{125} x^3 \, dx = \frac{3}{500} x^4 \Big|_0^5 = \frac{15}{4}.$

   $\text{Var}(x) = \int_0^5 \frac{3}{125} x^4 \, dx - \frac{225}{16} = \frac{3}{625} x^5 \Big|_0^5 - \frac{225}{16} = 15 - \frac{225}{16} = \frac{15}{16}.$

   $\sigma = \sqrt{\dfrac{15}{16}} = \dfrac{\sqrt{15}}{4} \approx 0.9682.$

5. $\mu = \int_1^5 \frac{3}{32} x(x-1)(5-x) \, dx = \frac{3}{32} \int_1^5 (-x^3 + 6x^2 - 5x) dx$

   $= \frac{3}{32} \left(-\frac{1}{4} x^4 + 2x^3 - \frac{5}{2} x^2 \right) \Big|_1^5 = \frac{3}{32} [(-\frac{625}{4} + 250 - \frac{125}{2}) - (-\frac{1}{4} + 2 - \frac{5}{2})] = 3.$

   $\text{Var}(x) = \int_1^5 \frac{3}{32} [-\frac{1}{4} x^4 + 6x^3 - 5x^2) dx - 9 = \frac{3}{32} (-\frac{1}{5} x^5 + \frac{3}{2} x^4 - \frac{5}{3} x^3) \Big|_1^5 - 9$

   $= \frac{3}{32} [(-625 + \frac{1875}{2} - \frac{625}{3}) - (-\frac{1}{5} + \frac{3}{2} - \frac{5}{3})] - 9 \approx 0.8..$

   $\sigma = \sqrt{0.8} \approx 0.8944.$

7. $\mu = \int_1^8 \frac{8}{7x} \, dx = \frac{8}{7} \ln x \Big|_1^8 = \frac{8}{7} \ln 9 \approx 2.376504619.$

$$\text{Var}(x) = \int_0^8 \frac{8}{7} dx - 5.64777 = \frac{8}{7} x \Big|_1^8 - 5.64777 = 8 - 5.647777 \approx 2.3522.$$

$$\sigma = \sqrt{2.3522} \approx 1.5337.$$

9. $\mu = \int_1^4 \frac{3}{14} x^{3/2} dx = \frac{3}{35} x^{5/2} \Big|_1^4 = \frac{3}{35}(32-1) = \frac{93}{35}.$

$$\text{Var}(x) = \int_1^4 \frac{3}{14} x^{5/2} dx - \frac{9}{25} = \frac{3}{49} x^{7/2} \Big|_1^4 - \frac{9}{25} = \frac{3}{49}(128-1) - \frac{9}{25} \approx 0.715102.$$

$$\sigma = \sqrt{7.4155} \approx 0.8456.$$

11. $\mu = \int_1^\infty \frac{3}{x^3} dx = \lim_{b \to \infty} \int_1^b 3x^{-3} dx = \lim_{b \to \infty} -\frac{3}{2x^2} \Big|_1^b = \lim_{b \to \infty} \left(-\frac{3}{2b^2} + \frac{3}{2} \right) = \frac{3}{2}.$

$$\text{Var}(x) = \int_1^\infty \frac{3}{x^2} dx - \frac{9}{4} = \lim_{b \to \infty} \int_1^b 3x^{-2} dx - \frac{9}{4} = \lim_{b \to \infty} -\frac{3}{x} \Big|_1^b - \frac{9}{4}$$

$$= \lim_{b \to \infty} \left(-\frac{3}{b} + 3 \right) - \frac{9}{4} = \frac{3}{4}.$$

$$\sigma = \sqrt{\frac{3}{4}} = \frac{\sqrt{3}}{2} \approx 0.8660.$$

13. $\mu = \int_0^\infty \frac{1}{4} xe^{-x/4} dx = \lim_{b \to \infty} \int_0^b \frac{1}{4} xe^{-x/4} dx = \lim_{b \to \infty} 4(-\frac{1}{4} - 1)e^{-x/4} \Big|_0^b$

$$= \lim_{b \to \infty} [4(-\tfrac{1}{4}b - 1)e^{-b/4} + 4] = 4.$$

$$\text{Var}(x) = \int_0^\infty \frac{1}{4} x^2 e^{-x/4} dx - 16 = \lim_{b \to \infty} \int_0^b \frac{1}{4} x^2 e^{-x/4} dx - 16$$

$$= \lim_{b \to \infty} x^2 e^{-x/4} - 32(-\tfrac{1}{4}x - 1)e^{-x/4} \Big|_0^b - 16$$

$$= \lim_{b \to \infty} \{ [b^2 e^{-b/4} - 32(-\tfrac{1}{4}b - 1)e^{-b/4}] + 32 \} - 16$$

$$= 16.$$

$$\sigma = \sqrt{16} = 4.$$

15. $\mu = \int_0^\infty x \cdot \frac{1}{100} e^{-x/100} dx = \lim_{b \to \infty} \int_0^b \frac{1}{100} xe^{-x/100} dx = \lim_{b \to \infty} [100 \left(-\frac{x}{100} - 1 \right) e^{-x/100} \Big|_0^b$

$$= \lim_{b \to \infty} \left[100 \left(-\frac{b}{100} - 1 \right) e^{-b/100} + 100 \right] = 100.$$

So a plant of this species is expected to live 100 days.

17. $\mu = \int_0^5 t \cdot \frac{2}{25} t \, dt = \frac{2}{25} \int_0^5 t^2 \, dt = \frac{2}{75} t^3 \Big|_0^5 = \frac{2}{75} (125) = 3\frac{1}{3}.$

So a shopper is expected to spend $3\frac{1}{3}$ minutes in the magazine section.

19. $\mu = \int_0^3 x \cdot \frac{2}{9} x(3-x) \, dx = \frac{2}{9} \int_0^3 (3x^2 - x^3) \, dx = \frac{2}{9} (x^3 - \frac{1}{4} x^4) \Big|_0^3$

$\qquad = \frac{2}{9} \left(27 - \frac{81}{4} \right) = 1.5.$

So the expected amount of snowfall is 1.5 ft.

21. $\mu = \int_0^5 x \cdot \frac{6}{125} x(5-x) \, dx = \frac{6}{125} \int_0^5 (5x^2 - x^3) \, dx = \frac{6}{125} \left(\frac{5}{3} x^3 - \frac{1}{4} x^4 \right) \Big|_0^5$

$\qquad = \frac{6}{125} \left(\frac{625}{3} - \frac{625}{4} \right) = 2.5.$

So the expected demand is 2500 lb.

23. $P(x \le m) = \int_2^m \frac{1}{6} \, dx = \frac{1}{6} x \Big|_2^m = \frac{1}{6} (m-2) = \frac{1}{2}.$

$\qquad m - 2 = 3,$ or $m = 5.$

25. $P(x \le m) = \int_0^m \frac{3}{16} x^{1/2} \, dx = \frac{1}{8} x^{3/2} \Big|_0^m = \frac{1}{8} m^{3/2}.$ Next, we solve

$\qquad \frac{1}{8} m^{3/2} = \frac{1}{2}, \ m^{3/2} = 4,$ or $m = 4^{2/3} \approx 2.5198.$

27. $P(x \le m) = \int_1^m 3x^{-2} \, dx = -\frac{3}{x} \Big|_1^m = -\frac{3}{m} + 3.$

$\qquad -\frac{3}{m} + 3 = \frac{1}{2}$ gives $-\frac{3}{m} = -\frac{5}{2}, \ -5m = -6,$ or $m = \frac{6}{5}.$

29. False. The expected value of x is $\int_a^b x f(x) \, dx.$

1. a.

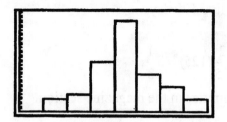

   b. μ = 4 and σ = 1.40

3. a. *X* gives the minimum age requirement for a regular driver's license.

   b.

x	15	16	17	18	19	21
P(X=x)	0.02	0.30	0.08	0.56	0.02	0.02

   c.

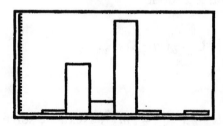

   d. μ = 17.34 and σ = 1.11

5. a. Let *X* denote the random variable that gives the weight of a carton of sugar.
   b. The probability distribution for the random variable *X* is

x	4.96	4.97	4.98	4.99	5.00	5.01	5.02	5.03
P(X = x)	$\frac{3}{30}$	$\frac{4}{30}$	$\frac{4}{30}$	$\frac{1}{30}$	$\frac{1}{30}$	$\frac{5}{30}$	$\frac{3}{30}$	$\frac{3}{30}$

x	5.04	5.05	5.06
$P(X = x)$	$\dfrac{4}{30}$	$\dfrac{1}{30}$	$\dfrac{1}{30}$

    c. $\mu = 5.00467 \approx 5.00$; $V(X) = 0.0009$; $\sigma = \sqrt{0.0009} = 0.03$

EXERCISES 10.3, page 765

1. $P(Z < 1.45) = 0.9265$.

3. $P(Z < -1.75) = 0.0401$.

5. $P(-1.32 < Z < 1.74) = P(Z < 1.74) - P(Z < -1.32)$
                                  $= 0.9591 - 0.0934 = 0.8657$.

7. $P(Z < 1.37) = 0.9147$.

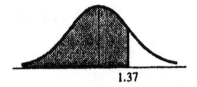

9. $P(Z < -0.65) = 0.2578$.

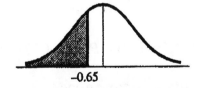

11. $P(Z > -1.25) = 1 - P(Z < -1.25) = 1 - 0.1056 = 0.8944$

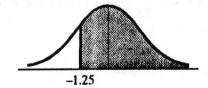

13. $P(0.68 < Z < 2.02) = P(Z < 2.02) - P(Z < 0.68)$
$$= 0.9783 - 0.7517 = 0.2266.$$

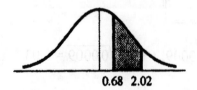

0.68 2.02

15. a. Referring to Table 4, we see that $P(Z < z) = 0.8907$ implies that $z = 1.23$.

b. Referring to Table 4, we see that $P(Z < z) = 0.2090$ implies that $z = -0.81$.

17. a. $P(Z > -z) = 1 - P(Z < -z) = 1 - 0.9713 = 0.0287$ implies $z = 1.9$.
b. $P(Z < -z) = 0.9713$ implies that $z = -1.9$.

19. a. $P(X < 60) = P(Z < \dfrac{60-50}{5}) = P(Z < 2) = 0.9772.$

b. $P(X > 43) = P(Z > \dfrac{43-50}{5}) = P(Z > -1.4) = P(Z < 1.4) = 0.9192.$

c. $P(46 < X < 58) = P(\dfrac{46-50}{5} < Z < \dfrac{58-50}{5}) = P(-0.8 < Z < 1.6)$
$$= P(Z < 1.6) - P(Z < -0.8)$$
$$= 0.9452 - 0.2119 = 0.7333.$$

21. $\mu = 20$ and $\sigma = 2.6$.

a. $P(X > 22) = P(Z > \dfrac{22-20}{2.6}) = P(Z > 0.77) = P(Z < -0.77) = 0.2206.$

b. $P(X < 18) = P(Z < \dfrac{18-20}{2.6}) = P(Z < -0.77) = 0.2206.$

c. $P(19 < X < 21) = P(\dfrac{19-20}{2.6} < Z < \dfrac{21-20}{2.6}) = P(-0.39 < Z < 0.39)$
$$= P(Z < 0.39) - P(Z < -0.39)$$

$$= 0.6517 - 0.3483 = 0.3034.$$

23. $\mu = 750$ and $\sigma = 75$.

a. $P(X > 900) = P(Z > \dfrac{900 - 750}{75}) = P(Z > 2) = P(Z < -2) = 0.0228.$

b. $P(X < 600) = P(Z < \dfrac{600 - 750}{75}) = P(Z < -2) = 0.0228.$

c. $P(750 < X < 900) = P(Z < \dfrac{750 - 750}{75} < Z < \dfrac{900 - 750}{75})$

$= P(0 < Z < 2) = P(Z < 2) - P(Z < 0)$

$= 0.9772 - 0.5000 = 0.4772.$

d. $P(600 < X < 800) = P(\dfrac{600 - 750}{75} < Z < \dfrac{800 - 750}{75})$

$= P(-2 < Z < .667) = P(Z < .667) - P(Z < -2)$

$= 0.7486 - 0.0228 = 0.7258.$

25. $\mu = 100$ and $\sigma = 15$.

a. $P(X > 140) = P(Z > \dfrac{140 - 100}{15}) = P(Z > 2.667) = P(Z < -2.667)$

$= 0.0038.$

b. $P(X > 120) = P(Z > \dfrac{120 - 100}{15}) = P(Z > 1.33) = P(Z < -1.33)$

$= 0.0918.$

c. $P(100 < X < 120) = P(\dfrac{100 - 100}{15} < Z < \dfrac{120 - 100}{15}) = P(0 < Z < 1.333)$

$= P(Z < 0) - P(Z < 1.333)$

$= 0.9082 - 0.5000 = 0.4082.$

d. $P(X < 90) = P(Z < \dfrac{90 - 100}{15}) = P(Z < -0.667) = 0.2514.$

27. Here $\mu = 475$ and $\sigma = 50$.

$P(450 < X < 550) = P(\dfrac{450 - 475}{50} < Z < \dfrac{550 - 475}{50}) = P(-0.5 < Z < 1.5)$

$= P(Z < 1.5) - P(Z < -0.5)$

$= 0.9332 - 0.3085 = 0.6247.$

29. Here $\mu = 22$ and $\sigma = 4$.

$$P(X < 12) = P\left(Z < \frac{12-22}{4}\right) = P(Z < -2.5) = 0.0062, \text{ or } 0.62 \text{ percent.}$$

31. $\mu = 70$ and $\sigma = 10$.

To find the cut-off point for an A, we solve $P(Y < y) = 0.85$ for y. Now

$$P(Y < y) = P\left(Z < \frac{y-70}{10}\right) = 0.85 \text{ implies } \frac{y-70}{10} = 1.04$$

or $y = 80.4 \approx 80$.

For a B: $P(Y < y) = P\left(Z < \frac{y-70}{10}\right) = 7.5$; $\frac{y-70}{10} = .67$, or $y \approx 77$.

For a C: $P(Y < y) = P\left(Z < \frac{y-70}{10}\right) = 0.60$ implies $\frac{y-70}{10} = 0.25$, or $y \approx 73$.

For a D: $P(Y < y) = P\left(Z < \frac{y-70}{10}\right) = 0.15$ implies $\frac{y-70}{10} = -0.84$ or $y \approx 62$.

For a F: $P(Y < y) = P\left(Z < \frac{y-70}{10}\right) = 0.05$ implies $\frac{y-70}{10} = -1.65$, or $y \approx 54$.

CHAPTER 10, REVIEW EXERCISES, page 770

1. First $f(x) = \frac{1}{28}(2x+3)$ is nonnegative on $[0,4]$. Next,

$$\int_0^4 \frac{1}{28}(2x+3)dx = \frac{1}{28}(x^2+3x)\Big|_0^4 = \frac{1}{28}(16+12) = 1.$$

3. First, $f(x) = \frac{1}{4} > 0$ on $[7,11]$. Next,

$$\int_7^{11} \frac{1}{4}dx = \frac{1}{4}x\Big|_7^{11} = \frac{1}{4}(11-7) = 1.$$

5. $\int_0^9 kx^2\,dx = \frac{k}{3}x^3\Big|_0^9 = \frac{k}{3}(729) = 1.$ Therefore, $k = \frac{1}{243}$.

7. $\int_1^3 kx^{-2}\,dx = \frac{2k}{3} = 1.$ Therefore, $k = \frac{3}{2}$.

9. a. $\int_2^4 \frac{2}{21}x\,dx = \frac{1}{21}x^2\Big|_2^4 = \frac{16}{21} - \frac{4}{21} = \frac{12}{21} = \frac{4}{7}$.

   b. $\int_4^4 \frac{2}{21}x\,dx = \frac{1}{21}x^2\Big|_4^4 = 0.$

b. $\int_4^4 \frac{2}{21} x \, dx = \frac{1}{21} x^2 \Big|_4^4 = 0.$

c. $\int_3^4 \frac{2}{21} x \, dx = \frac{1}{21} x^2 \Big|_3^4 = \frac{16}{21} - \frac{9}{21} = \frac{7}{21} = \frac{1}{3}.$

11. a. $P(1 \le x \le 3) = \frac{3}{16} \int_1^3 x^{1/2} \, dx = \frac{1}{8} x^{3/2} \Big|_1^3 = \frac{1}{8}(3\sqrt{3} - 1) \approx 0.52.$

b. $P(x \le 3) = \frac{3}{16} \int_0^3 x^{1/2} dx = \frac{1}{8} x^{3/2} \Big|_0^3 = \frac{1}{8}(3\sqrt{3} - 0) \approx 0.65.$

c. $P(x = 2) = \frac{3}{16} \int_2^2 x^{1/2} \, dx = 0.$

13. $\mu = \frac{1}{5} \int_2^7 x \, dx = \frac{1}{10} x^2 \Big|_2^7 = \frac{1}{10}(49 - 4) = \frac{9}{2}.$

$\mathrm{Var}(x) = \frac{1}{5} \int_2^7 x^2 \, dx - \left(\frac{9}{2}\right)^2 = \frac{1}{15} x^3 \Big|_2^7 - (4.5)^2 = \frac{1}{15}(343 - 8) - (4.5)^2$

$\approx 2.083.$

$\sigma = \sqrt{2.083} \approx 1.44.$

15. $\mu = \frac{1}{4} \int_{-1}^1 x(3x^2 + 1) dx = \frac{1}{4} \int_{-1}^1 (3x^3 + x) dx = \frac{1}{4}\left(\frac{3}{4} x^4 + \frac{1}{2} x^2\right)\Big|_{-1}^1 = 0.$

$\mathrm{Var}(x) = \int_{-1}^1 x^2(3x^2 - 1) \, dx - 0 = \frac{1}{4} \int_{-1}^1 (3x^4 + x^2) \, dx = \frac{1}{4}\left(\frac{3}{5} x^5 + \frac{1}{3} x^3\right)\Big|_{-1}^1$

$= \frac{7}{15}.$

$\sigma = \sqrt{\frac{7}{15}} \approx 0.6831.$

17. $P(Z < 2.24) = 0.9875.$

19. $P(0.24 \le Z \le 1.28) = P(Z \le 1.28) - P(Z \le 0.24) = 0.8997 - 0.5948$
$= 0.3049.$

21. a. $P(X \le 84) = P\left(Z \le \dfrac{84 - 80}{8}\right) = P(Z \le 0.5) = 0.6915.$

b. $P(X \ge 70) = P\left(\dfrac{70 - 80}{8}\right) = P(Z \ge -1.25) = P(Z \le 1.25) = 0.8944.$

c. $P(75 \le X \le 85) = P\left(\dfrac{75-80}{8} \le Z \le \dfrac{85-80}{8}\right) = P(-0.625 \le Z \le 0.625)$

$= P(Z \le 0.625) - P(Z \le -0.625) = 0.7341 - 0.2660$

$\approx 0.4681.$

23. a. $P(t > 6) = \displaystyle\int_{6}^{\infty} \tfrac{1}{4} e^{-t/4}\, dt = \lim_{b\to\infty} \int_{6}^{b} -e^{-t/4}\, dt = \lim_{b\to\infty} -e^{-t/4}\Big|_{6}^{b}$

$= \lim_{b\to\infty} -e^{-b/4} + e^{-6/4} = 0.22313.$

b. $P(t < 2) = \displaystyle\int_{0}^{2} \tfrac{1}{4} e^{-t/4}\, dt = -e^{-t/4}\Big|_{0}^{2} = -e^{-1/2} + e^{0} = -0.6065 + 1 = 0.39347.$

c. $\mu = \displaystyle\int_{0}^{\infty} \tfrac{1}{4} t e^{-t/4}\, dt = \lim_{b\to\infty} \int_{0}^{b} \tfrac{1}{4} t e^{-t/4}\, dt = \lim_{b\to\infty} 4(-\tfrac{1}{4} t - 1) e^{-t/4}\Big|_{0}^{b}$

$= 0 + 4 = 4.$

CHAPTER 11

EXERCISES 11.1, page 784

1. $f(x) = e^{-x}$, $f'(x) = -e^{-x}$, $f''(x) = e^{-x}$, and $f'''(x) = -e^{-x}$.

    So $f(0) = 1$, $f'(0) = -1$, $f''(0) = 1$, and $f'''(0) = -1$.

    Therefore,

    $P_1(x) = f(0) + f'(0)x = 1 - x$, or $P_1(x) = 1 - x$

    $P_2(x) = f(0) = f'(0)x + \dfrac{f''(0)}{2!}x^2 = 1 - x + \tfrac{1}{2}x^2$

    $P_3(x) = f(0) + f'(0)x + \dfrac{f''(0)}{2!}x^2 + \dfrac{f'''(0)}{3!}x^3 = 1 - x + \tfrac{1}{2}x^2 - \tfrac{1}{6}x^3.$

3. $f(x) = \dfrac{1}{x+1} = (x+1)^{-1}$, $f'(x) = -(x+1)^{-2}$, $f''(x) = 2(x+1)^{-3}$,

    and $f'''(x) = -6(x+1)^{-4}$. So

    $f(0) = 1$, $f'(0) = -1$, $f''(0) = 2$, and $f'''(0) = -6$. Therefore,

    $P_1(x) = 1 - x$

    $P_2(x) = 1 - x + \dfrac{2}{2!}x^2 = 1 - x + x^2$

    $P_3(x) = 1 - x + \dfrac{2}{2!}x^2 - \dfrac{6}{3!}x^3 = 1 - x + x^2 - x^3.$

5. $f(x) = \dfrac{1}{x} = x^{-2}$; $f'(x) = -x^{-2} = -\dfrac{1}{x^2}$, $f''(x) = (-1)(-2)x^{-3} = \dfrac{2}{x^3}$.

    $f'''(x) = -\dfrac{6}{x^4}.$

    So $f(1) = 1$, $f'(1) = -1$, $f''(1) = 2$, $f'''(1) = -6$. Therefore,

    $P_1(x) = f(1) + f'(x)(x-1) = 1 - (x-1)$, or $P_1 = 1 - (x-1)$

    $P_2(x) = f(1) + f'(1)(x-1) + \dfrac{f''(1)}{2!}(x-1)^2 = 1 - (x-1) + (x-1)^2$

    $P_3(x) = f(1) + f'(1)(x-1) + \dfrac{f''(1)}{2!}(x-1)^2 + \dfrac{f'''(1)}{3!}(x-1)^3$

    $= 1 - (x-1) + (x-1)^2 - (x-1)^3.$

369   *11 Taylor Polynomials*

7. $f(x) = (1-x)^{1/2}$, $f'(x) = -\dfrac{1}{2}(1-x)^{-1/2}$, $f''(x) = -\dfrac{1}{4}(1-x)^{-3/2}$,

$f'''(x) = -\dfrac{3}{8}(1-x)^{-5/2}$.

$f(0) = 1$, $f'(0) = -\dfrac{1}{2}$, $f''(0) = -\dfrac{1}{4}$, $f'''(0) = -\dfrac{3}{8}$.

$P_1 = 1 - \dfrac{1}{2}x$; $P_2(x) = 1 - \dfrac{1}{2}x + \dfrac{(-\frac{1}{4})}{2!}x^2 = 1 - \dfrac{1}{2}x - \dfrac{1}{8}x^2$

$P_3 = 1 - \dfrac{1}{2}x - \dfrac{1}{8}x^2 + \dfrac{(-\frac{3}{8})}{3!}x^3 = 1 - \dfrac{1}{2}x - \dfrac{1}{8}x^2 - \dfrac{1}{16}x^3$.

9. $f(x) = \ln(1-x)$, $f'(x) = -\dfrac{1}{1-x} = -(1-x)^{-1}$, $f''(x) = -(1-x)^{-2}$

$f'''(x) = -2(1-x)^{-3}$.

$f(0) = 0$, $f'(0) = -1$, $f''(0) = -1$, $f'''(0) = -2$.

$P_1(x) = -x$, $P_2(x) = -x + \dfrac{(-1)}{2!}x^2 = -x - \dfrac{1}{2}x^2$

$P_3(x) = -x - \dfrac{1}{2}x^2 + \dfrac{(-2)}{3!}x^3 = -x - \dfrac{1}{2}x^2 - \dfrac{1}{3}x^3$.

11. $f(x) = x^4$, $f'(x) = 4x^3$, $f''(x) = 12x^2$.

$f(2) = 16$, $f'(2) = 32$, $f''(2) = 48$

$P_2(x) = 16 + 32(x-2) + \dfrac{48}{2!}(x-2)^2 = 16 + 32(x-2) + 24(x-2)^2$.

13. $f(x) = \ln x$; $f'(x) = \dfrac{1}{x}$, $f''(x) = -\dfrac{1}{x^2}$, $f'''(x) = \dfrac{2}{x^3}$, $f^{(4)} = -\dfrac{6}{x^4}$.

$f(1) = 0$, $f'(1) = 1$, $f''(1) = -1$, $f'''(1) = 2$, $f^{(4)} = -6$.

$P_4(x) = (x-1) + \dfrac{(-1)}{2!}(x-1)^2 + \dfrac{2}{3!}(x-1)^3 + \dfrac{(-6)}{4!}(x-1)^4$

$= (x-1) - \dfrac{1}{2}(x-1)^2 + \dfrac{1}{3}(x-1)^3 - \dfrac{1}{4}(x-1)^4$.

15. $f(x) = e^x$, $f'(x) = f''(x) = f'''(x) = f^{(4)}(x) = e^x$

$$f(1) = f'(1) = f''(1) = f'''(x) = f^{(4)}(x) = e$$

$$P_4(x) = e + e(x-1) + \frac{e}{2!}(x-1)^2 + \frac{e}{3!}(x-1)^3 + \frac{e}{4!}(x-1)^4$$

17. $f(x) = (1-x)^{1/2}$, $f'(x) = -\frac{1}{2}(1-x)^{-1/2}$, $f''(x) = -\frac{1}{4}(1-x)^{-3/2}$,

$$f'''(x) = -\frac{3}{8}(1-x)^{-5/2}$$

$$f(0) = 1, \ f'(0) = -\frac{1}{2}, \ f''(0) = -\frac{1}{4}, \ f'''(0) = -\frac{3}{8}.$$

Therefore,

$$P_3(x) = 1 - \frac{1}{2}x + \frac{\left(-\frac{1}{4}\right)}{2!}x^2 + \frac{\left(-\frac{3}{8}\right)}{3!}x^3 = 1 - \frac{1}{2}x - \frac{1}{8}x^2 - \frac{1}{16}x^3.$$

19. $f(x) = \dfrac{1}{2x+3} = (2x+3)^{-1}$, $f'(x) = -2(2x+3)^{-2}$, $f''(x) = 8(2x+3)^{-3}$

$$f'''(x) = -48(2x+3)^{-4}.$$

So $f(0) = \dfrac{1}{3}$, $f'(0) = -\dfrac{2}{9}$, $f''(0) = \dfrac{8}{27}$, $f'''(0) = -\dfrac{16}{27}$. Therefore,

$$P_3(x) = \frac{1}{3} - \frac{2}{9}x + \frac{\left(\frac{8}{27}\right)}{2!}x^2 + \frac{\left(-\frac{16}{27}\right)}{3!}x^3 = \frac{1}{3} - \frac{2}{9}x + \frac{4}{27}x^2 - \frac{8}{81}x^3.$$

21. $f(x) = \dfrac{1}{1+x} = (1+x)^{-1}$, $f'(x) = -(1+x)^{-2}$, $f''(x) = 2(1+x)^{-3}$,

$f'''(x) = -6(1+x)^{-4}$, $f^{(4)}(x) = 24(1+x)^{-5}$. So

$f(0) = 1$, $f'(0) = -1$, $f''(0) = 2$, $f'''(0) = -6$, $f^{(4)} = 24$.

Therefore, $P_n(x) = 1 - x + x^2 - x^3 + \cdots + (-1)^n x^n$.

In particular, $P_4(x) = 1 - x + x^2 - x^3 + x^4$ and so

$P_4(0.1) = 1 - 0.1 + 0.01 - 0.001 + 0.0001 = 0.9091.$

$$f(0.1) = \frac{1}{1.1} \approx 0.909090\ldots \ .$$

23. $f(x) = e^{-x/2}$, $f'(x) = -\frac{1}{2}e^{-x/2}$, $f''(x) = \frac{1}{4}e^{-x/2}$, $f'''(x) = -\frac{1}{8}e^{-x/2}$

$f^{(4)}(x) = \frac{1}{16}e^{-x/2}.$

$f(0) = 1$, $f'(0) = -\frac{1}{2}$, $f''(0) = \frac{1}{4}$, $f'''(0) = -\frac{1}{8}$, $f^{(4)}(0) = \frac{1}{16}$.

Therefore, $P_4(x) = 1 - \frac{1}{2}x + \frac{1}{8}x^2 - \frac{1}{48}x^3 + \frac{1}{384}x^4$.

Since $f(x) = e^{-x/2}$, we let $x = 0.02$ to obtain $e^{-0.1}$. So

$e^{-0.1} \approx P(0.2) = 1 - \frac{1}{2}(0.2) + \frac{1}{8}(0.2)^2 - \frac{1}{48}(0.2)^3 + \frac{1}{384}(0.2)^4 \approx 0.90484$.

25. $f(x) = x^{1/2}$, $f'(x) = \frac{1}{2}x^{-1/2}$, $f''(x) = -\frac{1}{4}x^{-3/2}$. Therefore,

$\qquad f(16) = 4$, $f'(16) = \frac{1}{8}$, $f''(16) = -\frac{1}{256}$. Therefore,

$P_2(x) = 4 + \frac{1}{8}(x - 16) - \frac{1}{512}(x - 16)^2$

$\sqrt{15.6} \approx f(15.6) \approx 3.94969$.

27. $f(x) = \ln(x + 1)$.

We first obtain $P_3(x) = x - \frac{1}{2}x^2 + \frac{1}{3}x^3$.

Next, $\displaystyle\int_0^{1/2} \ln(x+1)\,dx = \int_0^{1/2} (x - \frac{1}{2}x^2 + \frac{1}{3}x^3)\,dx = \frac{1}{2}x^2 - \frac{1}{6}x^3 + \frac{1}{12}x^4 \Big|_0^{1/2}$

$\qquad\qquad\qquad = \frac{1}{8} - \frac{1}{48} + \frac{1}{192} \approx 0.109$.

Actual value: $\displaystyle\int_0^{1/2} \ln(x+1)\,dx = (x+1)\ln(x+1) - (x+1) \Big|_0^{1/2}$

$\qquad\qquad\qquad\qquad = \frac{3}{2}\ln\frac{3}{2} - \frac{3}{2} + 1 = \frac{3}{2}\ln\frac{3}{2} - \frac{1}{2} \approx 0.108$.

29. $f(x) = x^{1/2}$, $f'(x) = \frac{1}{2}x^{-1/2}$, $f''(x) = -\frac{1}{4}x^{-3/2}$, $f'''(x) = \frac{3}{8}x^{-5/2}$

$f(4) = 2$, $f'(4) = \frac{1}{4}$, $f''(4) = -\frac{1}{32}$.

$P_2(x) = 2 + \frac{1}{4}(x - 4) - \frac{1}{64}(x - 4)^2$.

$\sqrt{4.06} = f(4.06) = P_2(4.06) = 2 + \frac{1}{4}(0.06) - \frac{1}{64}(0.06)^2 = 2.01494375$.

To find a bound for the approximation, observe that $f'''(x) = \dfrac{3}{8x^{5/2}}$ is

decreasing on $[4, 4.06]$. Therefore, f''' attains the absolute maximum value at

$x = 4$. So $|f'''(x)| \le \dfrac{3}{8(4)^{5/2}} = \dfrac{3}{256}$. Therefore,

$\qquad\qquad |R_2(4.06)| \le \dfrac{\frac{3}{256}}{3!}(0.06)^3 \approx 0.000000421$.

31. $f(x) = (1-x)^{-1}$, $f'(x) = (1-x)^{-2}$, $f''(x) = 2(1-x)^{-3}$,

$\qquad f'''(x) = 6(1-x)^{-4}$, $f^{(4)}(x) = 24(1-x)^{-5}$, $f^{(5)} = 120(1-x)^{-6} > 0$

So $f(0) = 1$, $f'(0) = 1$, $f''(0) = 2$, $f'''(0) = 6$, $f^{(4)}(0) = 24$, and $P_3(x) = 1 + x + x^2 + x^3$. Therefore,

$$f(0.2) \approx P_3(0.2) = 1 + 0.2 + 0.04 + 0.008 = 1.248.$$

Since $f^{(5)}(x) > 0$, $f^{(4)}(x)$ is increasing and the maximum is

$$M = \frac{24}{(1-0.2)^5} = 73.24 < 74.$$

Then the error bound is $\dfrac{74}{4!}(0.2)^4 = 0.00493$.

The exact value of $f(0.2)$ is $f(0.2) = \dfrac{1}{1-0.2} = 1.25$.

33. $f(x) = \ln x$, $f'(x) = \dfrac{1}{x}$, $f''(x) = -\dfrac{1}{x^2}$, $f'''(x) = \dfrac{2}{x^3}$, $f^{(4)} = -\dfrac{6}{x^4}$.

$f(1) = 0$, $f'(1) = 1$, $f''(1) = -1$ and $P_2(x) = x - 1 - \frac{1}{2}(x-1)^2$.

Therefore, $f(1.1) = 1.1 - 1 - \frac{1}{2}(0.1)^2 = 0.095$.

Since $f^{(4)}(x) < 0$, $f^{(4)}(x)$ is decreasing and $M = \dfrac{2}{x^3}\bigg|_{x=1}$,

we see that $M = 2$. Therefore, the error bound is $\dfrac{2}{3!}(0.1)^3 = 0.00033$.

35. a. $f(x) = e^{-x}$, $f'(x) = -e^{-x}$, $f''(x) = e^{-x}$, $f'''(x) = -e^{-x}$, $f^{(4)}(x) = e^{-x}$.
Since $f^{(5)}(x) = -e^{-x} < 0$, $f^{(4)}(x)$ is decreasing and the maximum value of $f^{(4)}(x)$ is attained at $x = 0$. So, we may take

$$M = \max_{0 \le x \le 1} |f^{(4)}(x)| = e^0 = 1.$$

Therefore, a bound on the approximation of $f(x)$ by $P_3(x)$ is

$$\tfrac{1}{4!}(1-0)^4 = 0.04167.$$

b. $\displaystyle\int_0^1 e^{-x}\,dx \approx \int_0^1 P_3(x)\,dx = \int_0^1 (1 - x + \tfrac{1}{2}x^2 - \tfrac{1}{6}x^3)\,dx$ (See Problem 1 for $P_3(x)$)

$$= x - \tfrac{1}{2}x^2 + \tfrac{1}{6}x^3 - \tfrac{1}{24}x^4\bigg|_0^1 = 1 - \tfrac{1}{2} + \tfrac{1}{6} - \tfrac{1}{24} = 0.625.$$

c. The error bound on the approximation in (b) is

11 *Taylor Polynomials*

$$\int_0^1 0.04167\,dx = 0.04167x\Big|_0^1 = 0.04167 \qquad \text{(using the result of (a))}$$

d. Since

$$\int_0^1 e^{-x}\,dx = -e^{-x}\Big|_0^1 = -e^{-1}+1 \approx 0.632121,$$

the actual error is $0.632121 - 0.625 = 0.007121$.

37. The required area is $A = \displaystyle\int_0^{0.5} e^{-x^2/2}\,dx$.

Now $e^u = 1 + u + \frac{1}{2}u^2$ is the second Taylor polynomial at $u = 0$.

Therefore, $e^{-x^2/2} \approx 1 - \frac{1}{2}x^2 + \frac{1}{2}(-\frac{x^2}{2})^2 = 1 - \frac{1}{2}x^2 + \frac{1}{8}x^4$

is the fourth Taylor polynomial at $x = 0$. So

$$A = \int_0^{0.5} e^{-x^2/2}\,dx = \int_0^{0.5}\left(1 - \tfrac{1}{2}x^2 + \tfrac{1}{8}x^4\right)dx = \left(x - \tfrac{1}{6}x^3 + \tfrac{1}{40}x^5\right)\Big|_0^{0.5}$$

$$= 0.5 - \tfrac{1}{6}(0.5)^3 + \tfrac{1}{40}(0.5)^5 \approx 0.47995,$$

or approximately 0.48 square units.

39. The percentage of the nonfarm work force in the service industries t decades from now is given by $P(t) = \displaystyle\int 6e^{1/(2t+1)}\,dt$.

Let $u = \dfrac{1}{2t+1}$, so that $2t+1 = \dfrac{1}{u}$ and $t = \dfrac{1}{2}\left(\dfrac{1}{u}-1\right)$. Therefore, $dt = -\dfrac{1}{2u^2}\,du$.

So we have

$$\tilde{P}(u) = 6\int e^u\left(\frac{du}{-2u^2}\right) = -3\int \frac{e^u}{u^2}\,du = -3\int \frac{1}{u^2}\left(1 + u + \frac{u^2}{2} + \frac{u^3}{6} + \frac{u^4}{24}\right)du$$

$$= -3\int\left(\frac{1}{u^2} + \frac{1}{u} + \frac{1}{2} + \frac{u}{6} + \frac{u^2}{24}\right)du = -3\left(-\frac{1}{u} + \ln u + \frac{1}{2}u + \frac{u^2}{12} + \frac{u^3}{72}\right) + C.$$

Therefore,

$$P(t) = -3\left[-(2t+1) + \ln\left(\frac{1}{2t+1}\right) + \frac{1}{2(2t+1)} + \frac{1}{12(2t+1)^2} + \frac{1}{72(2t+1)^3}\right] + C.$$

Using the condition $P(0) = 30$, we find $P(0) = -3(-1 + \ln 1 + \frac{1}{2} + \frac{1}{12} + \frac{1}{72}) = 30$,

or $C = 28.79$. So

$$P(t) = -3\left[-2(2t+1) + \ln\left(\frac{1}{2t+1}\right) + \frac{1}{2(2t+1)} + \frac{1}{12(2t+1)^2} + \frac{1}{72(2t+1)^3}\right] + 28.79.$$

Two decades from now, the percentage of nonfarm workers will be

$$P(2) \approx -3\left[-5 + \ln\left(\tfrac{1}{5}\right) + \tfrac{1}{10} + \tfrac{1}{300} + \tfrac{1}{9000}\right] + 28.79 \approx 48.31,$$

or approximately 48 percent.

41. The average enrollment between $t = 0$ and $t = 2$ is given by

$$A = \frac{1}{2}\int_0^2\left[-\frac{20,000}{\sqrt{1+0.2t}} + 21,000\right]dt = -10,000\int_0^2(1+0.2t)^{-1/2}\,dt + \frac{1}{2}(21,000t)\Big|_0^2$$

$$= -10,000\int_0^2(1+0.2t)^{-1/2}\,dt + 21,000.$$

To evaluate the above integral, we approximate the integrand by the second Taylor polynomial at $t = 0$. We compute

$f(t) = (1+0.2t)^{-1/2}$, $f'(t) = -0.1(1+0.2t)^{-3/2}$,

$f''(t) = 0.03(1+0.2t)^{-5/2}$,

$f(0) = 1$, $f'(0) = -0.1$, $f''(0) = 0.03$.

$P_2(t) = 1 - 0.1t + 0.015t^2$.

Therefore,

$$A = -10,000\int_0^2(1-0.1t+0.015t^2)\,dt + 21,000$$

$$= -10,000(t - 0.05t^2 + 0.005t^3)\Big|_0^2 + 21,000$$

$$= -10,000(2 - 0.2 + 0.04) + 21,000 = 2600.$$

So the average enrollment between $t = 0$ and $t = 2$ is 2600.

43. False. The fourth Taylor polynomial $P_4(x)$ coincides with $f(x)$ for all values of x in this case. To see this, note that $R_4(x) = \dfrac{f^{(5)}(c)}{5!}(x-a)^5 = 0$ since $f^{(5)}(x) = 0$.

45. True. Since $f^{(n+1)}(x) = 0$ for all values of x, it follows that

$$R_n(x) = \frac{f^{(n+1)}(c)}{(n+1)!}(x-a)^{n+1} = 0$$

for all values of x.

1. 1, 2, 4, 8, 16.

3. $\dfrac{1-1}{1+1}, \dfrac{2-1}{2+1}, \dfrac{3-1}{3+1}, \dfrac{4-1}{4+1}, \dfrac{5-1}{5+1}$, or $0, \dfrac{1}{3}, \dfrac{2}{4}, \dfrac{3}{5}, \dfrac{4}{6}$.

5. $a_1 = \dfrac{2^0}{1!} = 1,\ a_2 = \dfrac{2^1}{2!} = 1,\ a_3 = \dfrac{2^2}{3!} = \dfrac{4}{6},\ a_4 = \dfrac{2^3}{4!} = \dfrac{8}{24},\ a_5 = \dfrac{2^4}{5!} = \dfrac{16}{120}$.

7. $a_1 = e,\ a_2 = \dfrac{e^2}{8},\ a_3 = \dfrac{e^3}{27},\ a_4 = \dfrac{e^4}{64},\ a_5 = \dfrac{e^5}{125}$.

9. $a_1 = \dfrac{3-1+1}{2+1} = 1,\ a_2 = \dfrac{12-2+1}{8+1} = \dfrac{11}{9},\ a_3 = \dfrac{27-3+1}{18+1} = \dfrac{25}{19}$,

$a_4 = \dfrac{48-4+1}{32+1} = \dfrac{45}{33} = \dfrac{15}{11},\ a_5 = \dfrac{75-5+1}{50+1} = \dfrac{71}{51}$.

11. $a_n = 3n - 2$ 13. $a_n = \dfrac{1}{n^3}$ 15. $a_n = 2\left(\dfrac{4}{5}\right)^{n-1} = \dfrac{2^{2n-1}}{5^{n-1}}$.

17. $a_n = \dfrac{(-1)^{n+1}}{2^{n-1}}$ 19. $a_n = \dfrac{n}{(n+1)(n+2)}$ 21. $a_n = \dfrac{e^{n-1}}{(n-1)!}$

23.   25.

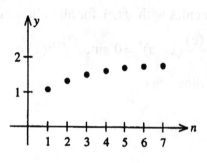   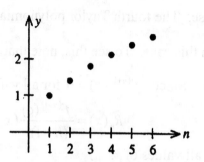

27.

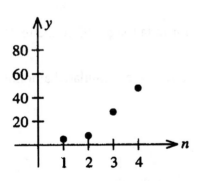

29.

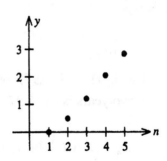

31. $\displaystyle\lim_{n\to\infty}\frac{n+1}{2n}=\lim_{n\to\infty}\frac{1+\frac{1}{n}}{2}=\frac{1}{2}.$

33. $\displaystyle\lim_{n\to\infty}\frac{(-1)^n}{\sqrt{n}}=0.$

35. $\displaystyle\lim_{n\to\infty}\frac{\sqrt{n}-1}{\sqrt{n}+1}=\lim_{n\to\infty}\frac{1-\dfrac{1}{\sqrt{n}}}{1+\dfrac{1}{\sqrt{n}}}=1.$

37. $\displaystyle\lim_{n\to\infty}\frac{2n^3-1}{n^3+2n+1}=\lim_{n\to\infty}\frac{2-\dfrac{1}{n^3}}{1+\dfrac{2}{n^2}+\dfrac{1}{n^3}}=2.$

39. $\displaystyle\lim_{n\to\infty}\left(2-\frac{1}{2^n}\right)=\lim_{n\to\infty}2-\lim_{n\to\infty}\frac{1}{2^n}=2-0=2.$

41. $\displaystyle\lim_{n\to\infty}\frac{2^n}{3^n}=\lim_{n\to\infty}\left(\frac{2}{3}\right)^n=0.$

43. $\displaystyle\lim_{n\to\infty}\frac{n}{\sqrt{2n^2+3}}=\lim_{n\to\infty}\frac{1}{\sqrt{2+\dfrac{3}{n^2}}}=\frac{1}{\sqrt{2}}=\frac{\sqrt{2}}{2}.$

45. a. $a_1=0.015,\ a_{10}=0.140,\ a_{100}=0.77939,\ a_{1000}=0.999999727$

b. $\lim_{n\to\infty} a_n = \lim_{n\to\infty}[1-(0.985)^n] = 1 - \lim_{n\to\infty}(0.985)^n = 1 - 0 = 1.$

47. a. The result follows from the compound interest formula (page 393 in the text).

b. $a_n = 100(1.01)^n$

c. $a_{24} = 100(1.01)^{24} = 126.97,$ or $126.97. Therefore, the accumulated amount at the end of two years is $126.97.

49. True. $\lim_{n\to\infty} a_n b_n = \left(\lim_{n\to\infty} a_n\right)\left(\lim_{n\to\infty} b_n\right) = (L)(0) = 0.$

51. False. Let $\lim_{n\to\infty} a_n = (-1)^n$ and $b_n = 1 + \dfrac{1}{n}.$ Then $|a_n| \le 1$ for all n and

$\lim_{n\to\infty} b_n = \lim_{n\to\infty}\left(1 + \dfrac{1}{n}\right) = 1.$ But $\lim_{n\to\infty} a_n b_n = \lim_{n\to\infty}(-1)^n\left(1 + \dfrac{1}{n}\right)$ does not exist

since if n is very large, $a_n b_n$ is close to 1 if n is even, but close to -1 if n is odd.

Therefore, $a_n b_n$ cannot approach a specific number as n approaches infinity.

EXERCISES 11.3, page 805

1. $S_1 = -2,\ S_2 = -2 + 4 = 2,\ S_3 = -2 + 4 - 8 = -6,\ S_4 = -6 + 16 = 10,$ and

so $\lim_{N\to\infty} S_N$ does not exist and the series $\displaystyle\sum_{n=1}^{\infty}(-2)^n$ is divergent.

3. $S_N = \displaystyle\sum_{n=1}^{N}\dfrac{1}{n^2 + 3n + 2} = \sum_{n=1}^{N}\left(\dfrac{1}{n+1} - \dfrac{1}{n+2}\right)$

$= \left(\dfrac{1}{2} - \dfrac{1}{3}\right) + \left(\dfrac{1}{3} - \dfrac{1}{4}\right) + \cdots + \left(\dfrac{1}{N} - \dfrac{1}{N+1}\right) + \left(\dfrac{1}{N+1} - \dfrac{1}{N+2}\right)$

$= \dfrac{1}{2} - \dfrac{1}{N+2}.$

So $\lim_{N\to\infty} S_N = \lim_{N\to\infty}\left(\dfrac{1}{2} - \dfrac{1}{N+2}\right) = \dfrac{1}{2}$ and $\displaystyle\sum_{n=1}^{\infty}\dfrac{1}{n^2 + 3n + 2} = \dfrac{1}{2}.$

5. $\displaystyle\sum_{n=0}^{\infty}\left(\dfrac{1}{3}\right)^n = 1 + \dfrac{1}{3} + \left(\dfrac{1}{3}\right)^2 + \cdots$

$$= \frac{1}{1-\frac{1}{3}} = \frac{1}{\frac{2}{3}} = \frac{3}{2}.$$

7. This is a geometric series with $r = 1.01 > 1$ and so it diverges.

9. $\displaystyle\sum_{n=0}^{\infty} \frac{(-2)^n}{3^n} = \sum_{n=0}^{\infty}\left(-\frac{2}{3}\right)^n$ is a geometric series with $|r| = \left|-\frac{2}{3}\right| = \frac{2}{3} < 1$,

and so it converges. The sum is $\displaystyle\frac{1}{1-\left(-\frac{2}{3}\right)} = \frac{1}{1+\frac{2}{3}} = \frac{3}{5}.$

11. $\displaystyle\sum_{n=0}^{\infty} \frac{2^n}{3^{n+2}} = \sum_{n=0}^{\infty}\frac{1}{9}\left(\frac{2}{3}\right)^n = \frac{1}{9}\cdot\frac{1}{1-\frac{2}{3}} = \left(\frac{1}{9}\right)(3) = \frac{1}{3}.$

13. $\displaystyle\sum_{n=0}^{\infty} e^{-0.2n} = \sum_{n=0}^{\infty}\left(e^{-0.2}\right)^n = \sum_{n=0}^{\infty}\left(\frac{1}{e^{0.2}}\right)^n = \frac{1}{1-\frac{1}{e^{0.2}}} \approx 5.52.$

15. Here $r = \left(-\dfrac{3}{\pi}\right)$ and since $|r| = \left|-\dfrac{3}{\pi}\right| < 1$, the series converges. In fact,

$$\sum_{n=0}^{\infty}\left(-\frac{3}{\pi}\right)^n = \frac{1}{1-\left(-\dfrac{3}{\pi}\right)} = \frac{1}{1+\dfrac{3}{\pi}} = \frac{1}{\dfrac{\pi+3}{\pi}} = \frac{\pi}{\pi+3}.$$

17. $8+4+\dfrac{1}{2}\left[1+\dfrac{1}{2}+\left(\dfrac{1}{2}\right)^2+\cdots\right] = 12 + \dfrac{1}{2}\left(\dfrac{1}{1-\frac{1}{2}}\right) = 13.$

19. $3-\dfrac{1}{3}\left[1-\dfrac{1}{3}+\left(-\dfrac{1}{3}\right)^2-\cdots\right] = 3-\dfrac{1}{3}\cdot\dfrac{1}{1+\dfrac{1}{3}} = 3-\dfrac{1}{3}\cdot\dfrac{3}{4} = 3-\dfrac{1}{4} = \dfrac{11}{4}.$

21. $\displaystyle\sum_{n=0}^{\infty} \frac{3+2^n}{3^n} = 3\sum_{n=0}^{\infty}\left(\frac{1}{3^n}\right) + \sum_{n=0}^{\infty}\left(\frac{2}{3}\right)^n = 3\cdot\frac{1}{1-\frac{1}{3}} + \frac{1}{1-\frac{2}{3}} = \frac{9}{2} + 3 = \frac{15}{2}.$

23. $\displaystyle\sum_{n=0}^{\infty} \frac{3 \cdot 2^n + 4^n}{3^n} = \sum_{n=0}^{\infty}\left[3\left(\tfrac{2}{3}\right)^n + \left(\tfrac{4}{3}\right)^n\right]$ diverges because the series

$\displaystyle\sum_{n=0}^{\infty}\left(\tfrac{4}{3}\right)^n$ is geometric with constant ratio $\tfrac{4}{3} > 1$.

25. Since $\dfrac{e}{\pi} < 1$, $\displaystyle\sum_{n=1}^{\infty}\left(\dfrac{e}{\pi}\right)^n$ converges. Also, $\dfrac{\pi}{e^2} < 1$ and so $\displaystyle\sum_{n=1}^{\infty}\left(\dfrac{\pi}{e^2}\right)^n$ converges.

Therefore, $\displaystyle\sum_{n=1}^{\infty}\left[\left(\dfrac{e}{\pi}\right)^n + \left(\dfrac{\pi}{e^2}\right)^n\right] = \sum_{n=1}^{\infty}\left(\dfrac{e}{\pi}\right)^n + \sum_{n=1}^{\infty}\left(\dfrac{\pi}{e^2}\right)^n$

$$= \frac{e}{\pi} \cdot \frac{1}{1 - \dfrac{e}{\pi}} + \frac{\pi}{e^2} \cdot \frac{1}{1 - \dfrac{\pi}{e^2}}$$

$$= \frac{e}{\pi} \cdot \frac{\pi}{\pi - e} + \frac{\pi}{e^2} \cdot \frac{e^2}{e^2 - \pi}$$

$$= \frac{e}{\pi - e} + \frac{\pi}{e^2 - \pi} = \frac{e^3 - \pi e + \pi^2 - \pi e}{(\pi - e)(e^2 - \pi)}$$

$$= \frac{e^3 - 2\pi e + \pi^2}{(\pi - e)(e^2 - \pi)}$$

27. $0.3333\ldots = 0.3 + 0.03 + 0.003 + \cdots = \dfrac{3}{10} + \dfrac{3}{10^2} + \dfrac{3}{10^3} + \cdots$

$$= \frac{3}{10}\sum_{n=0}^{\infty}\left(\frac{1}{10}\right)^n = \frac{\frac{3}{10}}{1 - \frac{1}{10}} = \frac{3}{10} \cdot \frac{10}{9} = \frac{1}{3}.$$

29. $1.213213213 = 1 + 0.213\left[1 + (0.0001) + (0.001)^2 + \cdots\right]$

$$= 1 + \frac{213}{1000} \cdot \frac{1}{1 - 0.001} = 1 + \frac{213}{1000} \cdot \frac{1000}{999} = 1 + \frac{213}{999} = \frac{1212}{999}$$

$$= \frac{404}{333}.$$

31. The given series is geometric with $r = (-x)^n$. Therefore, it converges if $|-x| < 1$, $|x| < 1$, or $-1 < x < 1$. Furthermore, $\displaystyle\sum_{n=0}^{\infty}(-x)^n = \dfrac{1}{1-(-x)} = \dfrac{1}{1+x}$.

33. The given series is geometric with $r = 2(x-1)$ and so it converges provided
$$|2(x-1)| < 1, |x-1| < \frac{1}{2}, \ -\frac{1}{2} < x-1 < \frac{1}{2}, \ \text{ or } \ \frac{1}{2} < x < \frac{3}{2}. \text{ Next,}$$
$$\sum_{n=1}^{\infty} 2^n (x-1)^n = 2(x-1)\cdot\frac{1}{1-2(x-1)} = 2(x-1)\cdot\frac{1}{3-2x} = \frac{2(x-1)}{3-2x}.$$

35. The additional spending generated by the proposed tax cut will be
$$(0.91)(30) + (0.91)^2(30) + (0.91)^3(30) + \cdots$$
$$= (0.91)(30)[1 + 0.91 + (0.91)^2 + (0.91)^3 + \cdots]$$
$$= 27.3\left[\frac{1}{1-0.91}\right] = 303.33, \ \text{ or approximately } \$303 \text{ billion.}$$

37. $p = \dfrac{1}{6} + \left(\dfrac{1}{6}\right)\left(\dfrac{5}{6}\right)^2 + \dfrac{1}{6}\left(\dfrac{5}{6}\right)^4 + \cdots$
$$= \frac{1}{6}\left\{1 + \left(\frac{5}{6}\right)^2 + \left[\left(\frac{5}{6}\right)^2\right]^2 + \cdots\right\} \qquad \left(r = \left(\frac{5}{6}\right)^2\right)$$
$$= \frac{1}{6}\cdot\frac{1}{1-\left(\dfrac{5}{6}\right)^2} = \frac{1}{6}\cdot\frac{1}{1-\dfrac{25}{36}}$$
$$= \frac{1}{6}\cdot\frac{36}{36-25} = \frac{6}{11}.$$

39. The required upper bound is no larger than
$$B = a_1 + aNa_1 + aNa_2 + aNa_3 + \cdots$$
$$= a_1 + aNa_1 + aN(ra_1) + aN(r^2 a_1) + \cdots. \qquad \text{(a geometric series)}$$

$$= a_1 + aNa_1(1+r+r^2+\cdots)$$
$$= a_1 + \frac{aNa_1}{1-r} = \frac{a_1 - a_1(1+aN-b)+aNa_1}{1-(1+aN-b)}$$
$$= \frac{a_1 b}{b-aN}.$$

41. $\quad A = Pe^{-r} + Pe^{-2r} + Pe^{-3r} + \cdots$
$$= Pe^{-r}(1+e^{-r}+e^{-2r}+\cdots)$$
$$= \frac{Pe^{-r}}{1-e^{-r}} = \frac{P}{e^{r}-1}.$$

43. b. $\quad C + R \leq S$
$$C + \frac{Ce^{-kt}}{1-e^{-kt}} \leq S, \quad \frac{C-Ce^{-kt}+Ce^{-kt}}{1-e^{-kt}} \leq S, \quad \frac{C}{1-e^{-kt}} \leq S$$
$$1-e^{-kt} \geq \frac{C}{S}, \quad e^{-kt} \leq 1 - \frac{C}{S} = \frac{S-C}{S}.$$
$$-kt \leq \ln\frac{S-C}{S}, \quad kt \geq -\ln\frac{S-C}{S} = \ln\frac{S}{S-C}$$
or $\quad t \geq \frac{1}{k}\ln\frac{S}{S-C}.$

Therefore, the minimum time between dosages should be
$$\frac{1}{k}\ln\frac{S}{S-C} \text{ hours.}$$

45. False. Take $a_n = -n$ and $b_n = n$. Then both $\displaystyle\sum_{n=0}^{\infty} a_n$ and $\displaystyle\sum_{n=0}^{\infty} b_n$ diverge but

$$\sum_{n=0}^{\infty}(a_n + b_n) = \sum_{n=0}^{\infty} 0 \quad \text{clearly converges to zero.}$$

47. True. This is a convergent geometric series with common ratio $|r|$. If $|r| < 1$, then

$$\sum_{n=0}^{\infty}|r|^n = \frac{1}{1-|r|}.$$

1. Here $a_n = \dfrac{n}{n+1}$ and since $\lim\limits_{n\to\infty} \dfrac{n}{n+1} = 1 \neq 0$, the series is divergent.

3. Here $a_n = \dfrac{2n}{3n+1}$ and $\lim\limits_{n\to\infty} \dfrac{2n}{3n+1} = \dfrac{2}{3} \neq 0$, and so the series diverges.

5. Here $a_n = 2(1.5)^n$ and $\lim\limits_{n\to\infty} 2(1.5)^n = \infty$, and so the series diverges.

7. Here $a_n = \dfrac{1}{2+3^{-n}}$, and $\lim\limits_{n\to\infty} \dfrac{1}{2+3^{-n}} = \dfrac{1}{2} \neq 0$, and so the series diverges.

9. Here $a_n = \left(-\dfrac{\pi}{3}\right)^n = (-1)^n \left(\dfrac{\pi}{3}\right)^n$. Because $\dfrac{\pi}{3} > 1$, we see that $\lim\limits_{n\to\infty} a_n$ does not exist, and so the series is divergent.

11. Take $f(x) = \dfrac{1}{x+1}$ and note that it is nonnegative and decreasing for $x \geq 1$. Next,

    $$\int_1^\infty \frac{1}{x+1}\,dx = \lim_{b\to\infty} \int_1^b \frac{1}{x+1}\,dx = \lim_{b\to\infty}\left[\ln(x+1)\big|_1^b\right] = \lim_{b\to\infty}\left[\ln(b+1) - \ln 2\right] = \infty$$

    and so the series is divergent.

13. First $f(x) = \dfrac{x}{2x^2+1}$ is nonnegative and decreasing for $x \geq 1$. Next,

    $$\int_1^\infty \frac{x}{2x^2+1}\,dx = \lim_{b\to\infty}\int_1^b \frac{x}{2x^2+1}\,dx = \lim_{b\to\infty}\left[\frac{1}{4}\ln(2x^2+1)\bigg|_1^b\right] = \lim_{b\to\infty}\left[\frac{1}{4}\ln(2b^2+1) - \frac{1}{4}\ln 3\right] = \infty$$

    and so the series is divergent.

15. Here $f(x) = xe^{-x}$ and nonnegative and decreasing for $x \geq 1$ (study its derivative).

    Next, $\displaystyle\int_1^\infty xe^{-x}\,dx = \lim_{b\to\infty}\int_1^b xe^{-x}\,dx = \lim_{b\to\infty}\left[-(x+1)e^{-x}\big|_1^b\right]$ (Integrate by parts)

    $$= \lim_{b\to\infty}\left[-(b+1)e^{-b} + 2e^{-b}\right] = 0$$

    [You can verify that be^{-b} approaches 0 by graphing it.]. So the integral converges.

17. Let $f(x) = \dfrac{x}{(x^2+1)^{3/2}}$. Then f is nonnegative and decreasing on $[1, \infty)$. Next,

$$\int_1^\infty \frac{x}{(x^2+1)^{3/2}}\,dx = \lim_{b\to\infty} \int_1^b x(x^2+1)^{-3/2}\,dx$$

$$= \lim_{b\to\infty}\left[\left(-\frac{1}{\sqrt{x^2+1}}\right)\Bigg|_1^b\right] \quad \text{[Use substitution.]}$$

$$= \lim_{b\to\infty}\left(-\frac{1}{\sqrt{b^2+1}}+\frac{1}{\sqrt{2}}\right) = \frac{1}{\sqrt{2}}$$

which converges and so the series converges.

19. Let $f(x) = \dfrac{1}{x\ln^3 x}$ which is nonnegative and decreasing on $[9,\infty)$. Next,

$$\int_9^\infty \frac{1}{x(\ln x)^3}\,dx = \lim_{b\to\infty}\int_9^b \frac{(\ln x)^{-3}}{x}\,dx$$

$$= \lim_{b\to\infty}\left[-\frac{1}{2(\ln x)^2}\Bigg|_9^b\right] \quad \text{[Use substitution with } u = \ln x.\text{]}$$

$$= \lim_{b\to\infty}\left[-\frac{1}{2(\ln b)^2}+\frac{1}{2(\ln 9)^2}\right] = \frac{1}{2(\ln 9)^2}$$

which is convergent, and so the series is convergent.

21. Here $p = 3 > 1$ and so the series is convergent.

23. Here $p = 1.01 > 1$, and so the series is convergent.

25. Here $p = \pi > 1$, and so the series is convergent.

27. Let $a_n = \dfrac{1}{2n^2+1}$. Then $a_n = \dfrac{1}{2n^2+1} < \dfrac{1}{2n^2} < \dfrac{1}{n^2} = b_n$. Since $\sum b_n$ is a convergent p-series, $\displaystyle\sum_{n=1}^\infty \frac{1}{2n^2+1}$ converges by the comparison test.

29. Let $a_n = \dfrac{1}{n-2}$. Then $a_n = \dfrac{1}{n-2} > \dfrac{1}{n} = b_n$.

Since $\sum b_n$ diverges, the comparison test implies that $\displaystyle\sum_{n=3}^\infty \frac{1}{n-2}$ diverges as well.

31. Let $a_n = \dfrac{1}{\sqrt{n^2-1}}$. Then $a_n = \dfrac{1}{\sqrt{n^2-1}} > \dfrac{1}{\sqrt{n^2}} = \dfrac{1}{n} = b_n$. Since the harmonic series $\sum b_n$ diverges, the comparison test shows that the given series also diverges.

33. Let $a_n = \dfrac{2^n}{3^n+1} < \dfrac{2^n}{3^n} = \left(\dfrac{2}{3}\right)^n = b_n$. Since $\displaystyle\sum_{n=0}^{\infty}\left(\dfrac{2}{3}\right)^n$ is a convergent geometric series, $\displaystyle\sum_{n=0}^{\infty} a_n$ converges by the comparison test.

35. Let $a_n = \dfrac{\ln n}{n}$. Since $\ln n > 1$ if $n > 3$, we see that $a_n = \dfrac{\ln n}{n} > \dfrac{1}{n} = b_n$. But $\sum b_n$ is the divergent harmonic series, and the comparison test implies that $\displaystyle\sum_{n=2}^{\infty}\dfrac{\ln n}{n}$ diverges as well.

37. We use the Integral Test with $f(x) = \dfrac{1}{\sqrt{x+1}}$ which is nonnegative and decreasing on $[0, \infty)$. We find

$$\int_0^{\infty}\dfrac{1}{\sqrt{x+1}}\,dx = \lim_{b\to\infty}\int_0^b (x+1)^{-1/2}\,dx = \lim_{b\to\infty}\left[2(x+1)^{1/2}\Big|_0^b\right]$$
$$= \lim_{b\to\infty}[2(b+1)^{1/2} - 2] = \infty$$

and so the series diverges.

39. Let $a_n = \dfrac{1}{n(\sqrt{n^2+1}}$. Observe that if n is large, n^2+1 behaves like n^2 and this suggests we compare $\sum a_n$ with $\sum b_n$ where $b_n = \dfrac{1}{n\sqrt{n^2}} = \dfrac{1}{n^2}$. Now $\sum b_n$ is a convergent p-series with $p = 2$. Since $0 < \dfrac{1}{n\sqrt{n^2+1}} < \dfrac{1}{n^2}$ the comparison test implies that the given series converges.

41. $\displaystyle\sum_{n=1}^{\infty}\dfrac{1}{n\sqrt{n}} = \sum_{n=1}^{\infty}\dfrac{1}{n^{3/2}}$ is a convergent p-series with $p = \dfrac{3}{2}$. Next, $\displaystyle\sum_{n=1}^{\infty}\dfrac{2}{n^2}$ is a

convergent p-series with $p = 2$. Therefore, $\displaystyle\sum_{n=1}^{\infty}\left(\frac{1}{n\sqrt{n}} + \frac{2}{n^2}\right)$ is convergent.

43. We know that $\ln n > 1$ if $n > 3$ and so $a_n = \dfrac{\ln n}{\sqrt{n}} > \dfrac{1}{\sqrt{n}} = b_n$ if $n > 3$. But

$\displaystyle\sum_{n=2}^{\infty} b_n = \sum_{n=2}^{\infty} \frac{1}{n^{1/2}}$ is a divergent p-series with $p = \dfrac{1}{2}$. Therefore, by the comparison

test, the given series is divergent.

45. We use the Integral Test, letting $f(x) = \dfrac{1}{x(\ln x)^2}$. Observe that f is nonnegative

and decreasing on $[2, \infty)$. Next, we compute

$$\int_2^{\infty} \frac{1}{x(\ln x)^2}\,dx = \lim_{b\to\infty}\int_2^b \frac{(\ln x)^{-2}}{x}\,dx$$

$$= \lim_{b\to\infty}\left[-\frac{1}{\ln x}\Big|_2^b\right] \qquad \text{(Use substitution with } u = \ln x)$$

$$= \lim_{b\to\infty}\left(-\frac{1}{\ln b} + \frac{1}{\ln 2}\right) = \frac{1}{\ln 2},$$

which converges, and so the given series converges.

47. We use the comparison test. Since $a_n = \dfrac{1}{\sqrt{n}+4} > \dfrac{1}{3\sqrt{n}} = b_n$ and $\displaystyle\sum_{n=1}^{\infty} b_n$ is

divergent, we conclude that $\displaystyle\sum_{n=1}^{\infty}\frac{1}{\sqrt{n}+4}$ is also divergent.

49. We use the Integral Test with $f(x) = \dfrac{1}{x(\ln x)^p}$. Observe that f is nonnegative and

decreasing on $[2, \infty)$. Next, we compute

$$\int_2^{\infty}\frac{1}{x(\ln x)^p}\,dx = \lim_{b\to\infty}\int_2^b \frac{(\ln x)^{-p}}{x}\,dx$$

$$= \lim_{b \to \infty} \left[\frac{(\ln x)^{-p+1}}{1-p} \Big|_2^b \right] \qquad \text{(Use substitution with } u = \ln x)$$

$$= \lim_{b \to \infty} \left[\frac{(\ln b)^{1-p}}{1-p} - \frac{(\ln 2)^{1-p}}{1-p} \right]$$

$$= \frac{(\ln 2)^{1-p}}{p-1} \qquad \text{if } p > 1.$$

If $p \le 1$, then the improper integral diverges. Therefore, $\displaystyle\sum_{n=2}^{\infty} \frac{1}{n(\ln n)^p}$ converges for $p > 1$.

51. Denoting the Nth partial sum of the series by S_N, we have

$$S_N = \sum_{n=1}^{N} \left(\frac{a}{n+1} - \frac{1}{n+2} \right) = \left(\frac{a}{2} - \frac{1}{3} \right) + \left(\frac{a}{3} - \frac{1}{4} \right) + \cdots + \left(\frac{a}{N+1} - \frac{1}{N+2} \right)$$

$$= \frac{a}{2} + \frac{a-1}{3} + \frac{a-1}{4} + \cdots + \frac{a-1}{N+1} - \frac{1}{N+2}.$$

If $a = 1$, then $S_N = \dfrac{1}{2} - \dfrac{1}{N+2}$ is the Nth partial sum of a telescoping series that converges to ½. If $a \ne 0$, then S_N is the Nth partial sum of a series akin to the harmonic series $\displaystyle\sum_{n=1}^{\infty} \frac{1}{n}$ and in this case, the series diverges. Therefore, the series converges only for $a = 1$.

53. $\displaystyle\int_1^{\infty} \frac{1}{x^p} dx = \lim_{b \to \infty} \int_1^b x^{-p} dx = \lim_{b \to \infty} \left[\frac{x^{1-p}}{1-p} \Big|_1^b \right] = \lim_{b \to \infty} \left[\frac{b^{1-p}}{1-p} - \frac{1}{1-p} \right] = \frac{1}{p-1}$ if $p > 1$

and diverges to infinity if $p < 1$. If $p = 1$, then we have

$$\int_1^{\infty} \frac{1}{x} dx = \lim_{b \to \infty} \int_1^b \frac{1}{x} dx = \lim_{b \to \infty} \left[\ln |x| \Big|_1^b \right] = \lim_{b \to \infty} (\ln b - \ln 1) = \infty$$

and so the integral diverges in this case as well.

55. True. Compare it with the divergent harmonic series $\displaystyle\sum_{n=1}^{\infty} \frac{1}{n}$. In fact, if $\displaystyle\sum_{n=1}^{\infty} \frac{x}{n}$

converges for $x \neq 0$. Then $\sum\limits_{n=1}^{\infty} \dfrac{x}{n} = x \sum\limits_{n=1}^{\infty} \dfrac{1}{n}$ converges, a contradiction.

57. False. The harmonic series $\sum\limits_{n=1}^{\infty} a_n$ with $a_n = \dfrac{1}{n}$ $(n = 1, 2, 3, \ldots)$ is divergent, but

$\lim\limits_{n \to \infty} a_n = 0$.

59. False. Let $a_n = \dfrac{1}{n^2}$ and $b_n = \dfrac{2}{n^2}$. Clearly $b_n \geq a_n$ $(n = 1, 2, 3, \ldots)$ but both

$\sum a_n$ and $\sum b_n$ converge.

EXERCISES 11.5, page 827

1. $R = \lim\limits_{n \to \infty} \left| \dfrac{a_n}{a_{n+1}} \right| = \lim\limits_{n \to \infty} 1 = 1$.   Therefore, $R = 1$ and the interval of convergence is $(0, 2)$.

3. $R = \lim\limits_{n \to \infty} \left| \dfrac{a_n}{a_{n+1}} \right| = \lim\limits_{n \to \infty} \dfrac{n^2}{(n+1)^2} = \lim\limits_{n \to \infty} \dfrac{n^2}{n^2 + 2n + 1} = 1; \ (-1,1)$

5. $R = \lim\limits_{n \to \infty} \left| \dfrac{a_n}{a_{n+1}} \right| = \lim\limits_{n \to \infty} \dfrac{\dfrac{1}{4^n}}{\dfrac{1}{4^{n+1}}} = \lim\limits_{n \to \infty} 4 = 4; \quad (-4,4)$

7. $R = \lim\limits_{n \to \infty} \left| \dfrac{a_n}{a_{n+1}} \right| = \lim\limits_{n \to \infty} \dfrac{1}{n!2^n}(n+1)!2^{n+1} = \lim\limits_{n \to \infty} 2(n+1) = \infty.$

    Therefore, $R = \infty$ and the interval of convergence is $(-\infty, \infty)$.

9. $R = \lim\limits_{n \to \infty} \left| \dfrac{a_n}{a_{n+1}} \right| = \lim\limits_{n \to \infty} \dfrac{\dfrac{n!}{2^n}}{\dfrac{(n+1)!}{2^{n+1}}} = \lim\limits_{n \to \infty} \dfrac{2}{n+1} = 0; \quad x = -2$

11. $R = \lim_{n \to \infty} \left| \dfrac{a_n}{a_{n+1}} \right| = \lim_{n \to \infty} \dfrac{\dfrac{1}{(n+1)^2}}{\dfrac{1}{(n+2)^2}} = \lim_{n \to \infty} \dfrac{n^2 + 4n + 4}{n^2 + 2n + 1} = 1; \quad (-4, -2)$

13. $R = \lim_{n \to \infty} \left| \dfrac{a_n}{a_{n+1}} \right| = \lim_{n \to \infty} \dfrac{2n(n+2)!}{(n+1)!(2n+2)} = \infty; \quad (-\infty, \infty)$

15. $\quad R = \lim_{n \to \infty} \left| \dfrac{a_n}{a_{n+1}} \right| = \lim_{n \to \infty} \dfrac{\dfrac{n}{n+1}}{\dfrac{n+1}{n+2}} = \lim_{n \to \infty} \dfrac{n(n+2)}{(n+1)^2} = \lim_{n \to \infty} \dfrac{n^2 + 2n}{n^2 + 2n + 1} = 1; \quad (-\tfrac{1}{2}, \tfrac{1}{2})$

17. $\quad R = \lim_{n \to \infty} \left| \dfrac{a_n}{a_{n+1}} \right| = \lim_{n \to \infty} \dfrac{\dfrac{n!}{3^n}}{\dfrac{(n+1)!}{3^{n+1}}} = \lim_{n \to \infty} \dfrac{3}{n+1} = 0; \quad x = -1$

19. $\quad R = \lim_{n \to \infty} \left| \dfrac{a_n}{a_{n+1}} \right| = \lim_{n \to \infty} \dfrac{n^3 (3^{n+1})}{3^n (n+1)^3} = \lim_{n \to \infty} \dfrac{3}{\left(\dfrac{n+1}{n} \right)^3} = 3; \quad (0, 6)$

21. $f(x) = x^{-1}, \; f'(x) = -x^{-2}, \; f''(x) = 2x^{-3}, \; f'''(x) = -3 \cdot 2x^{-4}, \; ...,$
$f^{(n)}(x) = (-1)^n n! x^{-n-1}.$
Therefore,
$$f(1) = 1, \; f'(1) = -1, \; f''(x) = 2, \; f'''(1) = -3!, \; ..., \; f^{(n)}(1) = (-1)^n n!,$$

and so
$$f(x) = 1 - (x-1) + \frac{2}{2!}(x-1)^2 + \cdots + \frac{(-1)^n n!}{n!}(x-1)^n + \cdots$$
$$= \sum_{n=0}^{\infty} (-1)^n (x-1)^n; \; R = 1$$
and the interval of convergence is $(0, 2)$.

23. $f(x) = (x+1)^{-1}, \; f'(x) = -(x+1)^{-2}, \; f''(x) = 2(x+1)^{-3}, f'''(x) = -3!(x+1)^{-4}, \; ...,$

$f^{(n)}(x) = (-1)^n n!(x+1)^{-n-1}$.

Therefore, $f(2) = \dfrac{1}{3}$, $f'(2) = -\dfrac{1}{3^2}$, $f''(2) = \dfrac{2}{3^3}$, ..., $f^{(n)} = \dfrac{(-1)^n n!}{3^{n+1}}$,

and so

$$f(x) = \frac{1}{3} - \frac{1}{3^2}(x-2) + \frac{1}{3^3}(x-2)^2 + \cdots + \frac{(-1)^n}{3^{n+1}}(x-2)^n + \cdots$$

$$= \sum_{n=0}^{\infty}(-1)^n \frac{(x-2)^n}{3^{n+1}}.$$

$R = 3$ and the interval of convergence is (-1,5).

25. $f(x) = (1-x)^{-1}$, $f'(x) = (1-x)^{-2}$, $f''(x) = 2(1-x)^{-3}$, $f'(x) = 6(1-x)^{-4}$, ...,
$f^{(n)}(x) = n!(1-x)^{-n-1}$.

So
$$f(2) = -1,\ f'(2) = 1,\ f''(2) = -2,\ f'''(2) = 6, ...,$$
$$f^{(n)}(2) = (-1)^{n+1} n!$$

Therefore,
$$f(x) = -1 + (x-2) - (x-2)^2 + \cdots + (-1)^{n+1}(x-2)^n + \cdots$$

$$= \sum_{n=0}^{\infty}(-1)^{n+1}(x-2)^n.$$

$R = 1$ and the interval of convergence is (1,3).

27. $f(x) = x^{1/2}$, $f'(x) = \dfrac{1}{2}x^{-1/2}$, $f''(x) = -\dfrac{1}{4}x^{-3/2}$, $f'''(x) = \dfrac{3}{8}x^{-5/2}$,

$f^{(4)}(x) = -\dfrac{3\cdot 5}{16}x^{-7/2}$,... . So

$f(1) = 1$, $f'(1) = \dfrac{1}{2}$, $f''(1) = -\dfrac{1}{2^2}$, $f'''(1) = \dfrac{1\cdot 3}{2^3}$, ...,

$f^{(n)}(1) = (-1)^{n+1}\dfrac{1\cdot 3\cdot 5\cdots(2n-3)}{2^n}$ $(n \geq 2)$

So $f(x) = 1 + \dfrac{1}{2}(x-1) + \sum_{n=2}^{\infty}(-1)^{n+1}\dfrac{1\cdot 3\cdot 5\cdots(2n-3)}{n!2^n}(x-1)^n$

$R = 1$ and the interval of convergence is (0,2).

29. $f(x) = e^{2x}$, $f'(x) = 2e^{2x}$, $f''(x) = 2^2 e^{2x}$, ..., $f^{(n)}(x) = 2^n e^{2x}$.

$f(0) = 1,\ f'(0) = 2,\ f''(0) = 2^2,\ ...,\ f^{(n)}(0) = 2^n.$

Therefore,

$$f(x) = 1 + 2x + 2^2 x^2 + \cdots + \frac{2^n}{n!} x^n + \cdots = \sum_{n=0}^{\infty} \frac{2^n}{n!} x^n.$$

$R = \infty$ and the interval of convergence is $(-\infty, \infty)$.

31. $f(x) = (x+1)^{-1/2},\ f'(x) = -\frac{1}{2}(x+1)^{-3/2},\ f''(x) = \frac{3}{4}(x+1)^{-5/2}.$

$f'''(x) = -\dfrac{1 \cdot 3 \cdot 5}{2^3} x^{-7/2},\ ...,\ f^{(n)}(x) = (-1)^n \dfrac{1 \cdot 3 \cdot 5 \cdots (2n-1)}{2^n} x^{-(2n+1)/2}.$

So $\displaystyle\sum_{n=0}^{\infty} (-1)^n \dfrac{1 \cdot 3 \cdot 5 \cdots (2n-1)}{n! 2^n} x^n.$

$R = 1$ and the interval of convergence is $(-1,1)$.

33. For a Taylor series, $S_n(x) = P_n(x)$. Furthermore, $P_n(x) = f(x) - R_n(x)$. Therefore,

$$\lim_{n \to \infty} S_n(x) = \lim_{n \to \infty} P_n(x) = \lim_{n \to \infty}[f(x) - R_n(x)] = f(x) - \lim_{n \to \infty} R_n(x).$$

In other words for fixed x, the sequence of partial sums of the Taylor series of f converges to f if and only if $\lim_{n \to \infty} R_n(x) = 0$.

35. True. This follows from Theorem 9.

EXERCISES 11.6, page 837

1. $f(x) = \dfrac{1}{1-x} = \dfrac{1}{-1-(x-2)} = -\dfrac{1}{1+(x-2)}.$

Now, use the fact that

$$\frac{1}{1+u} = 1 - u + u^2 - u^3 + \cdots = \sum_{n=0}^{\infty} (-1)^n u^n \qquad |u| < 1$$

with $u = (x - 2)$ to obtain

$$f(x) = \frac{1}{1-x} = \sum_{n=0}^{\infty} (-1)^{n+1}(x-2)^n$$

with an interval of convergence of $(1,3)$.

3. Let $u = 3x$ in the series

$$\frac{1}{1+u} = 1 - u + u^2 - u^3 + \cdots = \sum_{n=0}^{\infty} (-1)^n u^n \qquad |u| < 1$$

and we have

$$f(x) = \frac{1}{1+3x} = 1 - 3x + (3x)^2 - (3x)^3 + \cdots = \sum_{n=0}^{\infty} (-1)^n 3^n x^n \qquad |3x| < 1$$

with an interval of convergence of $(-\frac{1}{3}, \frac{1}{3})$.

5. $f(x) = \dfrac{1}{4 - 3x} = \dfrac{1}{4(1 - \frac{3x}{4})} = \dfrac{1}{4} \sum_{n=0}^{\infty} \left(\dfrac{3x}{4}\right)^n = \sum_{n=0}^{\infty} \dfrac{3^n}{4^{n+1}} x^n \qquad \left|\dfrac{3x}{4}\right| < 1$

with an interval of convergence of $(-\frac{4}{3}, \frac{4}{3})$.

7. $f(x) = \dfrac{1}{1 - x^2} = 1 + (x^2) + (x^2)^2 + (x^2)^3 + \cdots = 1 + x^2 + x^4 + x^6 + \cdots$

$$= \sum_{n=0}^{\infty} x^{2n}.$$

The interval of convergence is $(-1, 1)$.

9. $f(x) = e^{-x} = 1 + (-x) + \dfrac{(-x)^2}{2!} + \dfrac{(-x)^3}{3!} + \cdots$

$$= 1 - x + \frac{x^2}{2!} - \frac{x^3}{3!} + \cdots = \sum_{n=0}^{\infty} (-1)^n \frac{x^n}{n!}.$$

The interval of convergence is $(-\infty, \infty)$.

11. $f(x) = xe^{-x^2} = x\left[1 - (-x^2) + \dfrac{(-x^2)^2}{2!} + \dfrac{(-x^2)^3}{3!} + \cdots \right]$

$$= x\left(1 - x^2 + \frac{x^4}{2!} - \frac{x^6}{3!} + \cdots \right) = \sum_{n=0}^{\infty} (-1)^n \frac{x^{2n+1}}{n!}.$$

The interval of convergence is $(-\infty, \infty)$.

13. $f(x) = \dfrac{1}{2}(e^x + e^{-x}) = \dfrac{1}{2}\left[\left(1 + x + \dfrac{x^2}{2!} + \dfrac{x^3}{3!} + \cdots\right) + \left(1 - x + \dfrac{x^2}{2!} - \dfrac{x^3}{3!} + \cdots\right) \right]$

$$= 1 + \frac{x^2}{2!} + \frac{x^4}{4!} + \frac{x^6}{6!} + \cdots + \frac{x^{2n}}{2n!} + \cdots$$

$$= \sum_{n=0}^{\infty} \frac{x^{2n}}{2n!}$$

The interval of convergence is $(-\infty, \infty)$.

15. $\ln x = (x-1) - \frac{1}{2}(x-1)^2 + \frac{1}{3}(x-1)^3 + \cdots$

Replace x by $1 + 2x$ to obtain

$$f(x) = \ln(1+2x) = 2x - \frac{1}{2}(2x)^2 + \frac{1}{3}(2x)^3 - \cdots = \sum_{n=1}^{\infty} (-1)^{n-1} \cdot \frac{2^n x^n}{n}.$$

We must have $0 < 1 + 2x \le 2$ so $-1 < 2x \le 1$, or $-\frac{1}{2} < x \le \frac{1}{2}$.

So the interval of convergence is $(-\frac{1}{2}, \frac{1}{2}]$.

17. $f(x) = \ln(1+x^2) = x^2 - \frac{1}{2}x^4 + \frac{1}{3}x^6 - \cdots$

$$f(x) = x^2 - \frac{1}{2}x^4 + \frac{1}{3}x^6 + \cdots + \frac{(-1)^{n+1}}{n}x^{2n} + \cdots = \sum_{n=1}^{\infty} (-1)^{n+1} \frac{x^{2n}}{n};$$

The interval of convergence is $(-1,1)$.

19. Replace x by $x + 1$ in the formula for $\ln x$ in Table 11.1, giving

$$\ln(x+1) = x - \frac{1}{2}x^2 + \frac{1}{3}x^3 - \cdots \qquad (-1 < x \le 1).$$

Next, observe that $x = x - 1 + 2 = 2\left(1 + \frac{x-2}{2}\right)$. Therefore,

$$\ln x = \ln 2\left(1 + \frac{x-2}{2}\right) = \ln 2 + \ln\left(1 + \frac{x-2}{2}\right).$$

If we replace x in the expression for $\ln(x+1)$ by $\frac{x-2}{2}$, we obtain

$$\ln x = \ln 2 + \frac{x-2}{2} - \frac{1}{2}\left(\frac{x-2}{2}\right)^2 + \frac{1}{3}\left(\frac{x-2}{2}\right)^3 - \cdots$$

$$= \ln 2 + \sum_{n=1}^{\infty} (-1)^{n-1}\left(\frac{1}{n \cdot 2^n}\right)(x-2)^n.$$

Therefore, $f(x) = (x-2)\ln x = (x-2)\ln 2 + \sum_{n=1}^{\infty} (-1)^{n-1}\left(\frac{1}{n \cdot 2^n}\right)(x-2)^n$.

To find the interval of convergence, observe that x must satisfy

$$-1 < \frac{x-2}{2} < 1, \ -2 < x-2 \le 2, \text{ or } 0 < x \le 4.$$

So the interval of convergence is $(0,4]$.

21. Replacing x in the formula for $\ln x$ in Table 11.1 by $1+x$, we obtain

$$\ln(1+x) = x - \frac{1}{2}x^2 + \frac{1}{3}x^3 - \frac{1}{4}x^4 + \cdots$$

Differentiating, we obtain

$$\frac{1}{1+x} = 1 - x + x^2 - x^3 + \cdots + (-1)^n x^n + \cdots = \sum_{n=0}^{\infty} (-1)^n x^n.$$

23. $\dfrac{1}{1+x} = 1 - x + x^2 - x^3 + \cdots$

$$\int \frac{1}{1+x}dx = x - \frac{1}{2}x^2 + \frac{1}{3}x^3 - \frac{1}{4}x^4 + \cdots + \frac{(-1)^{n+1}}{n} + \cdots$$

Therefore,

$$f(x) = \ln(1+x) = x - \frac{1}{2}x^2 + \frac{1}{3}x^3 - \frac{1}{4}x^4 + \cdots + \frac{(-1)^{n+1}}{n}x^n + \cdots .$$

25. $\displaystyle\int_0^{0.5} \frac{1}{\sqrt{1+x^2}}dx = \int_0^{0.5}\left(1 - \frac{1}{2}x^2 + \frac{3}{8}x^4 - \frac{5}{16}x^6\right)dx$

$$= x - \frac{1}{6}x^3 + \frac{3}{40}x^5 - \frac{5}{112}x^7\Big|_0^{0.5} \approx 0.4812.$$

27. $\displaystyle\int_0^1 e^{-x^2}dx = \int_0^1\left(1 - x^2 + \frac{x^4}{2} - \frac{x^6}{6} + \frac{x^8}{24}\right)dx$

$$= x - \frac{1}{3}x^3 + \frac{1}{10}x^5 - \frac{1}{42}x^7 + \frac{1}{216}x^9\Big|_0^1 \approx 0.7475.$$

29. $\pi = 4\displaystyle\int_0^1 \frac{dx}{1+x^2} = 4\int_0^1 (1 - x^2 + x^4 - x^6 + x^8)dx$

$$= 4\left(x - \frac{1}{3}x^3 + \frac{1}{5}x^5 - \frac{1}{7}x^7 + \frac{1}{9}x^9\right)\Big|_0^1 = 4(1 - 0.333333 + 0.2 - 0.1428571 + 0.1111111)$$

$$\approx 3.34.$$

31. $P(28 \le x \le 32) = \dfrac{1}{10\sqrt{2\pi}} \displaystyle\int_{28}^{32} e^{(-1/2)[(x-30)/10]^2} \, dx$

$$= \frac{1}{10\sqrt{2\pi}} \int_{28}^{32}\left[1 - \frac{1}{2}\left(\frac{x-30}{10}\right)^2 + \frac{1}{8}\left(\frac{x-30}{10}\right)^4 - \frac{1}{48}\left(\frac{x-30}{10}\right)^6\right] dx$$

$$= \frac{1}{10\sqrt{2\pi}}\left[1 - \frac{1}{6}\left(\frac{x-30}{10}\right)^3 + \frac{1}{40}\left(\frac{x-30}{10}\right)^5 - \frac{1}{336}\left(\frac{x-30}{10}\right)^7\right]_{28}^{32}$$

$$\approx 15.85, \text{ or } 15.85 \text{ percent.}$$

EXERCISES 11.7, page 846

1. Take $f(x) = x^2 - 3$ so that $f'(x) = 2x$. We have

$$x_{n+1} = x_n - \frac{x_n^2 - 3}{2x_n} = \frac{x_n^2 + 3}{2x_n}.$$

With $x_0 = 1.5$, we find

$x_1 = 1.75$, $x_2 = 1.7321429$, $x_3 = 1.7320508$ and so $\sqrt{3} \approx 1.732051$.

3. $x_{n+1} = x_n - \dfrac{x_n^2 - 7}{2x_n} = \dfrac{x_n^2 + 7}{2x_n}$

$x_0 = 2.5$, $x_1 = 2.65$, $x_2 = 2.645755$, $x_3 = 2.645751$

Therefore, $\sqrt{7} \approx 2.64575$.

5. $x_{n+1} = x_n - \dfrac{x_n^3 - 14}{3x_n^2} = \dfrac{2x_n^3 + 14}{3x_n^2}$

$x_0 = 2.5$, $x_1 = 2.413333$, $x_2 = 2.410146$, $x_3 = 2.410142264$

Therefore, $\sqrt[3]{14} \approx 2.410142$.

7. $f(x) = x^2 - x - 3$, $f'(x) = 2x - 1$

$$x_{n+1} = x_n - \frac{x_n^2 - x_n - 3}{2x_n - 1} = \frac{2x_n^2 - x_n - x_n^2 + x_n + 3}{2x_n - 1} = \frac{x_n^2 + 3}{2x_n - 1},$$

and $x_0 = 2$, $x_1 = 2.33333$, $x_2 = 2.30303$, $x_3 = 2.30278$, and $x_4 = 2.30278$.

9. $x_{n+1} = x_n - \dfrac{x_n^3 + 2x_n^2 + x_n - 5}{3x_n^2 + 4x_n + 1} = \dfrac{2x_n^3 + 2x_n^2 + 5}{3x_n^3 + 4x_n + 1}$

$x_0 = 1$, $x_1 = 1.53333$, $x_2 = 1.19213$, $x_3 = 1.11949$, $x_4 = 1.11635$, $x_5 = 1.11634$,
$x_6 = 1.11634$.

11. $x_{n+1} = x_n - \dfrac{\sqrt{x_n + 1} - x_n}{\frac{1}{2}(x_n + 1)^{-1/2} - 1} = \dfrac{x_n + 2}{2\sqrt{x_n + 1} - 1}$ (upon simplification)

$x_0 = 1$, $x_2 = 1.640764$, $x_2 = 1.618056$, $x_3 = 1.618034$.
Therefore, the zero is approximately 1.61803.

13. $x_{n+1} = x_n - \dfrac{e^{x_n} - \dfrac{1}{x_n}}{e^{x_n} + \dfrac{1}{x_n^2}} = \dfrac{x_n^2 e^{x_n}(x_n - 1) + 2x_n}{x_n^2 e^{x_n} + 1}$ (upon simplification)

$x_0 = 0.5$, $x_1 = 0.562187$, $x_2 = 0.56712$, $x_3 = 0.567143$.
Therefore, the zero is approximatley 0.5671.

15. a. $f(0) = -2$ and $f(1) = 3$. Since $f(x)$ is a polynomial, it is continuous.
Furthermore, since f changes sign between $x = 0$ and $x = 1$, we conclude that f has a root in the interval $(0,1)$.
b. $f(x) = 2x^3 - 9x^2 + 12x - 2$ and $f'(x) = 6x^2 - 18x + 12$.

$$x_{n+1} = x_n - \frac{2x_n^3 - 9x_n^2 + 12x_n - 2}{6x_n^2 - 18x_n + 12} = \frac{4x_n^3 - 9x_n^2 + 2}{6x_n^2 - 18x_n + 12}$$

$x_0 = 0.5$, $x_1 = \dfrac{0.25}{4.5} = 0.055556$, $x_2 = \dfrac{1.972908}{11.0185108} \approx 0.193556$.

$$x_3 = \frac{1.734419}{8.969390} \approx 0.193371, \quad x_4 = \frac{1.692391}{0.193371} \approx 0.179054.$$

$$x_5 = \frac{1.691830}{0.193556} \approx 0.193556, \text{ or } 0.19356.$$

17. a. $f(1) = -3$ and $f(2) = 1.$ Since $f(x)$ has opposite signs at $x = 1$ and $x = 2$, we see that the continuous function f has at least one zero in the interval $(1,2)$.

   b. $f'(x) = 3x^2 - 3,$ and the required iteration formula is

   $$x_{n+1} = x_n - \frac{x_n^3 - 3x_n - 1}{3x_n^2 - 3} = \frac{3x_n^3 - 3x_n - x_n^3 + 3x_n + 1}{3x_n^2 - 3} = \frac{2x_n^3 + 1}{3x_n^2 - 3}.$$

   With $x_0 = 1.5,$ we find

   $x_1 = 2.066667,$ $x_2 = 1.900876,$ $x_3 = 1.879720,$ $x_4 = 1.879385,$
   $x_5 = 1.879385,$

   and the required root is 1.87939.

19. Let $F(x) = 2\sqrt{x+3} - 2x + 1.$ To solve $F(x) = 0$, we use the iteration

   $$x_{n+1} = x_n - \frac{2\sqrt{x_n + 3} - 2x_n + 1}{(x_n + 3)^{-1/2} - 2}$$

   $x_0 = 3,$ $x_1 = 2.93654,$ $x_2 = 2.93649,$ and $x_3 = 2.9365.$

21. $F(x) = e^{-x} - x + 1,$ $F'(x) = -e^{-x} - 1.$

   Therefore, $x_{n+1} = x_n + \dfrac{e^{-x_n} - x_n + 1}{e^{-x_n} + 1}$

   $x_0 = 1.2,$ $x_1 = 1.2 + \dfrac{0.1011942}{1.3011942} \approx 1.2777703,$ $x_2 = 1.277703 + \dfrac{0.0000955}{1.2786767} \approx 1.2784499,$

   or approximately 1.2785.

   $x_3 \approx 1.2784499 + \dfrac{0.0000187}{1.2784686} \approx 1.2784645,$ or approximately 1.2785.

23. To solve $F(x) = e^{-x} - \sqrt{x} = 0,$ we use the iteration

   $$x_{n+1} = x_n - \frac{e^{-x_n} - x_n^{1/2}}{-e^{-x_n} - \frac{1}{2}x^{-1/2}} = x_n + \frac{2\sqrt{x_n}\,e^{-x_n} - 2x_n}{2\sqrt{x_n}\,e^{-x_n} + 1}$$

   $x_0 = 1,$ $x_1 = 0.271649,$ $x_2 = 0.411602,$ $x_3 = 0.426303,$ $x_4 = 0.426303.$

25. The daily average cost is $\overline{C}(x) = \dfrac{C(x)}{x} = 0.0002x^2 - 0.06x + 120 + \dfrac{5000}{x}$.

To find the minimum of $\overline{C}(x)$, we set

$$\overline{C}'(x) = 0.0004x - 0.06 - \frac{5000}{x^2} = 0$$

obtaining $0.0004x^3 - 0.06x^2 - 5000 = 0$, or $x^3 - 150x^2 - 12,500,000 = 0$.
Write $f(x) = x^3 - 150x^2 - 12,500,000$ and observe that $f(0) < 0$, whereas $f(500) > 0$. So the root (critical point of $\overline{C}$) lies between $x = 0$ and $x = 500$.
Take $x_0 = 250$ and use the iteration

$$x_{n+1} = x_n - \frac{x_n^3 - 150x_n^2 - 12,500,000}{3x_n^2 - 300x_n} = \frac{2x_n^3 - 150x_n^2 + 12,500,000}{3x_n^2 - 300x_n}.$$

We find

$x_0 = 250$, $x_1 = 305.556$, $x_2 = 294.818$, $x_3 = 294.312$, and $x_4 = 294.311$.

You can show that $\overline{C}(x)$ is concave upward on $(0, \infty)$. So the level of production that minimizes $\overline{C}(x)$ is 294 units/day.

27. We solve the equation $f(t) = 0.05t^3 - 0.4t^2 - 3.8t - 15.6 = 0$.
Use the iteration

$$t_{n+1} = t_n - \frac{0.05t_n^3 - 0.4t_n^2 - 3.8t_n - 15.6}{0.15t_n^2 - 0.8t_n - 3.8} = \frac{0.1t_n^3 - 0.4t_n^2 + 15.6}{0.15t_n^2 - 0.8t_n - 3.8}$$

with $t_0 = 14$, obtaining $t_1 = 14.6944$, $t_2 = 14.6447$, $t_3 = 14.6445$.
So the temperature is $0°F$ at 8:39 P.M.

29. Here $f(x) = 120,000x^3 - 80,000x^2 - 60,000x - 40,000$
$\qquad f'(x) = 360,000x^2 - 160,000x - 60,000$.
Therefore,

$$x_{n+1} = x_n - \frac{120,000x_n^3 - 80,000x_n^2 - 60,000x_n - 40,000}{360,000x_n^2 - 160,000x_n - 60,000}$$

$$= x_n - \frac{6x_n^3 - 4x_n^2 - 3x_n - 2}{18x_n^2 - 8x_n - 3} = \frac{12x_n^3 - 4x_n^2 + 2}{18x_n^2 - 8x_n - 3}.$$

Then, $x_0 = 1.13$, $x_1 = 1.2981455$, $x_2 = 1.2692040$, $x_3 = 1.2681893$.
$\qquad x_4 = 1.2681880$, $x_5 = 1.2681880$, or $r = 26.819$.
Therefore, the rate of return is approximately 26.82 percent per year.

31. Here $C = 100{,}000$, $R = 1053$, $N = (12)(25) = 300$. Therefore,

$$r_{n+1} = r_n - \frac{100{,}000r_n + 1053\left[(1+r_n)^{-300} - 1\right]}{100{,}000 - 315{,}900(1+r_n)^{-301}}.$$

With $r_0 = 0.1$, we find

$r_1 = 0.0153$, $r_2 = 0.0100043$, $r_3 = 0.00999747$, and $r_4 = 0.00999747$.

Therefore, r is approximately 0.01 and the interest rate is approximately $12(0.01) = 0.12$ or 12 percent per year.

33. Here $C = 6000$ (75% of 8000), $N = 12(4) = 48$, and $R = 152.18$.

$$r_{n+1} = r_n - \frac{6000r_n + 152.18[(1+r_n)^{-48} - 1]}{6000 - 7304.64(1+r_n)^{-49}}.$$

With $r_0 = 0.01$, we find

$$r_1 = 0.01 - \frac{6000(0.01) + 152.18[(1.01)^{-48} - 1]}{6000 - 7304.64(1.01)^{-49}} \approx 0.00853956,$$

$r_2 \approx 0.0083388$, $r_3 \approx 0.0083346$.

So r is approximately $12(0.008335) = 0.1000$, that is, 10 percent per year.

35. We are required to solve the equation $s(x) = d(x)$ or

$$0.1x + 20 = \frac{50}{0.01x^2 + 1}; \quad 0.001x^3 + 0.2x^2 + 0.1x + 20 = 50,$$

$$0.001x^3 + 0.2x^2 + 0.1x - 30 = 0;$$

that is, the equation

$$F(x) = x^3 + 200x^3 + 100x - 30{,}000 = 0.$$

We use the iteration

$$x_{n+1} = x_n - \frac{x_n^3 + 200x_n^2 + 100x_n - 30{,}000}{3x_n^2 + 400x_n + 100} = \frac{2x_n^3 + 200x_n^2 + 30{,}000}{3x_n^2 + 400x_n + 100}.$$

With $x_0 = 10$, we find

$x_1 = 11.8182$, $x_2 = 11.6721$, $x_3 = 11.6711$, $x_4 = 11.6711$.

Therefore, the equilibrium quantity is approximately 11,671 units and the equilibrium price is $p = 0.1(11.671) + 20$, or $21.17/unit.

37. a. Let $f(x) = x^n - a$. We want to solve the equation $f(x) = 0$, $f'(x) = nx^{n-1}$.
Therefore, we have the iteration

$$x_{i+1} = x_i - \frac{f(x_i)}{f'(x_i)} = x_i - \frac{x_i^n - a}{nx_i^{n-1}}$$

$$= x_i - \frac{x_i^n}{nx_i^{n-1}} + \frac{a}{nx_i^{n-1}} = x_i - \frac{1}{n}x_i + \frac{a}{nx_i^{n-1}}$$

$$= \left(1 - \frac{1}{n}\right)x_i + \frac{a}{nx_i^{n-1}} = \left(\frac{n-1}{n}\right)x_i + \frac{a}{nx_i^{n-1}}.$$

b. Use part (a) with $n = 4$ and $a = 42$ and initial guess $x_0 = 2$ (x_0 is not unique!).

We find
$$x_{i+1} = \left(\frac{3}{4}\right)x_i + \frac{42}{4x_i^3}$$

and so

$$x_1 = \left(\frac{3}{4}\right)(2) + \frac{42}{4(2^3)} = 2.8125$$

$$x_2 = (.75)(2.8125) + \frac{42}{4(2.8125)^3} \approx 2.5813$$

$$x_3 \approx (.75)(2.5813) + \frac{42}{4(2.5813)^3} \approx 2.5464.$$

CHAPTER 11 REVIEW EXERCISES, page 850

1. $f(x) = \dfrac{1}{x+2} = \dfrac{1}{(x+1)+2-1} = \dfrac{1}{1+(x+1)}.$

    $f(x) = 1 - (x+1) + (x+1)^2 - (x+1)^3 + (x+1)^4.$

3. Observe that $\ln(1+x) = x - \dfrac{1}{2}x^2 + \dfrac{1}{3}x^3 - \dfrac{1}{4}x^4$. Therefore,

    $\ln(1+x^2) = x^2 - \frac{1}{2}(x^2)^2 + \frac{1}{3}(x^2)^3 - \frac{1}{4}(x^2)^4$

    $= x^2 - \frac{1}{2}x^4.$

5. $f(x) = x^{1/3}$, $f'(x) = \frac{1}{3}x^{-2/3}$, $f''(x) = -\frac{2}{9}x^{-5/3}$. So

    $f(8) = 2$, $f'(8) = \dfrac{1}{12}$, $f''(8) = -\dfrac{1}{144}$. Therefore,

    $f(x) = f(8) + f'(8)(x-8) + \dfrac{f''(8)}{2!}(x-8)^2 = 2 + \dfrac{1}{12}(x-8) - \dfrac{1}{288}(x-8)^2$

$$\sqrt[3]{7.8} = f(7.8) \approx 2 + \frac{1}{12}(-0.2) - \frac{1}{288}(-0.2)^2 \approx 1.9832.$$

7. $f(x) = x^{1/3}$, $f'(x) = \frac{1}{3}x^{-2/3}$, $f''(x) = -\frac{2}{9}x^{-5/3}$, $f'''(x) = \frac{10}{27}x^{-8/3}$

$$f(27) = 3,\ f'(27) = \frac{1}{27},\ f'(27) = -\frac{2}{2187}.$$

$$f(x) = f(27) + f'(27)(x - 27) + \frac{f''(27)}{2!}(x - 27)^2$$

$$= 3 + \frac{1}{27}(x - 27) - \frac{1}{2187}(x - 27)^2$$

$$\sqrt[3]{26.98} = f(26.98) = 3 + \frac{1}{27}(-0.02) - \frac{1}{2187}(-0.02)^2 \approx 2.9992591.$$

The error is less than $\dfrac{M}{3!}|x - 27|^3$ where M is a bound for

$f'''(x) = \dfrac{10}{27x^{7/3}}$ on $[26.98, 26]$. Observe that $f'''(x)$ is decreasing on the interval

and so

$$|f'''(x)| \le \frac{10}{27(26.98)^{8/3}} < 0.00006.$$

So the error is less than

$$\frac{0.0002}{6}(0.02)^3 < 8 \times 10^{-11}.$$

9. $e^x = 1 + x + \dfrac{x^2}{2!} + \dfrac{x^3}{3!} + \cdots$. Therefore,

$$e^{-1} \approx 1 - 1 + \frac{1}{2} - \frac{1}{6} + \frac{1}{24} - \frac{1}{120} + \cdots \approx 0.367.$$

11. $\displaystyle \lim_{n \to \infty} \frac{2n^2 + 1}{3n^2 - 1} = \lim_{n \to \infty} \frac{2 + \frac{1}{n^2}}{3 - \frac{1}{n^2}} = \frac{2}{3}.$

13. $\displaystyle \lim_{n \to \infty} a_n = \lim_{n \to \infty}\left(1 - \frac{1}{2^n}\right) = \lim_{n \to \infty} 1 - \lim_{n \to \infty} \frac{1}{2^n} = 1 - 0 = 1.$

15. $\displaystyle\sum_{n=1}^{\infty}\frac{2^n}{3^n}=\sum_{n=1}^{\infty}\left(\frac{2}{3}\right)^n=\frac{2}{3}\left(\frac{1}{1-\frac{2}{3}}\right)=\frac{2}{3}\cdot 3=2.$

17. $\displaystyle\sum_{n=1}^{\infty}(-1)^{n-1}\left(\frac{1}{\sqrt{2}}\right)^n=\frac{1}{\sqrt{2}}-\left(\frac{1}{\sqrt{2}}\right)^2+\cdots=\frac{\dfrac{1}{\sqrt{2}}}{1-\left(-\dfrac{1}{\sqrt{2}}\right)}$

$$=\frac{1}{\sqrt{2}}\cdot\frac{1}{1+\dfrac{1}{\sqrt{2}}}=\frac{1}{1+\sqrt{2}}.$$

19. $1.424242\ldots=1+\dfrac{42}{10^2}+\dfrac{42}{10^4}+\cdots$

$$=1+\frac{42}{100}\left[1+\frac{1}{100}+\left(\frac{1}{100}\right)^2+\cdots\right]$$

$$=1+\frac{42}{100}\left(\frac{1}{1-\frac{1}{100}}\right)=1+\frac{42}{100}\left(\frac{100}{99}\right)=1+\frac{42}{99}=\frac{141}{99}.$$

21. Let $a_n=\dfrac{n^2+1}{2n^2-1}$. Since $\displaystyle\lim_{n\to\infty}\frac{n^2+1}{2n^2-1}=\lim_{n\to\infty}\frac{1+\dfrac{1}{n^2}}{2-\dfrac{1}{n^2}}=\frac{1}{2}\neq 0$, the divergence test

implies that the series $\displaystyle\sum_{n=1}^{\infty}a_n$ diverges.

23. $\displaystyle\sum_{n=1}^{\infty}\left(\frac{1}{n}\right)^{1.1}=\sum_{n=1}^{\infty}\frac{1}{n^{1.1}}$ is a convergent p-series with $p=1.1>1$.

25. $R=\displaystyle\lim_{n\to\infty}\left|\frac{a_n}{a_{n+1}}\right|=\lim_{n\to\infty}\frac{\dfrac{1}{n^2+2}}{\dfrac{1}{(n-1)^2+2}}=\lim_{n\to\infty}\frac{(n+1)^2+2}{n^2+2}$

$$=\lim_{n\to\infty}\frac{n^2+2n+3}{n^2+2}=1.$$

The interval of convergence is (-1,1).

27. $R = \lim\limits_{n \to \infty} \left| \dfrac{a_n}{a_{n+1}} \right| = \lim\limits_{n \to \infty} \dfrac{(n+1)(n+2)}{n(n+1)} = 1.$

So $R = 1$ and the interval of convergence is (0,2).

29. We have $\dfrac{1}{1+x} = 1 + x + x^2 + x^3 + \cdots$ $(-1 < x < 1)$

Replacing x by $2x$ in the expression, we find

$$f(x) = \dfrac{1}{2x-1} = -\dfrac{1}{1-2x} = -[1 + 2x + (2x)^2 + (2x)^3 + \cdots]$$

$$= -1 - 2x - 4x^2 - 8x^3 - \cdots - 2^n x^n - \cdots. \qquad (-\tfrac{1}{2} < x < \tfrac{1}{2}).$$

31. We know that

$$\ln(1+x) = x - \dfrac{1}{2}x^2 + \dfrac{1}{3}x^3 - \dfrac{1}{4}x^4 + \cdots + (-1)^{n+1}\dfrac{x^n}{n} \qquad (-1 < x < 1).$$

Replace x by $2x$, to obtain

$$f(x) = \ln(1+2x) = 2x - 2x^2 + \dfrac{8}{3}x^3 - \cdots + (-1)^{n+1}\dfrac{2^n x^n}{n}$$

$$= \sum_{n=0}^{\infty} (-1)^n \dfrac{2^n}{n} x^n.$$

The interval of convergence is $(-\tfrac{1}{2}, \tfrac{1}{2}]$..

33. $f(x) = x^3 - 12, \ f'(x) = 3x^2$

Therefore, $x_{n+1} = x_n - \dfrac{x_n^3 - 12}{3x_n^2} = \dfrac{3x_n^3 - x_n^3 + 12}{3x_n^2} = \dfrac{2x_n^3 + 12}{3x_n^2}.$

Using $x_0 = 2$, we have
$x_1 = 2.3333333, \ x_1 = 2.2902491, \ x_3 = 2.2894277$, and
the root is approximately 2.28943.

35. We solve the equation $F(x) = 2x - e^{-x}$.

$F'(x) = 2 + e^{-x}$. So the iteration is

$$x_{n+1} = x_n - \dfrac{2x_n - e^{-x_n}}{2 + e^{-x_n}} = \dfrac{(x_n + 1)e^{-x_n}}{2 + e^{-x_n}}.$$

Taking $x_0 = 0.5$, we find $x_1 = 0.349045$, $x_2 = 0.351733$, $x_3 = 0.351733$.
So the point of intersection is approximately $(0.35173, 0.70346)$.

37. The amount required

$$A = 10,000[e^{-0.09} + e^{-0.09(2)} + \cdots] = \frac{10,000e^{-0.09}}{1 - e^{-0.09}} = 106,186.10,$$

or $106,186.10.

39. We compute

$$P(63.5 \le x \le 65.5) = \frac{1}{2.5\sqrt{2\pi}} \int_{63.5}^{65.5} e^{-1/2[(x-64.5)/2.5]^2}.$$

Replacing x with $-\frac{1}{2}[(x-64.5)/2.5]^2$ in the expression

$$e^x = 1 + x + \frac{1}{2!}x^2 + \frac{1}{3!}x^3,$$

we obtain

$$e^{-1/2[(x-64.5)/2.5]^2} \approx 1 - \tfrac{1}{2}[(x-64.5)/2.5]^2 + \tfrac{1}{2!}\left\{-\tfrac{1}{2}[(x-64.5)/2.5]^2\right\}^2]$$

$$+ \tfrac{1}{3!}\left\{-\tfrac{1}{2}[(x-64.5)/2.5]^2\right\}^3$$

$$= 1 - \tfrac{1}{12.5}(x-64.5)^2 + \tfrac{1}{312.5}(x-64.5)^4 - \tfrac{1}{11718.75}(x-64.5)^6.$$

Therefore,
$P(63.5 \le x \le 65.5) \approx$

$$\frac{1}{2.5\sqrt{2\pi}} \int_{63.5}^{65.5} \left[1 - \frac{1}{12.5}(x-64.5)^2 + \frac{1}{312.5}(x-64.5)^4 - \frac{1}{11718.75}(x-64.5)^6\right] dx$$

$$= \frac{1}{2.5\sqrt{2\pi}} \left[x - \frac{1}{37.5}(x-64.5)^3 + \frac{1}{1607.5}(x-64.5)^5 - \frac{1}{82031.25}(x-64.5)^7\right]_{63.5}^{65.5}$$

$$= \frac{1}{2.5\sqrt{2\pi}}[(65.5 - 0.02667 + 0.00062 - 0.00001)$$

$$- (63.5 + 0.02667 - 0.00062 + 0.00001)]$$

$$\approx 0.3108,$$

or approximately 31.08%.

CHAPTER 12

EXERCISES 12.1, page 858

1. $450° = \dfrac{450}{180}\pi = \dfrac{5\pi}{2}$ rad.

3. $-270° = -\dfrac{270}{180}\pi = -\dfrac{3\pi}{2}$ rad.

5. a. III   b. III   c. II   d. I

7. $f(x) = \dfrac{\pi}{180}x$ radians, $f(75) = \dfrac{\pi}{180}(75)$ radians $= \dfrac{5\pi}{12}$ radians

9. $f(x) = \dfrac{\pi}{180}x$ radians, $f(160) = \dfrac{\pi}{180}(160)$ radians $= \dfrac{8\pi}{9}$ radians

11. $f(630) = \dfrac{\pi}{180}(630)$ radians $= \dfrac{7\pi}{2}$ radians

13. $g\left(\dfrac{2\pi}{3}\right) = \left(\dfrac{180}{\pi}\right)\left(\dfrac{2\pi}{3}\right) = 120°.$

15. $g\left(-\dfrac{3\pi}{2}\right) = \left(\dfrac{180}{\pi}\right)\left(-\dfrac{3\pi}{2}\right) = -270°.$

17. $g\left(\dfrac{22\pi}{18}\right) = \left(\dfrac{180}{\pi}\right)\left(\dfrac{22\pi}{18}\right) = 220°.$

19.

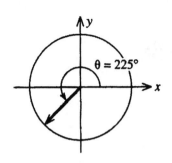

21.

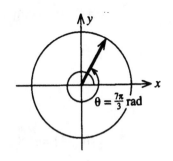

23. $\dfrac{5\pi}{6}$ rad $= 150°$; coterminal angle is $-210°$.

25. $-\dfrac{\pi}{4}$ rad $= -45°$; coterminal angle: $360° - 45° = 315°$

27. True. $3630 = (360)(10) + 30$ and the result is evident.

29. True. Adding $n(360)$ degrees to θ revolves the angle θ $|n|$ revolutions, clockwise if n is positive and counter-clockwise if n is negative.

EXERCISES 12.2, page 868

1. $\sin 3\pi = 0$.

3. $\sin\left(\dfrac{9\pi}{2}\right) = 1$.

5. $\sin\left(-\dfrac{4\pi}{3}\right) = \sin\left(\dfrac{\pi}{3}\right) = \dfrac{\sqrt{3}}{2}$.

7. $\tan\dfrac{\pi}{6} = \dfrac{\sqrt{3}}{3}$.

9. $\sec\left(-\dfrac{5\pi}{8}\right) = \sec\left(\dfrac{5\pi}{8}\right) = -2.6131$.

11. $\sin\dfrac{\pi}{2} = 1$, $\cos\dfrac{\pi}{2} = 0$, $\tan\dfrac{\pi}{2}$ is undefined, $\sec\dfrac{\pi}{2}$ is undefined, $\csc\dfrac{\pi}{2} = 1$, $\cot\dfrac{\pi}{2} = 0$.

13. $\sin\left(\dfrac{5\pi}{3}\right) = -\dfrac{\sqrt{3}}{2}$, $\cos\left(\dfrac{5\pi}{3}\right) = \dfrac{1}{2}$, $\tan\left(\dfrac{5\pi}{3}\right) = -\sqrt{3}$, $\csc\left(\dfrac{5\pi}{3}\right) = -\dfrac{2\sqrt{3}}{2}$, $\sec\left(\dfrac{5\pi}{3}\right) = 2$, $\cot\left(\dfrac{5\pi}{3}\right) = -\dfrac{\sqrt{3}}{3}$.

15. $\theta = \dfrac{7\pi}{6}$ or $\dfrac{11\pi}{6}$.

17. $\theta = \dfrac{5\pi}{6}$ or $\dfrac{11\pi}{6}$

19. $\theta = \pi$

21. $\sin\theta = \sin\left(-\dfrac{4\pi}{3}\right) = \sin\left(\dfrac{2\pi}{3}\right) = \sin\left(\dfrac{\pi}{3}\right)$, so $\theta = \dfrac{2\pi}{3}$ or $\dfrac{\pi}{3}$.

23.

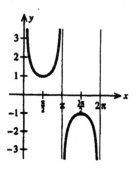

25.

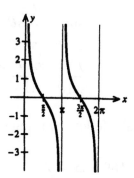

27.

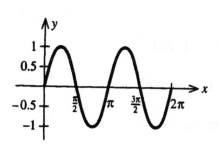

29.

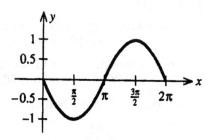

31. $\cos^2\theta - \sin^2\theta = \cos^2\theta - (1-\cos^2\theta) = 2\cos^2\theta - 1.$

33. $(\sec\theta + \tan\theta)(1-\sin\theta) = \sec\theta + \tan\theta - \sin\theta\sec\theta - \tan\theta\sin\theta$
$$= \sec\theta + \tan\theta - \tan\theta - \tan\theta\sin\theta$$
$$= \frac{1}{\cos\theta} - \frac{\sin^2\theta}{\cos\theta} = \frac{1}{\cos\theta}(1-\sin^2\theta)$$
$$= \frac{\cos^2\theta}{\cos\theta} = \cos\theta.$$

35. $(1+\cot^2\theta)\tan^2\theta = \csc^2\theta\tan^2\theta = \frac{1}{\sin^2\theta}\cdot\frac{\sin^2\theta}{\cos^2\theta} = \frac{1}{\cos^2\theta}$
$$= \sec^2\theta.$$

37.
$$\frac{\csc\theta}{\tan\theta+\cot\theta}=\frac{\dfrac{1}{\sin\theta}}{\dfrac{\sin\theta}{\cos\theta}+\dfrac{\cos\theta}{\sin\theta}}=\frac{\dfrac{1}{\sin\theta}}{\dfrac{\sin^2\theta+\cos^2\theta}{\cos\theta\sin\theta}}$$

$$=\frac{1}{\sin\theta}\cdot\frac{\cos\theta}{1}\cdot\frac{\sin\theta}{1}=\cos\theta.$$

39. The results follow by using similar triangles.

41. $|AB|=\sqrt{169-25}=12.$

$\sin\theta=\dfrac{5}{13},\ \cos\theta=\dfrac{12}{13},\ \tan\theta=\dfrac{5}{12},\ \csc\theta=\dfrac{13}{5},\ \sec\theta=\dfrac{13}{12},\cot\theta=\dfrac{12}{5}.$

43. a. $P(t)=100+20\sin 6t.$

The maximum value of P occurs when $\sin t=1$, and
$$P(t)=100+20=120.$$

The minimum value of P occurs when $\sin 6t=-1$ and
$$P(t)=100-20=80.$$

b. $\sin 6t=1$ implies that $6t=\frac{\pi}{2},\frac{3\pi}{2},...$; that is,
$$t=\frac{\pi(4n+1)}{12}\qquad (n=0,1,2,\ ...).$$

$\sin 6t=-1$ implies that $6t=\frac{3\pi}{2},\frac{7\pi}{2}$; that is,
$$t=\frac{\pi(4n+3)}{12}\qquad (n=0,1,2,\ ...).$$

45. False. In fact, $\sin\theta=-\dfrac{\sqrt{3}}{2}$ if $\theta=\dfrac{4\pi}{3}$ or $\dfrac{5\pi}{3}$.

47. True. $\cos 2\theta=\cos^2\theta-\sin^2\theta=\cos^2\theta-(1-\cos^2\theta)=2\cos^2\theta-1.$

EXERCISES 12.3, page 881

1. $f(x)=\cos 3x;\ f'(x)=-3\sin 3x.$

3. $f(x)=2\cos\pi x;\ f'(x)=(-2\sin\pi x)\dfrac{d}{dx}(\pi x)=-2\pi\sin\pi x.$

5. $f(x) = \sin(x^2 + 1); f'(x) = \cos(x^2 + 1)\dfrac{d}{dx}(x^2 + 1) = 2x\cos(x^2 + 1).$

7. $f(x) = \tan 2x^2;\ f'(x) = (\sec^2 2x^2)\dfrac{d}{dx}(2x^2) = 4x\sec^2 2x^2.$

9. $f(x) = x\sin x\ ;\ \ f'(x) = \sin x + x\cos x.$

11. $f(x) = 2\sin 3x + 3\cos 2x;$
    $f'(x) = (2\cos 3x)(3) - (3\sin 2x)(2) = 6(\cos 3x - \sin 2x).$

13. $f(x) = x^2\cos 2x;$
    $f'(x) = 2x\cos 2x + x^2(-\sin 2x)(2) = 2x(\cos 2x - x\sin 2x).$

15. $f(x) = \sin\sqrt{x^2 - 1} = \sin(x^2 - 1)^{1/2}$
    $$f'(x) = \cos\sqrt{x^2 - 1}\,\dfrac{d}{dx}(x^2 - 1)^{1/2} = (\cos\sqrt{x^2 - 1})(\tfrac{1}{2})(x^2 - 1)^{-1/2}(2x)$$
    $$= \dfrac{x\cos\sqrt{x^2 - 1}}{\sqrt{x^2 - 1}}.$$

17. $f(x) = e^x\sec x$
    $f'(x) = e^x\sec x + e^x(\sec x\tan x) = e^x\sec x(1 + \tan x).$

19. $f(x) = x\cos\dfrac{1}{x};$
    $$f'(x) = \cos\dfrac{1}{x} + x(-\sin\tfrac{1}{x})\dfrac{d}{dx}\left(\dfrac{1}{x}\right) = \cos\dfrac{1}{x} - x\left(\dfrac{1}{x^2}\right)\sin\dfrac{1}{x}$$
    $$= \dfrac{1}{x}\left(\sin\dfrac{1}{x} + x\cos\dfrac{1}{x}\right)\ \text{or}\ \ \cos\dfrac{1}{x} + \dfrac{1}{x}\sin\dfrac{1}{x}.$$

21. $f(x) = \dfrac{x - \sin x}{1 + \cos x}$

12 Trigonometric Functions

$$f'(x) = \frac{(1+\cos x)(1-\cos x) - (x - \sin x)(-\sin x)}{(1+\cos x)^2}$$

$$= \frac{1 - \cos^2 x + x \sin x - \sin^2 x}{(1+\cos x)^2} = \frac{x \sin x}{(1+\cos x)^2}.$$

23. $f(x) = \sqrt{\tan x}$; $f'(x) = \frac{1}{2}(\tan x)^{-1/2} \sec^2 x = \frac{\sec^2 x}{2\sqrt{\tan x}}$.

25. $f(x) = \frac{\sin x}{x}$; $f'(x) = \frac{x \cos x - \sin x}{x^2}$.

27. $f(x) = \tan^2 x$. $f'(x) = 2 \tan x \sec^2 x$.

29. $f(x) = e^{\cot x}$; $f'(x) = e^{\cot x}(-\csc^2 x) = -\csc^2 x \cdot e^{\cot x}$.

31. $f(x) = \cot 2x$.
    $f'(x) = -2 \csc^2 2x$.
    Therefore, $f'(\frac{\pi}{4}) = -\frac{2}{\sin^2(\frac{\pi}{2})} = -2$.
    Then $y - 0 = -2(x - \frac{\pi}{4})$, or $y = -2x + \frac{\pi}{2}$.

33. $f(x) = e^x \cos x$.
    $f'(x) = e^x \cos x + e^x(-\sin x) = e^x(\cos x - \sin x)$.
    Setting $f'(x) = 0$ gives $\cos x - \sin x = 0$, or $\tan x = 1$; that is, when
    $x = \frac{\pi}{4}$, or $\frac{5\pi}{4}$. From the following sign diagram for f'

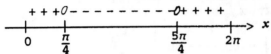

we see that f is increasing on $(0, \frac{\pi}{4}) \cup (\frac{5\pi}{4}, 2\pi)$ and decreasing on $(\frac{\pi}{4}, \frac{5\pi}{4})$.

35.

37.

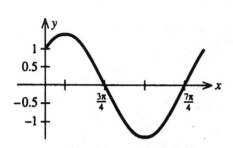

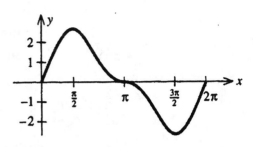

39. a. $f(x) = \sin x$, $f'(x) = \cos x$, $f''(x) = -\sin x$, $f'''(x) = -\cos x$,

$f^{(4)}(x) = \sin x$, ...

$f(0) = 0$, $f'(0) = 1$, $f''(0) = 0$, $f'''(0) = -1$, $f^{(4)}(0) = 0$, ...,

Therefore,

$$f(x) = \sin x = x - \frac{x^3}{3!} + \frac{x^5}{5!} - \frac{x^7}{7!} + \cdots + (-1)^{n+1}\frac{x^{2n+1}}{(2n+1)!} + \cdots$$

b. $\displaystyle\lim_{x \to 0}\frac{\sin x}{x} = \lim_{x \to 0}(1 - \frac{x^2}{3!} + \frac{x^4}{5!} - \cdots) = 1.$

41. $P(t) = 8000 + 1000\sin\left(\frac{\pi t}{24}\right)$.

$P'(t) = [1000\cos\left(\frac{\pi t}{24}\right)](\frac{\pi}{24})$ and $P''(12) = \dfrac{1000\pi}{24}\cos\dfrac{\pi}{2} = 0$; that is, the

wolf population is not changing during the twelfth month.

$$p(t) = 40{,}000 + 12{,}000\cos\left(\frac{\pi t}{24}\right)$$

$$p'(t) = [-12{,}000\sin\left(\frac{\pi t}{24}\right)](\frac{\pi}{24})$$

and $p'(12) = -500\sin\frac{\pi(12)}{24} = -500\pi$; that is, the caribou population is decreasing

at the rate of 1571/month.

43. $f(t) = 3\sin\dfrac{2\pi}{365}(t-79) + 12.$ $f'(t) = 3\cos\dfrac{2\pi}{365}(t-79) \cdot \dfrac{2\pi}{365}$

Therefore, $f'(79) = \dfrac{6\pi}{365} \approx 0.05164.$

The number of hours of daylight is increasing at the rate of 0.05 hrs per day on March 21.

45. $T = 62 - 18\cos\dfrac{2\pi(t-23)}{365}.$ $T' = \dfrac{36\pi}{365}\sin\left(\dfrac{-46\pi + 2\pi t}{365}\right).$

Setting $T' = 0$, we obtain
$$\dfrac{-46\pi + 2\pi t}{365} = 0,\ -46\pi = -2\pi t,\ \text{or } t = 23,$$

and $\quad\dfrac{-46\pi + 2\pi t}{365} = \pi,\ -46\pi + 2\pi t = 365\pi,\ 2\pi t = 411\pi,\ \text{or } t = 205.5.$

From the sign diagram shown below,

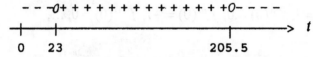

we see that a minimum occurs at $t = 23$ and a maximum occurs at $t = 205.5$. We conclude that the warmest day is July 25th and the coldest day is January 23rd.

47. $R(t) = 2(5 - 4\cos\frac{\pi t}{6}).$ $R'(t) = \frac{4\pi}{3}\sin(\frac{\pi t}{6}).$

Setting $R' = 0$, we obtain $\frac{\pi t}{6} = 0$, and conclude that $t = 0$, $t = 6$, and $t = 12$. From the sign diagram for R

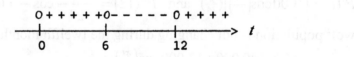

we conclude that a maximum occurs at $t = 6$.

49. From Problem 46, we have
$$V'(t) = \frac{6}{5\pi}\left(\sin\frac{\pi t}{2}\right)\left(\frac{\pi}{2}\right) = \frac{3}{5}\sin\frac{\pi t}{2}.$$

Then $\quad V''(t) = \frac{3}{5}\cos\frac{\pi t}{2}\left(\frac{\pi}{2}\right) = \frac{3\pi}{10}\cos\frac{\pi t}{2}.$

Setting $V'(t) = 0$ gives $t = 1, 3, 5, 7, 9, 11, 13, 15, \ldots$, as critical points. Evaluating $V''(t)$ at each of these points, we see that the rate of flow of air is at a maximum when $t = 1, 5, 9, 13, \ldots$, and it is at a minimum when $t = 3, 7, 11, 15, \ldots$.

51. $\tan\theta = \dfrac{y}{20}$. Therefore, $\sec^2\theta \cdot \dfrac{d\theta}{dt} = \dfrac{1}{20}\cdot\dfrac{dy}{dt}$.

We want to find $\dfrac{dy}{dt}$ when $z = 30$. But when $z = 30$,

$y = \sqrt{900-400} = \sqrt{500} = 10\sqrt{5}$, and $\sec\theta = \dfrac{30}{20} = \dfrac{3}{2}$. Therefore, with

$\dfrac{d\theta}{dt} = \dfrac{\pi}{2}$ rad/sec, we find

$$\dfrac{dy}{dt} = 20\sec^2\theta\cdot\dfrac{d\theta}{dt} = 20\left(\dfrac{3}{2}\right)^2\cdot\dfrac{\pi}{2} = \dfrac{45\pi}{2} \approx 70.7$$

or 70.7 ft/sec.

53. The area of the cross-section is
$$A = (2)(\tfrac{1}{2})(5\cos\theta)(5\sin\theta) + 5(5\sin\theta)$$
$$= 25(\cos\theta\sin\theta + \sin\theta).$$
$$\dfrac{dA}{d\theta} = 25(-\sin^2\theta + \cos^2\theta + \cos\theta)$$
$$= 25(\cos^2\theta - 1 + \cos^2\theta + \cos\theta)$$
$$= 25(2\cos^2\theta + \cos\theta - 1)$$
$$= 25(2\cos\theta - 1)(\cos\theta + 1).$$

Setting $\dfrac{dA}{d\theta} = 0$ gives $\cos\theta = \tfrac{1}{2}$ or $\cos\theta = -1$, or $\theta = \tfrac{\pi}{3}$ or π. The absolute

maximum of A occurs at $\theta = \tfrac{\pi}{3}$ as can be seen from the following table:

θ	A
0	0
$\frac{\pi}{3}$	$\frac{75\sqrt{3}}{4}$
π	0

Therefore, the angle shoud be $60°$.

55. Let $f(\theta) = \theta - 0.5\sin\theta - 1$. then $f'(\theta) = 1 - 0.5\cos\theta$ and so Newton's Method leads to the iteration.

$$\theta_{i+1} = \theta_i - \frac{f(\theta_i)}{f'(\theta_i)} = \theta_i - \frac{\theta_i - 0.5\sin\theta_i - 1}{1 - 0.5\cos\theta_i} = \frac{\theta_i - 0.5\theta_i\cos\theta_i - \theta_i + 0.5\sin\theta_i + 1}{1 - 0.5\cos\theta_i}$$

$$= \frac{1 + 0.5(\sin\theta_i - \theta_i\cos\theta_i)}{1 - 0.5\cos\theta_i}.$$

With $\theta_i = 1.5$, we find

$$\theta_1 = \frac{1 + 0.5(\sin 1.5 - 1.5\cos 1.5)}{1 - 0.5\cos 1.5} = \frac{1.4456946}{0.9646314} \approx 1.4987.$$

$$\theta_2 = \frac{1 + 0.5(\sin 1.4987 - 1.4987\cos 1.4987)}{1 - 0.5\cos 1.4987} = \frac{1.4447225}{0.9639831} \approx 1.4987.$$

57. True. Let $f(x) = \sin x$. Then

$$f'(0) = \lim_{x \to 0} \frac{\sin x - \sin 0}{x} = \lim_{x \to 0} \frac{\sin x}{x}.$$

But $f'(0) = \cos 0 = 1$. So $\lim_{x \to 0} \frac{\sin x}{x} = 1$.

59. False. Take $x = \pi$. Then f has a relative minimum if $x = \pi$, but g does not have a relative maximum at $x = \pi$.

61. $h(x) = \csc f(x) = \dfrac{1}{\sin f(x)}$.

$$h'(x) = \frac{-\cos f(x) \cdot f'(x)}{\sin^2 f(x)} = -\left(\csc f(x)\cot f(x)\right)f'(x).$$

63. $h(x) = \cot f(x) = \dfrac{\cos f(x)}{\sin f(x)}$.

$$h'(x) = \frac{\sin f(x)[-\sin f(x)]f'(x) - (\cos f(x))[\cos f(x)]f'(x)}{\sin^2 f(x)}$$

$$= \frac{\left[-\sin^2 f(x) - \cos^2 f(x)\right]f'(x)}{\sin^2 f(x)} = \frac{-1 \cdot f'(x)}{\sin^2 f(x)}$$

$$= -\csc^2 f(x)f'(x).$$

1. 1.2038 3. 0.7762 5. -0.2368

7. 0.8415; -0.2172 9. 1.1271; 0.2013

11. a. b. ≈ $0.63 c. ≈ $27.79

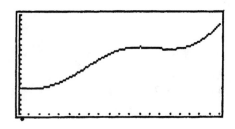

13. a. 15. ≈ 0.006 ft.

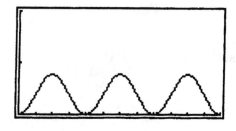

EXERCISES 12.4, page 891

1. $\int \sin 3x \, dx = -\frac{1}{3}\cos 3x + C.$

3. $\int (3\sin x + 4\cos x)\, dx = -3\cos x + 4\sin x + C.$

5. Let $u = 2x$ so that $du = 2dx$. Then
$$\int \sec^2 2x \, dx = \frac{1}{2}\int \sec^2 u \, du = \frac{1}{2}\tan u + C = \frac{1}{2}\tan 2x + C.$$

7. Let $u = x^2$ so that $du = 2x\,dx$, or $x\,dx = \frac{1}{2}du$. Then

$$\int x\cos x^2 dx = \frac{1}{2}\int \cos u\,du = \frac{1}{2}\sin u + C = \frac{1}{2}\sin x^2 + C.$$

9. Let $u = \pi x$ so that $du = \pi\,dx$, or $dx = \frac{1}{\pi}du$. Then

$$\int \csc \pi x \cot \pi x\,dx = \frac{1}{\pi}\int \csc u \cot u\,du = -\frac{1}{\pi}\csc u + C$$

$$= -\frac{1}{\pi}\csc \pi x + C.$$

11. $\displaystyle\int_{-\pi/2}^{\pi/2}(\sin x + \cos x)\,dx = -\cos x + \sin x\Big|_{-\pi/2}^{\pi/2} = 1 - (-1) = 2.$

13. $\displaystyle\int_0^{\pi/6} \tan 2x\,dx = -\frac{1}{2}\ln|\cos 2x|\Big|_0^{\pi/6} = -\frac{1}{2}\ln\frac{1}{2}.$

15. Let $u = \sin x$ so that $du = \cos x\,dx$. Then

$$\int \sin^3 x \cos x\,dx = \int u^3\,du = \frac{1}{4}u^4 + C = \frac{1}{4}\sin^4 x + C.$$

17. Let $u = \pi x$ so that $du = \pi\,dx = \frac{1}{\pi}du$. Then

$$\int \sec \pi x\,dx = \frac{1}{\pi}\int \sec u\,du = \frac{1}{\pi}\ln|\sec u + \tan u| + C$$

$$= \frac{1}{\pi}\ln|\sec \pi x + \tan \pi x| + C.$$

19. Let $u = 3x$ so that $du = 3\,dx$, or $du = \frac{1}{3}du$. If $x = 0$, then $u = 0$ and if $x = \frac{\pi}{12}$, then $u = \frac{\pi}{4}$. Then

$$\int_0^{\pi/12} \sec 3x\,dx = \frac{1}{3}\int_0^{\pi/4}\sec u\,du = \frac{1}{3}\ln|\sec u + \tan u|\Big|_0^{\pi/4}$$

$$= \frac{1}{3}[\ln(\sqrt{2} + 1) - \ln 1] = \frac{1}{3}\ln(\sqrt{2} + 1).$$

21. Let $u = \cos x$ so that $du = -\sin x\, dx$. Then
$$\int \sqrt{\cos x}\, \sin x\, dx = -\int \sqrt{u}\, du = -\tfrac{2}{3}u^{3/2} + C = -\tfrac{2}{3}(\cos x)^{3/2} + C.$$

23. $\int \cos 3x(1 - 2\sin 3x)^{1/2}\, dx$. Put $u = \sin 3x$, then $du = 3\cos 3x\, dx$. Therefore,
$$I = \tfrac{1}{3}\int (1 - 2u)^{1/2}\, du = -\tfrac{1}{6}\cdot\tfrac{2}{3}(1 - 2u)^{3/2} + C$$
$$= -\tfrac{1}{9}(1 - 2\sin 3x)^{3/2} + C.$$

25. $\int \tan^3 x \sec^2 x\, dx = \tfrac{1}{4}\tan^4 x + C.$

27. Let $u = \cot x - 1$ so that $du = -\csc^2 x\, dx$. Then
$$\int \csc^2 x(\cot x - 1)^3\, dx = -\int u^3\, du = -\tfrac{1}{4}u^4 + C = -\tfrac{1}{4}(\cot x - 1)^4 + C.$$

29. Put $u = \ln x$, $du = \dfrac{1}{x}dx$. Therefore,
$$I = \int \sin u\, du = -\cos u + C = -\cos(\ln x) + C$$
and $\displaystyle \int_1^{e^\pi} \frac{\sin(\ln x)}{x}\, dx = -\cos(\ln x)\Big|_1^{e^\pi} = -\cos\pi + \cos 0 = 2.$

31. $\int \sin(\ln x)\, dx$. Let $I = \int \sin(\ln x)\, dx$. Put $u = \sin(\ln x)$, $dv = dx$.

Then $\quad du = \dfrac{\cos(\ln x)}{x}$, $v = x$. Integrating by parts, we have
$$I = x\sin(\ln x) - \int \cos(\ln x)\, dx.$$

Let $J = \int \cos(\ln x)\, dx$ and $u = \cos(\ln x)\, dx$. Then $u = \cos(\ln x), dv = dx$, and
$$du = -\frac{\sin(\ln x)}{x}\, dx, \text{ and } v = x,$$

and $\quad J = x\cos(\ln x) + \int \sin(\ln x)\, dx = x\cos(\ln x) + I.$

So $\quad I = x[\sin(\ln x) - \cos(\ln x)] - I,$

or $2I = x[\sin(\ln x) - \cos(\ln x)]$.   Therefore,

$$\int \sin(\ln x)\, dx = \tfrac{1}{2}x[\sin(\ln x) - \cos(\ln x)] + C.$$

33. Area $= \int_0^{\pi} \cos\frac{x}{4}\, dx = 4\sin\frac{x}{4}\Big|_0^{\pi} = 4\left(\frac{\sqrt{2}}{2}\right) = 2\sqrt{2}$ sq units.

35. $A = \int_0^{\pi/4} \tan x\, dx = -\ln|\cos x|\Big|_0^{\pi/4} = -\ln\frac{\sqrt{2}}{2} = -\ln\frac{1}{\sqrt{2}}$

$\quad = \ln\sqrt{2} = \tfrac{1}{2}\ln 2$ sq units.

37. $A = \int_0^{\pi}(x - \sin x)\, dx = \tfrac{1}{2}x^2 + \cos x\Big|_0^{\pi} = \tfrac{1}{2}\pi^2 - 1 - 1 = \tfrac{1}{2}\pi^2 - 2 = \tfrac{1}{2}(\pi^2 - 4)$ sq units.

39. The average is $\quad A = \tfrac{1}{15}\int_0^{15}(80 + 3t\cos\frac{\pi}{6})\, dt = 80 + \tfrac{1}{5}\int_0^{15} t\cos\frac{\pi}{6}\, dt.$

Integrating by parts, we find $\quad \int t\cos\frac{\pi}{6}\, dt = \left(\frac{6}{\pi}\right)^2\left[\cos\frac{\pi}{6} + \frac{\pi}{6}\sin\frac{\pi}{6}\right].$

So $\quad A = 80 + \tfrac{1}{5}\left(\frac{6}{\pi}\right)^2\left[\cos\frac{\pi}{6} + \frac{\pi}{6}\sin\frac{\pi}{6}\right]_0^{15} = 80 + \tfrac{1}{5}\left(\frac{6}{\pi}\right)^2\left(\frac{15\pi}{6} - 1\right)$

$\quad \approx 85$, or approximately \$85 per share.

41. $R = \int_0^{12} 2(5 - 4\cos\frac{\pi t}{6})\, dt = 10t - 8\cdot\left(\frac{6}{\pi}\right)\sin\frac{\pi t}{6}\Big|_0^{12} = 120$, or \$120,000,

43. The required volume is

$$V = \int R(t)\, dt = 0.6\int \sin\frac{\pi t}{2}\, dt = (0.6)\left(\frac{2}{\pi}\right)\cos\frac{\pi t}{2} + C = -\frac{1.2}{\pi}\cos\frac{\pi t}{2} + C.$$

When $t = 0$, $V = 0$, and so $0 = -\frac{1.2}{\pi} + C$, and $C = \frac{1.2}{\pi}$. Therefore,

$$V = \frac{1.2}{\pi}\left(1 - \cos\frac{\pi t}{2}\right).$$

45. $\dfrac{dQ}{dt} = 0.0001(4 + 5\cos 2t)Q(400 - Q).$

The differential equation can be written in the form

$$\frac{dQ}{Q(400 - Q)} = 0.0001(4 + 5\cos 2t)\, dt \text{ or}$$

$$\frac{1}{400}\left[\frac{1}{Q}+\frac{1}{400-Q}\right]dQ = 0.0001(4+5\cos 2t)dt \quad \text{. So}$$

$$\int\left(\frac{1}{Q}+\frac{1}{400-Q}\right)dQ = 0.04(4+5\cos 2t)dt$$

$$\ln Q - \ln|400-Q| = 0.04(4t+2.5\sin 2t)+C_1$$

$$\ln\left|\frac{Q}{400-Q}\right| = 0.04(4t+2.5\sin 2t)+C_1$$

$$\frac{Q}{400-Q} = Ce^{0.04(4t+2.5\sin 2t)}.$$

Using the condition $Q(0)=10$, we have $\dfrac{10}{390}=C.$

So $\qquad \dfrac{Q}{400-Q} = \dfrac{10}{390}e^{0.04(4t+25\sin 2t)}.$

When $t=20$, we have

$$\frac{Q(20)}{400-Q(20)} = \frac{10}{390}e^{0.04(80+2.5\sin 40)} \approx 0.6777.$$

$$Q(20) = 0.6777[400-Q(20)] = 271.08 - 0.6777Q(20)$$

$$1.6777Q(20) \approx 271.08, \text{ and } Q(20) \approx 161.578,$$

or approximately 162 flies.

47. $\displaystyle\int_0^1 \cos t^2\, dt = \int_0^1\left(1-\frac{t^4}{2}+\frac{t^8}{24}-\frac{t^{12}}{720}\right)dt = t - \frac{t^5}{10}+\frac{t^9}{216}-\frac{t^{13}}{9360}\Big|_0^1$

$$= 1 - \frac{1}{10}+\frac{1}{216}-\frac{1}{9360}$$

$$\approx 1 - 0.1 + 0.0046296 - 0.0001068 \approx 0.9.$$

49. True.

$$\int_0^{b+2\pi} \cos x\, dx = \sin x\Big|_a^{b+2\pi} = \sin(b+2\pi)-\sin a$$

$$= \sin b\cos 2\pi + \cos b\sin 2\pi - \sin a$$

$$= \sin b - \sin a = \int_a^b \cos x\, dx$$

51. True. $\displaystyle\int_{-\pi/2}^{\pi/2}\left|\sin x\right|dx = 2\int_{0}^{\pi/2}\sin x\,dx = -2\cos x\Big|_{0}^{\pi/2} = 2$

$\displaystyle\int_{-\pi/2}^{\pi/2}\left|\cos x\right|dx = 2\int_{0}^{\pi/2}\cos x\,dx = -2\sin x\Big|_{0}^{\pi/2} = 2.$

USING TECHNOLOGY EXERCISES 12.4, page 893

1. 0.5419

3. 0.7544

5. 0.2231

7. 0.6587

9. -0.2032

11. 0.9045

13. a.

b. 2.2687 sq units

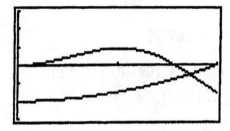

15. a.

b. 1.8239 sq units

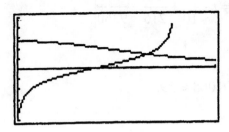

17. a.

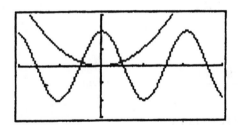

b. 1.2484 sq units

19. a.

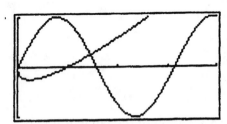

b. 1.0983 sq units

21. 7.6 ft

CHAPTER 12, REVIEW EXERCISES, page 897

1. $120° = \frac{2\pi}{3}$ rad.

3. $-225° = -\frac{225}{180}\pi = -\frac{5\pi}{4}$ rad.

5. $-\frac{5\pi}{2}$ rad $= -\frac{5}{2}(180) = -450°$.

7. $\cos\theta = \frac{1}{2}$ implies that $\theta = \frac{\pi}{3}$ or $\frac{5\pi}{3}$.

9. $f(x) = \sin 3x;\ f'(x) = 3\cos 3x$.

11. $f(x) = 2\sin x - 3\cos 2x$.
$f'(x) = 2\cos x + 3(\sin 2x)(2) = 2(\cos x + 3\sin 2x)$.

13. $f(x) = e^{-x}\tan 3x$.
$f'(x) = -e^{-x}\tan 3x + e^{-x}(\sec^2 3x)(3) = e^{-x}(3\sec^2 3x - \tan 3x)$.

15. $f(x) = 4\sin x \cos x$.
$f'(x) = 4[\sin x(-\sin x) + (\cos x)(\cos x)] = 4(\cos^2 x - \sin^2 x) = 4\cos 2x$.

17. $f(x) = \dfrac{1 - \tan x}{1 - \cot x}$.

$$f'(x) = \frac{(1 - \cot x)(-\sec^2 x) - (1 - \tan x)(\csc^2 x)}{(1 - \cot x)^2}$$

$$= \frac{-\sec^2 x + \sec^2 x \cot x - \csc^2 x + \csc^2 x \tan x}{(1 - \cot x)^2}$$

$$= \frac{(\cot x - 1)\sec^2 x - (1 - \tan x)\csc^2 x}{(1 - \cot x)^2}.$$

19. $f(x) = \sin(\sin x);\ f'(x) = \cos(\sin x) \cdot \cos x.$

21. $f(x) = \tan^2 x.\ f'(x) = 2\tan x \sec^2 x.$

So the slope is $f'(\frac{\pi}{4}) = 2(1)(\sqrt{2})^2 = 4$, and equation of the tangent line is

$$y - 1 = 4(x - \tfrac{\pi}{4}),\ \text{or}\ y = 4x + 1 - \pi.$$

23. Let $u = \frac{2}{3}x$ so that $du = \frac{2}{3}dx$, or $dx = \frac{3}{2}du$. So

$$\int \cos\tfrac{2}{3}x\,dx = \tfrac{3}{2}\int \cos u\,du = \tfrac{3}{2}\sin u + C = \tfrac{3}{2}\sin\tfrac{2}{3}x + C.$$

25. $\displaystyle\int x\csc x^2 \cot x^2\,dx.$ Put $u = x^2$, then $du = 2x\,dx.$ Then

$$\int x\csc x^2 \cot x^2\,dx = \tfrac{1}{2}\int \csc u \cot u\,du$$

$$= -\tfrac{1}{2}\csc u + C = -\tfrac{1}{2}\csc x^2 + C.$$

27. Let $u = \sin x$ so that $du = \cos x\,dx.$ Then

$$\int \sin^2 x \cos x\,dx = \int u^2\,du = \tfrac{1}{3}\sin^3 x + C.$$

29. Let $u = \sin x$ so that $du = \cos x\,dx.$ Then

$$\int \frac{\cos x}{\sin^2 x}\,dx = \int \frac{du}{u^2} = \int u^{-2}\,du = -\frac{1}{u} + C = -\frac{1}{\sin x} + C = -\csc x + C.$$

31. $\displaystyle\int_{\pi/6}^{\pi/2} \frac{\cos x}{1-\cos^2 x}\,dx = \int_{\pi/6}^{\pi/2} \frac{\cos x}{\sin^2 x}\,dx = \int_{\pi/6}^{\pi/2} \sin^{-2} x \cos x\,dx$

$$= -\frac{1}{\sin x}\Big|_{\pi/6}^{\pi/2} = -1+2 = 1.$$

33. $\displaystyle A = \int_{\pi/4}^{5\pi/4} (\sin x - \cos x)\,dx = -\cos x - \sin x \Big|_{\pi/4}^{5\pi/4}$

$$= \left(\tfrac{\sqrt{2}}{2} + \tfrac{\sqrt{2}}{2}\right) - \left(-\tfrac{\sqrt{2}}{2} - \tfrac{\sqrt{2}}{2}\right) = 2\sqrt{2} \quad \text{sq units.}$$

35. $R(t) = 60 + 37\sin^2\left(\frac{\pi t}{12}\right).$

$R'(t) = 37(2)\sin\left(\frac{\pi t}{12}\right)\cdot\cos\left(\frac{\pi t}{12}\right)\left(\frac{\pi}{12}\right) = \frac{37\pi}{6}\sin\left(\frac{\pi t}{6}\right)$ $\qquad$ [$\sin 2\theta = 2\sin\theta\cos\theta$]

Setting $R'(t) = 0$ gives $\frac{\pi t}{6} = 0,\ \pi,\ 2\pi,\ \ldots\,$. So $t = 0, 6, 12, \ldots\,$.

Therefore, $t = 6$ is a critical point in $(0,12)$. Now

$$R'(t) = \frac{37\pi}{6}\cos\left(\frac{\pi t}{6}\right)\cdot\frac{\pi}{6} = \frac{37\pi^2}{36}\cos\left(\frac{\pi t}{6}\right).$$

Since $R''(6) = -\frac{37\pi^2}{36} < 0$, the Second Derivative Test implies that the occupancy rate is highest when $t = 6$ (the beginning of December). Next, we set $R''(t) = 0$

giving $\frac{\pi t}{6} = \frac{\pi}{2}$, or $t = 3$.

You can show that $R'''(3) < 0$ and so $R'(t)$ is maximized at $t = 3$. So the occupancy rate is increasing most rapidly at the beginning of September.